Braiding Technology for Textiles

The Textile Institute and Woodhead Publishing

The Textile Institute is a unique organisation in textiles, clothing and footwear. Incorporated in England by a Royal Charter granted in 1925, the Institute has individual and corporate members in over 90 countries. The aim of the Institute is to facilitate learning, recognise achievement, reward excellence and disseminate information within the global textiles, clothing and footwear industries.

Historically, The Textile Institute has published books of interest to its members and the textile industry. To maintain this policy, the Institute has entered into partnership with Woodhead Publishing Limited to ensure that Institute members and the textile industry continue to have access to high calibre titles on textile science and technology.

Most Woodhead titles on textiles are now published in collaboration with The Textile Institute. Through this arrangement, the Institute provides an Editorial Board which advises Woodhead on appropriate titles for future publication and suggests possible editors and authors for these books. Each book published under this arrangement carries the Institute's logo.

Woodhead books published in collaboration with The Textile Institute are offered to Textile Institute members at a substantial discount. These books, together with those published by The Textile Institute that are still in print, are offered on the Elsevier website at: http://store.elsevier.com. Textile Institute books still in print are also available directly from the Institute's website at: www.textileinstitutebooks.com.

A list of Woodhead books on textiles science and technology, most of which have been published in collaboration with The Textile Institute, can be found towards the end of the contents pages.

Woodhead Publishing Series in Textiles: Number 158

Braiding Technology for Textiles

Y. Kyosev

ELSEVIER

AMSTERDAM • BOSTON • CAMBRIDGE • HEIDELBERG
LONDON • NEW YORK • OXFORD • PARIS • SAN DIEGO
SAN FRANCISCO • SINGAPORE • SYDNEY • TOKYO
Woodhead Publishing is an imprint of Elsevier

WP WOODHEAD PUBLISHING

Published by Woodhead Publishing in association with The Textile Institute
Woodhead Publishing is an imprint of Elsevier
80 High Street, Sawston, Cambridge, CB22 3HJ, UK
225 Wyman Street, Waltham, MA 02451, USA
Langford Lane, Kidlington, OX5 1GB, UK

British Library Cataloguing in Publication Data
A catalogue record for this book is available from the British Library.

Library of Congress Control Number: 2014944411

ISBN 978-0-08-101329-8 (print)
ISBN 978-0-85709-921-1 (online)

Typeset by Newgen Knowledge Works Pvt Ltd, India

Contents

Woodhead Publishing Series in Textiles

1 **Watson's textile design and colour Seventh edition**
Edited by Z. Grosicki
2 **Watson's advanced textile design**
Edited by Z. Grosicki
3 **Weaving Second edition**
P. R. Lord and M. H. Mohamed
4 **Handbook of textile fibres Volume 1: Natural fibres**
J. Gordon Cook
5 **Handbook of textile fibres Volume 2: Man-made fibres**
J. Gordon Cook
6 **Recycling textile and plastic waste**
Edited by A. R. Horrocks
7 **New fibers Second edition**
T. Hongu and G. O. Phillips
8 **Atlas of fibre fracture and damage to textiles Second edition**
J. W. S. Hearle, B. Lomas and W. D. Cooke
9 **Ecotextile '98**
Edited by A. R. Horrocks
10 **Physical testing of textiles**
B. P. Saville
11 **Geometric symmetry in patterns and tilings**
C. E. Horne
12 **Handbook of technical textiles**
Edited by A. R. Horrocks and S. C. Anand
13 **Textiles in automotive engineering**
W. Fung and J. M. Hardcastle
14 **Handbook of textile design**
J. Wilson
15 **High-performance fibres**
Edited by J. W. S. Hearle
16 **Knitting technology Third edition**
D. J. Spencer
17 **Medical textiles**
Edited by S. C. Anand
18 **Regenerated cellulose fibres**
Edited by C. Woodings

19 **Silk, mohair, cashmere and other luxury fibres**
Edited by R. R. Franck
20 **Smart fibres, fabrics and clothing**
Edited by X. M. Tao
21 **Yarn texturing technology**
J. W. S. Hearle, L. Hollick and D. K. Wilson
22 **Encyclopedia of textile finishing**
H-K. Rouette
23 **Coated and laminated textiles**
W. Fung
24 **Fancy yarns**
R. H. Gong and R. M. Wright
25 **Wool: Science and technology**
Edited by W. S. Simpson and G. Crawshaw
26 **Dictionary of textile finishing**
H-K. Rouette
27 **Environmental impact of textiles**
K. Slater
28 **Handbook of yarn production**
P. R. Lord
29 **Textile processing with enzymes**
Edited by A. Cavaco-Paulo and G. Gübitz
30 **The China and Hong Kong denim industry**
Y. Li, L. Yao and K. W. Yeung
31 **The World Trade Organization and international denim trading**
Y. Li, Y. Shen, L. Yao and E. Newton
32 **Chemical finishing of textiles**
W. D. Schindler and P. J. Hauser
33 **Clothing appearance and fit**
J. Fan, W. Yu and L. Hunter
34 **Handbook of fibre rope technology**
H. A. McKenna, J. W. S. Hearle and N. O'Hear
35 **Structure and mechanics of woven fabrics**
J. Hu
36 **Synthetic fibres: nylon, polyester, acrylic, polyolefin**
Edited by J. E. McIntyre
37 **Woollen and worsted woven fabric design**
E. G. Gilligan
38 **Analytical electrochemistry in textiles**
P. Westbroek, G. Priniotakis and P. Kiekens
39 **Bast and other plant fibres**
R. R. Franck
40 **Chemical testing of textiles**
Edited by Q. Fan

41 **Design and manufacture of textile composites**
Edited by A. C. Long
42 **Effect of mechanical and physical properties on fabric hand**
Edited by H. M. Behery
43 **New millennium fibers**
T. Hongu, M. Takigami and G. O. Phillips
44 **Textiles for protection**
Edited by R. A. Scott
45 **Textiles in sport**
Edited by R. Shishoo
46 **Wearable electronics and photonics**
Edited by X. M. Tao
47 **Biodegradable and sustainable fibres**
Edited by R. S. Blackburn
48 **Medical textiles and biomaterials for healthcare**
Edited by S. C. Anand, M. Miraftab, S. Rajendran and J. F. Kennedy
49 **Total colour management in textiles**
Edited by J. Xin
50 **Recycling in textiles**
Edited by Y. Wang
51 **Clothing biosensory engineering**
Y. Li and A. S. W. Wong
52 **Biomechanical engineering of textiles and clothing**
Edited by Y. Li and D. X-Q. Dai
53 **Digital printing of textiles**
Edited by H. Ujiie
54 **Intelligent textiles and clothing**
Edited by H. R. Mattila
55 **Innovation and technology of women's intimate apparel**
W. Yu, J. Fan, S. C. Harlock and S. P. Ng
56 **Thermal and moisture transport in fibrous materials**
Edited by N. Pan and P. Gibson
57 **Geosynthetics in civil engineering**
Edited by R. W. Sarsby
58 **Handbook of nonwovens**
Edited by S. Russell
59 **Cotton: Science and technology**
Edited by S. Gordon and Y-L. Hsieh
60 **Ecotextiles**
Edited by M. Miraftab and A. R. Horrocks
61 **Composite forming technologies**
Edited by A. C. Long
62 **Plasma technology for textiles**
Edited by R. Shishoo

63 **Smart textiles for medicine and healthcare**
Edited by L. Van Langenhove
64 **Sizing in clothing**
Edited by S. Ashdown
65 **Shape memory polymers and textiles**
J. Hu
66 **Environmental aspects of textile dyeing**
Edited by R. Christie
67 **Nanofibers and nanotechnology in textiles**
Edited by P. Brown and K. Stevens
68 **Physical properties of textile fibres Fourth edition**
W. E. Morton and J. W. S. Hearle
69 **Advances in apparel production**
Edited by C. Fairhurst
70 **Advances in fire retardant materials**
Edited by A. R. Horrocks and D. Price
71 **Polyesters and polyamides**
Edited by B. L. Deopura, R. Alagirusamy, M. Joshi and B. S. Gupta
72 **Advances in wool technology**
Edited by N. A. G. Johnson and I. Russell
73 **Military textiles**
Edited by E. Wilusz
74 **3D fibrous assemblies: Properties, applications and modelling of three-dimensional textile structures**
J. Hu
75 **Medical and healthcare textiles**
Edited by S. C. Anand, J. F. Kennedy, M. Miraftab and S. Rajendran
76 **Fabric testing**
Edited by J. Hu
77 **Biologically inspired textiles**
Edited by A. Abbott and M. Ellison
78 **Friction in textile materials**
Edited by B. S. Gupta
79 **Textile advances in the automotive industry**
Edited by R. Shishoo
80 **Structure and mechanics of textile fibre assemblies**
Edited by P. Schwartz
81 **Engineering textiles: Integrating the design and manufacture of textile products**
Edited by Y. E. El-Mogahzy
82 **Polyolefin fibres: Industrial and medical applications**
Edited by S. C. O. Ugbolue
83 **Smart clothes and wearable technology**
Edited by J. McCann and D. Bryson

84 **Identification of textile fibres**
Edited by M. Houck
85 **Advanced textiles for wound care**
Edited by S. Rajendran
86 **Fatigue failure of textile fibres**
Edited by M. Miraftab
87 **Advances in carpet technology**
Edited by K. Goswami
88 **Handbook of textile fibre structure Volume 1 and Volume 2**
Edited by S. J. Eichhorn, J. W. S. Hearle, M. Jaffe and T. Kikutani
89 **Advances in knitting technology**
Edited by K-F. Au
90 **Smart textile coatings and laminates**
Edited by W. C. Smith
91 **Handbook of tensile properties of textile and technical fibres**
Edited by A. R. Bunsell
92 **Interior textiles: Design and developments**
Edited by T. Rowe
93 **Textiles for cold weather apparel**
Edited by J. T. Williams
94 **Modelling and predicting textile behaviour**
Edited by X. Chen
95 **Textiles, polymers and composites for buildings**
Edited by G. Pohl
96 **Engineering apparel fabrics and garments**
J. Fan and L. Hunter
97 **Surface modification of textiles**
Edited by Q. Wei
98 **Sustainable textiles**
Edited by R. S. Blackburn
99 **Advances in yarn spinning technology**
Edited by C. A. Lawrence
100 **Handbook of medical textiles**
Edited by V. T. Bartels
101 **Technical textile yarns**
Edited by R. Alagirusamy and A. Das
102 **Applications of nonwovens in technical textiles**
Edited by R. A. Chapman
103 **Colour measurement: Principles, advances and industrial applications**
Edited by M. L. Gulrajani
104 **Fibrous and composite materials for civil engineering applications**
Edited by R. Fangueiro
105 **New product development in textiles: Innovation and production**
Edited by L.Horne

106 **Improving comfort in clothing**
Edited by G. Song
107 **Advances in textile biotechnology**
Edited by V. A. Nierstrasz and A. Cavaco-Paulo
108 **Textiles for hygiene and infection control**
Edited by B. McCarthy
109 **Nanofunctional textiles**
Edited by Y. Li
110 **Joining textiles: Principles and applications**
Edited by I. Jones and G. Stylios
111 **Soft computing in textile engineering**
Edited by A. Majumdar
112 **Textile design**
Edited by A. Briggs-Goode and K. Townsend
113 **Biotextiles as medical implants**
Edited by M. W. King, B. S. Gupta and R. Guidoin
114 **Textile thermal bioengineering**
Edited by Y. Li
115 **Woven textile structure**
B. K. Behera and P. K. Hari
116 **Handbook of textile and industrial dyeing. Volume 1: Principles, processes and types of dyes**
Edited by M. Clark
117 **Handbook of textile and industrial dyeing. Volume 2: Applications of dyes**
Edited by M. Clark
118 **Handbook of natural fibres. Volume 1: Types, properties and factors affecting breeding and cultivation**
Edited by R. Kozłowski
119 **Handbook of natural fibres. Volume 2: Processing and applications**
Edited by R. Kozłowski
120 **Functional textiles for improved performance, protection and health**
Edited by N. Pan and G. Sun
121 **Computer technology for textiles and apparel**
Edited by J. Hu
122 **Advances in military textiles and personal equipment**
Edited by E. Sparks
123 **Specialist yarn and fabric structures**
Edited by R. H. Gong
124 **Handbook of sustainable textile production**
M. I. Tobler-Rohr
125 **Woven textiles: Principles, developments and applications**
Edited by K. Gandhi
126 **Textiles and fashion: Materials design and technology**
Edited by R. Sinclair

127 **Industrial cutting of textile materials**
I. Viļumsone-Nemes
128 **Colour design: Theories and applications**
Edited by J. Best
129 **False twist textured yarns**
C. Atkinson
130 **Modelling, simulation and control of the dyeing process**
R. Shamey and X. Zhao
131 **Process control in textile manufacturing**
Edited by A. Majumdar, A. Das, R. Alagirusamy and V. K. Kothari
132 **Understanding and improving the durability of textiles**
Edited by P. A. Annis
133 **Smart textiles for protection**
Edited by R. A. Chapman
134 **Functional nanofibers and applications**
Edited by Q. Wei
135 **The global textile and clothing industry: Technological advances and future challenges**
Edited by R. Shishoo
136 **Simulation in textile technology: Theory and applications**
Edited by D. Veit
137 **Pattern cutting for clothing using CAD: How to use Lectra Modaris pattern cutting software**
M. Stott
138 **Advances in the dyeing and finishing of technical textiles**
M. L. Gulrajani
139 **Multidisciplinary know-how for smart textiles developers**
Edited by T. Kirstein
140 **Handbook of fire resistant textiles**
Edited by F. Selcen Kilinc
141 **Handbook of footwear design and manufacture**
Edited by A. Luximon
142 **Textile-led design for the active ageing population**
Edited by J. McCann and D. Bryson
143 **Optimizing decision making in the apparel supply chain using artificial intelligence (AI): From production to retail**
Edited by W. K. Wong, Z. X. Guo and S. Y. S. Leung
144 **Mechanisms of flat weaving technology**
V. V. Choogin, P. Bandara and E. V. Chepelyuk
145 **Innovative jacquard textile design using digital technologies**
F. Ng and J. Zhou
146 **Advances in shape memory polymers**
J. Hu

147 **Design of clothing manufacturing processes: A systematic approach to planning, scheduling and control**
J. Gersak
148 **Anthropometry, apparel sizing and design**
D. Gupta and N. Zakaria
149 **Silk: Processing, properties and applications**
Edited by K. Murugesh Babu
150 **Advances in filament yarn spinning of textiles and polymers**
Edited by D. Zhang
151 **Designing apparel for consumers: The impact of body shape and size**
Edited by M.-E. Faust and S. Carrier
152 **Fashion supply chain management using radio frequency identification (RFID) technologies**
Edited by W. K. Wong and Z. X. Guo
153 **High performance textiles and their applications**
Edited by C. A. Lawrence
154 **Protective clothing: Managing thermal stress**
Edited by F. Wang and C. Gao
155 **Composite nonwoven materials**
Edited by D. Das and B. Pourdeyhimi
156 **Functional finishes for textiles: Improving comfort, performance and protection**
Edited by R. Paul
157 **Assessing the environmental impact of textiles and the clothing supply chain**
S. S. Muthu
158 **Braiding technology for textiles**
Y. Kyosev
159 **Principles of colour appearance and measurement**
Volume 1: Object appearance, colour perception and instrumental measurement
A. K. R. Choudhury
160 **Principles of colour appearance and measurement**
Volume 2: Visual measurement of colour, colour comparison and management
A. K. R. Choudhury

Preface

Braiding is a simple but very good technique for making a range of products. As I started teaching braiding to textile engineering students, I realized that there was a large gap between the current state of the art and the available literature on braiding. I had not expected the gap in the literature available in English to be so much greater than that for German. My predecessors, Professor Bernhard Lepperhoff, Professor Friedhelm Goseberg and Professor Dr Harald Engels, all wrote excellent books on braiding in German. Since year 2000, the Niederrhein University has had Bachelor and Masters Degree courses both in German and English and a significant number of international students. The simple question then followed: how to teach braiding technology in English if all the available texts were out of date? What reading text could I give to my students?

The answer to these questions is now before you, and I hope you will enjoy the book and find it useful. My goal was to write a book which would help the beginner to understand the basics of braiding machines, carriers, patterning, and the forces during the braiding, i.e. the information required to engineer and manufacture high quality products. This explains why there are relatively few references at the end of each chapter. There are several papers about braiding and braided products, but most of these are based on investigation of a particular product or technique. What is missing is information on the background to the process and, in this area, only the older books were really helpful.

Where are the advanced topics, such as 3D braiding, lace braiding, composites, rope tests, tensile properties? These topics belong to a planned second book, where they will be covered in collaboration with my dear colleagues around the world who are the experts in each of these areas. However, such advances can only be understood from a solid background so, step by step, picture by picture, let us start braiding.

Yordan Kyosev
Mönchengladbach

This book would not have been possible without the help of one great teacher: Prof. Dr. Dr.h.c. Nicolas Schulmann.

Acknowledgements

First of all, I would like to express my profound thanks to my first teacher in braiding and narrow fabrics in general – Professor Dr Harald Engels. With his enthusiasm and fervour for braided and narrow woven fabrics, he educated several generations of German and international students at the Niederrhein University of Applied Sciences in Mönchengladbach, Germany. Thanks to his motivation and energy, and with the indispensable support of the faculty dean at the time, Professor Dr Marie Louise Klotz, it was possible to create a temporary endowed professorship for five years in the area of narrow fabrics. With this endowed professorship and its following transition into a permanent position, the continuity in knowledge and research in the subject of narrow fabrics at the university level was guaranteed.

The endowed professorship would not have been possible without the support of the companies Jakob Müller AG, Frick, Switzerland; August Herzog & Ko. KG, Oldenburg; Körting Nach. Steeger GmbH & Ko KG, Wuppertal; supporting companies through the German Industry Association for Technical Textiles (Deutsche Verband der Band und Flechtindustrie, currently IVGT – Industrieverband Veredlung – Garne – Gewebe – Technische Textilien e.V.) and the 'Deutsche Stiftungsverband'.

My personal thanks go to Dr Janpeter Horn, Guido Grave and Herbert Schneider from August Herzog Maschinenfabrik GmbH & Co. KG, Oldenburg and to Bernd Rehrmann and Axel Ludwig from Körting Nachfolger Wilhelm Steeger GmbH & Co. KG, Wuppertal, for support with pictures, information and valuable conversations about braiding machines and techniques.

My thanks to Dr Karin Birkefeld and Tjark von Reden, IFB Stuttgart, for valuable discussions about the open and closed braids and our collaboration in 2009, and also to Dr Amit Rawal for further discussions and collaboration in this area. I would like to express special thanks to Mr Frank Heimlich, my right-hand man in our braiding laboratory, for his continuous support and patience during the all the work with the students.

Several former students helped me indirectly during the work, for to which I would also like to express my gratitude: Svenia Hartmann for extensive help on patent research; Cristoph Fahrenbrach for managing the patent titles; Juila Biessmann for the experimental investigation of the geometry of tubular-braided structures; Franziska Becher and Ouail El Brini for investigations on the Vario Flechter; Priscilla Reiners for her work with spiral-braiding machines; Uttam Mittal and Lawrence Msalwiva for the testing and mechanical investigations of double-braided ropes; Mezgy Li for testing braiding software and the extended pattern for the Vario Flechter.

Two companies supported the writing of this book with software: Matworks Inc., with their great book author program, which allowed the use of Matlab® for figure generation and calculations, and TexMind UG for the software for 3D pictures of the braids.

Finally, this book would not have been possible without the endless patience and support of my family.

Introduction: the main types of braided structure using maypole braiding technology

1

1.1 Introduction

Braiding is a process of interlacing three or more threads diagonally to the product axis (parallel to the longest dimension of the resulting product) in order to obtain a thicker, wider or stronger product or in order to cover (overbraid) some profile. Braids can be linear products (ropes), curved or plane shell or solid structures (one, two or three-dimensional fabrics) with constant or variable cross-section, and of closed or open appearance.

Interlacing diagonally means that the yarns make an angle α with the product axis, which can be between 1° and 89° but is usually in the range of 30°–80° (Fig. 1.1c). This angle is called the braiding angle and is the most important geometrical parameter of braided structures.

The definition of what constitutes a one-, two- or three-dimensional fabric is not straightforward (Büsgen, 1993; Schneider, 2000; Hearle and Chen, 2013). The use of any dimension in such a classification requires information about the scale in which this term is used – micro-, meso-, or macro-scale – or equivalently, in textile terminology, fibre, yarn or product level.

From an engineering point of view, each braided rope or shoe lace can be classified as a one-dimensional product, because the length of each is significantly greater than the cross-section. If, however, we consider yarn interlacement, each rope is a three-dimensional product because of its curved outer structure. Hence, the definition of braids employed above.

1.2 Maypole braiding

The classical and mainly used braiding machines are known as maypole braiding machines. This name will always be used where these machines have to be distinguished from the other (spiral, lace) braiding machines. This section introduces the working principle of these machines, the types of structures and machines and their modifications in order to give an overview of the technology. The topics from this section are presented with more details in several chapters in the book.

Braiding Technology for Textiles. http://dx.doi.org/10.1533/9780857099211.1

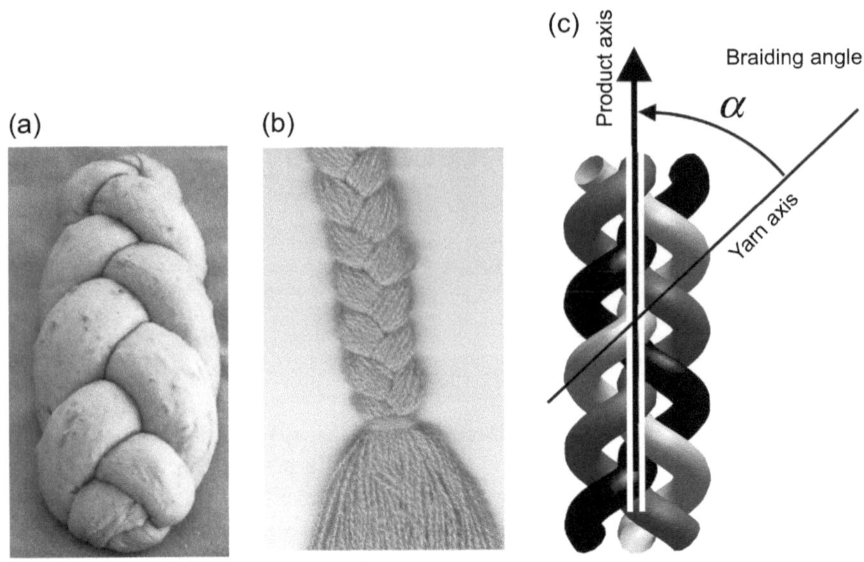

Figure 1.1 Braided cake (a), lace (b) and idealized representation (c).

1.2.1 Basic principle

In its basic form, braiding has been used for many years in a wide range of application areas, including the braiding of yarns, flowers, hair, etc. (Fig. 1.1), and the braiding process allows for the addition or replacement of yarns during each step (Fig. 1.2). The application areas for braided products are broad, ranging from the braiding of chirurgical sutures and other medical items, to such products as electric cables, ropes, laces and, at the largest scale, the huge ropes and tubes used in the marine oilfield sector. Also, a growing area of use of braids is in the production of fibre-reinforced composites.

All the products discussed above are manufactured in the same way, that is, by the interlacing of yarns at an angle of about 40°–60° to the main product axis. Figure 1.3 demonstrates the principle of hand braiding for a simple braid of three yarns (braided pigtail). The interlacement can be described in two steps:

- interlace the left two yarns, where the outer left yarn goes over the next one (Fig. 1.3b);
- interlace the right two yarns, where the right yarn goes over the left next nearest yarn (Fig. 1.3c), and so on.

1.2.2 Flat braiding

The above described process can be extended to more yarns, for examples 5, 7, etc. as demonstrated in Fig. 1.4. In this case, at the first step all left pairs have to be

Figure 1.2 The braiding allows adding new elements to the braid at each step: (a) garlic and (b) hair.

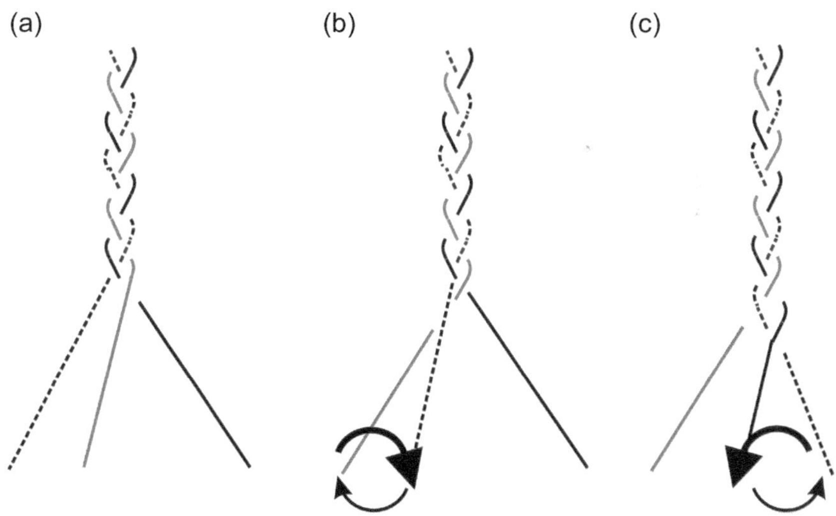

Figure 1.3 Principle of hand braiding with three yarns. (a) Initial configuration; (b) first step: interlacing the left yarn over the middle one; (c) second step: interlacing the right yarn over the new middle yarn.

interlaced, so that the left goes over the right, and in the second step, all right pairs have to be interlaced, so that the right yarn goes over the left.

The same principles are implemented on certain machines for braiding and the resulting product is a flat cross-sectioned braid, known as "flat braid". Flat braiding is possible for both odd and even numbers of yarns. Figure 1.5 demonstrates the

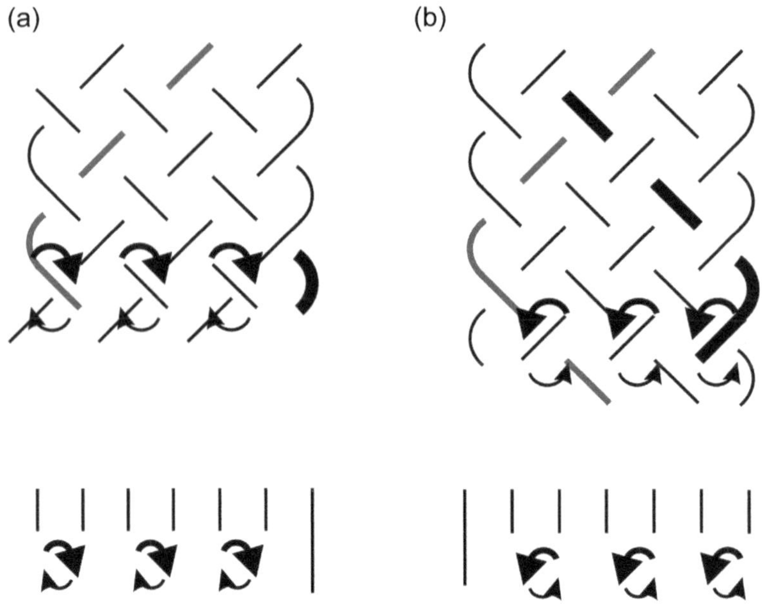

Figure 1.4 Hand braiding of odd (here seven) yarns. (a) Interlace each left yarn over each right in the pairs; (b) interlace each right yarn over the left one, but shifting the starting position on one yarn.

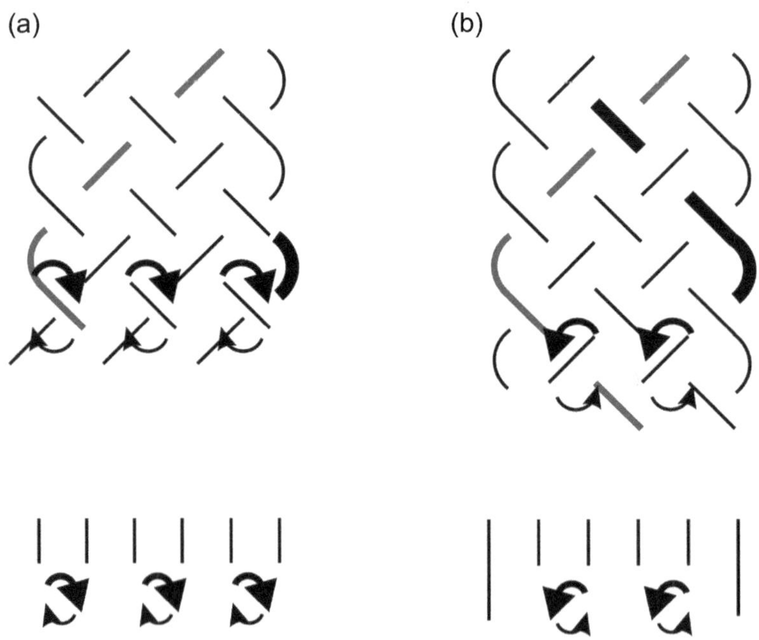

Figure 1.5 Hand braiding of even (here six) yarns. (a) Interlace each left yarn over each right in the pairs; (b) interlace each right yarn over the left one, but shifting the starting position on one yarn.

sequence in the case of six yarns: at every second step, yarns from both the left and right sides stay and wait, while when using an odd number of yarns, only one yarn per cycle stays unused during the interlacement.

1.2.3 Tubular braiding

Using an even number of yarns arranged around a circle allows production of tubular braids or ropes (Fig. 1.6). The sequence of steps in this case is the same as for the flat yarn. The easiest way to produce tubular braids by hand is through the use of a special table with a hole in the middle, so that the yarns can be kept on the circumference and arranged as desired (Fig. 1.7).

In order to produce a regular structure, all yarns have to be kept under constant tension during the braiding process. The simplest way of doing this is to connect weights to the yarn ends (Fig. 1.7). For hand braiding of short lengths, the most practical approach is to use a small bag filled with sand, because the braider can adjust the weight of each bag easily by adding or removing sand depending on the requirements.

The end of the braid can also be fixed or connected to a weight, which allows easy movement and adjustment of the braiding area. In order to keep the braid and the

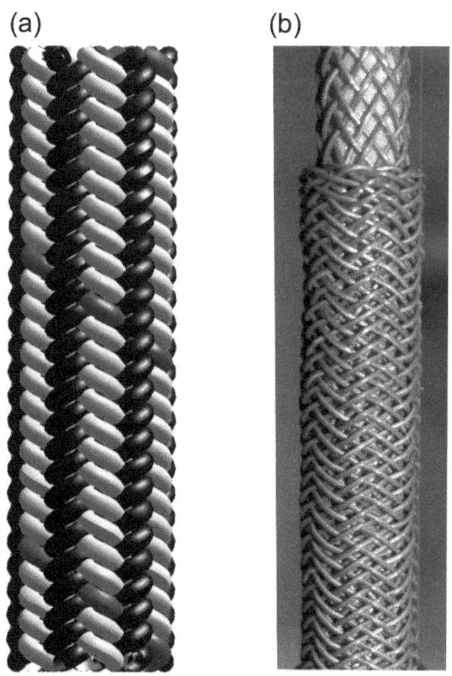

Figure 1.6 Tubular braids (rope). (a) Geometrical model and (b) single (upper part) and tripple (over-) braided part.

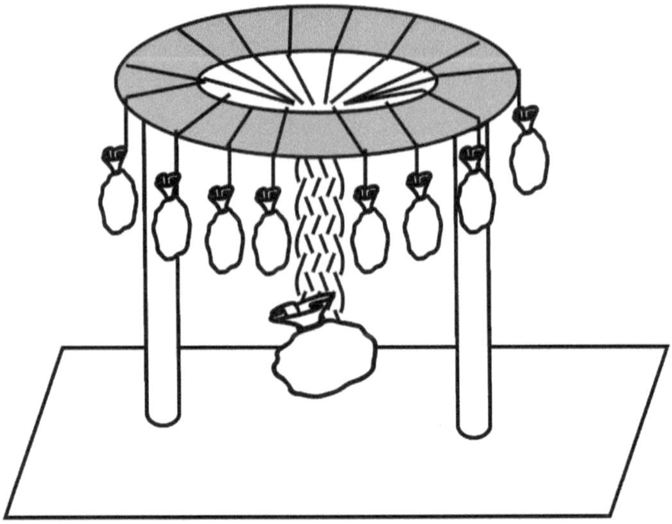

Figure 1.7 Table for hand braiding with weights for keeping the yarns under tension.

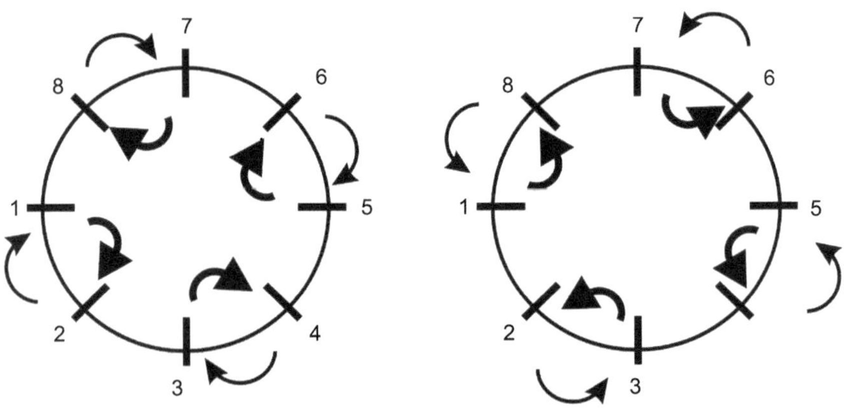

Figure 1.8 Sequence for tubular braiding.

yarns in equilibrium, the weight which is fixed on the braid M_{Braid} has to be equal to the sum of all weights at the yarns m_i.

$$M_{\text{Braid}} = m_1 + m_2 + \ldots + m_n = \sum_{i=1}^{n} m_i \quad [1.1]$$

The interlacing sequence in the case of tubular braiding is identical to that of flat braiding, except that all the yarn ends are located around a circle (Fig. 1.8). The process can be described as:

Figure 1.9 Maypole dance (Beard and Beard, 1893).

- interlacement of the left yarn over the right yarn in each pair (the yarn at position 1 goes over the yarn at position 2 and the yarn from position 2 goes to position 1, but interlacing under the first yarn; the pairs are 1–2; 3–4; 5–6; 7–8)
- after shifting the positions (the pairs become 8–1; 2–3; 4–5; 6–7) the right yarn goes over the left yarn in each pair.

This kind of sequence is similar to the maypole dance found in a number of European countries (Fig. 1.9). In this tradition, each dancer holds one yarn end and dances around a pole, at the top of which the yarn is fixed. Half of the dancers dance in a clockwise direction, and the other half counter-clockwise.

The beam is slowly covered from the top down. In braiding terminology, the beam would be called the core (if it had a regular form) or the mandrel (if it had a more complex geometry), and this principle is used today for overbraiding of profiles with carbon or glass fibre composites for aerospace, cables, high pressure ropes, etc. for automotive and other applications.

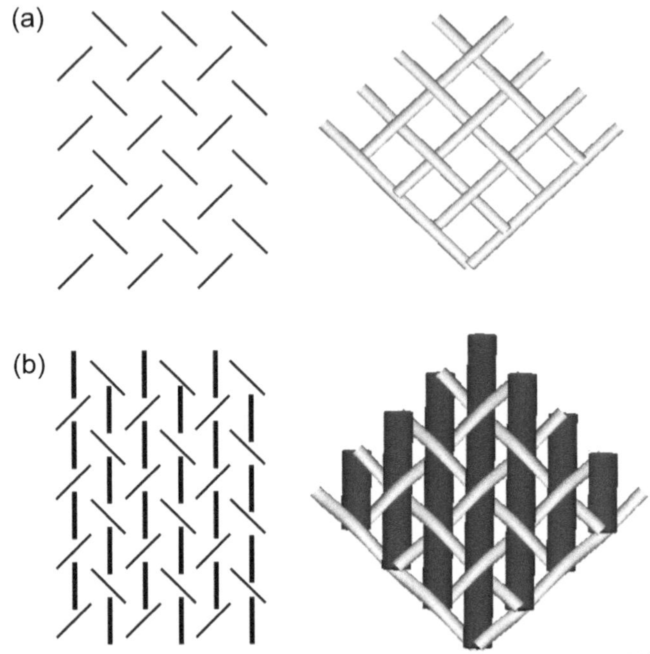

Figure 1.10 Normal (biaxial) braids (a) and braids with inlay yarns (triaxial) (b) as a draft and as 3D simulation. The right pictures are generated with WiseTex (Lomov *et al.*, 2012).

1.2.4 Biaxial and triaxial braids, inlay yarns

It is also possible to put inlay yarns between the core and the yarns. These are named differently in different application areas. For example, in the production of elastic tapes for underwear and other clothing accessories, they are often named "middle ends," because they stay at the middle between the braiding yarns, but for composites production, they are named inlays, and the braids with inlays are named triaxial braids, because of their three material axes (Fig. 1.10). Inlay yarns can be inserted into all kinds of braided structures; however, the core or mandrel requires a hollow structure and is mainly used in tubular braids (Fig. 1.11).

1.2.5 Industrial maypole braiding

The path of one dancer during the maypole dance is illustrated in Fig. 1.12. With industrial braiding, the machine element equivalent to this dancer is called a carrier, and the path of the carrier is called the track. If the motion of the dancers/carriers is analysed carefully, it can be seen that all dancers/carriers in one direction are following the same track, while the dancers/carriers in the opposite direction follow the opposite track. Thus, for tubular braiding, there are two tracks and two systems of carriers.

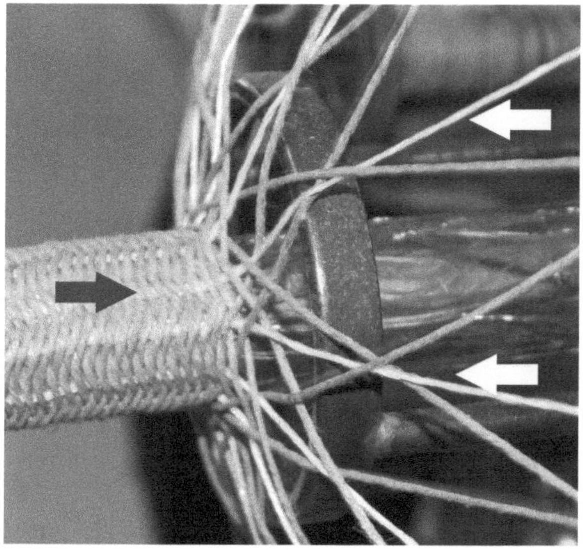

Figure 1.11 Tubular braid with core and inlay yarns (triaxial). The arrows point to the inlay yarns.

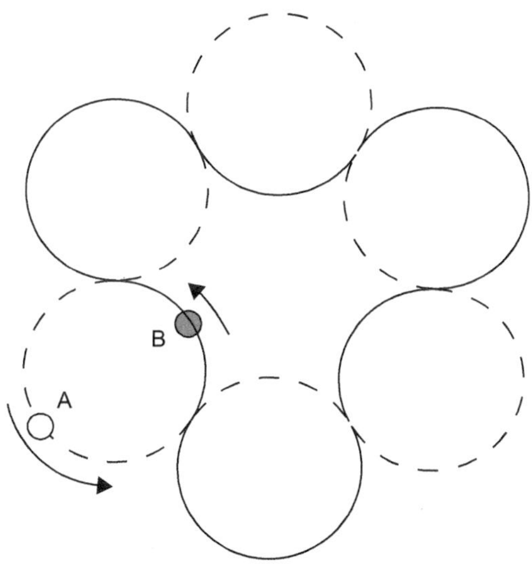

Figure 1.12 Track for tubular braid (the path of the dancers respectively carriers A and B).

The case of the flat braiding is a little different in that there is only one closed track and the carriers and their respective yarns change direction when they reach the boundary (Fig. 1.13).

The track determines the path of the motion, but the carriers are moved forwards by the horn gears (Fig. 1.14). Every *horn gear* is a disc with several cuts (slots) where the carrier foot fits. Each horn gear is mounted onto a shaft which has a corresponding

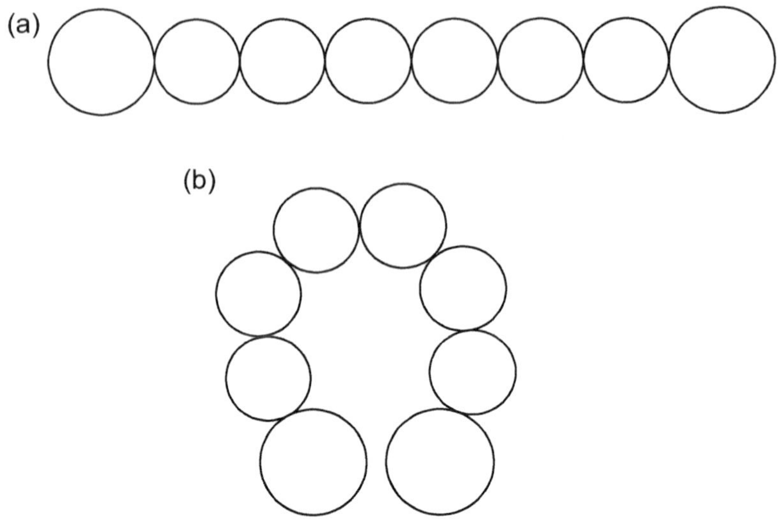

Figure 1.13 Possible tracks for flat braiding: (a) in one line, used for a smaller number of carriers; (b) arrangement around a circle – used for larger machines.

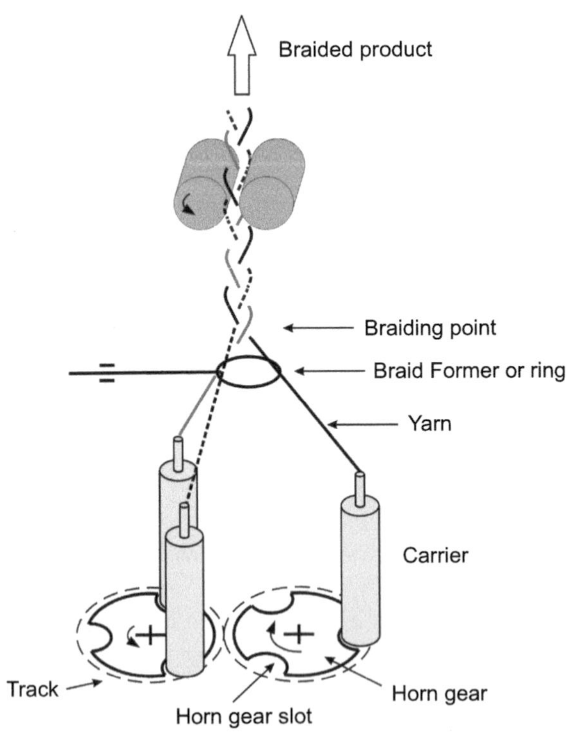

Figure 1.14 Principal construction of maypole braiding machine.

gear. This transmits the motion from the motor to the mounted horn gear and the neighbouring gear. At the beginning of the braiding, all outer ends of the yarns are connected together at the braiding point. Once the carriers start their motions, the yarns interlace together and build the next piece of braid at the braiding point. The braid produced is finally pulled out by a take-off system. The distance between each carrier and the braiding point does not remain constant during the braiding process and, as a result, the required length of the yarn between the carrier and the braiding point changes. The differences in the required yarn length should normally not be compensated for by yarn elongation (in most cases yarn elongation is in any case insufficient for such compensations) and a yarn length compensation mechanism has to be used on each carrier.

1.2.6 Carriers

As mentioned above, the yarn tension has to be kept constant during the braiding process and compensation should be made for any yarn length differences. Using weights on each yarn end (Fig. 1.7) solves these problems, but it is not helpful for braiding products more than half a metre in length. In order to produce longer braids, initially wooden carriers were used, which had two functions:

- to maintain yarn tension
- to hold extra yarn wound around it.

The yarn is kept in place by a simple knot.

Machine braiding was not possible with such carriers and, because the distance between the carrier and the braiding point during braiding varied, an additional functionality had to be built into their construction, as they have to be able to compensate for the yarn length differences.

During the maypole dance, the dancer can do this by moving the hands up and down. The carrier has a lever or guide which acts in a similar manner. Figure 1.15 shows historical industrial carriers (August Herzog Maschinenfabrik, approx. 1940). Current carriers have significantly less weight and smoother compensation, but the basic principles are the same. There is a lever or slider which pulls the yarn into the compensation area, or releases it if more yarn length is required. If the yarn length is insufficient the bobbin break is released and free yarn is unwound. For metal yarn a horizontal bobbin and special rollers with a larger radius of curvature are used in order to minimize damage to the braid. The functionality of the different type of carriers is presented in Chapter 7 (Carriers for braiding machines).

1.2.7 Form braiding and 3D braiding

For several applications thicker braided products of square or other cross-sectional type are required. These products are usually used for the manufacture of gaskets or fibre-reinforced composites. The braiding process in this case is named packing braiding, or 2.5D braiding (2.5D means more than the two dimensions of length

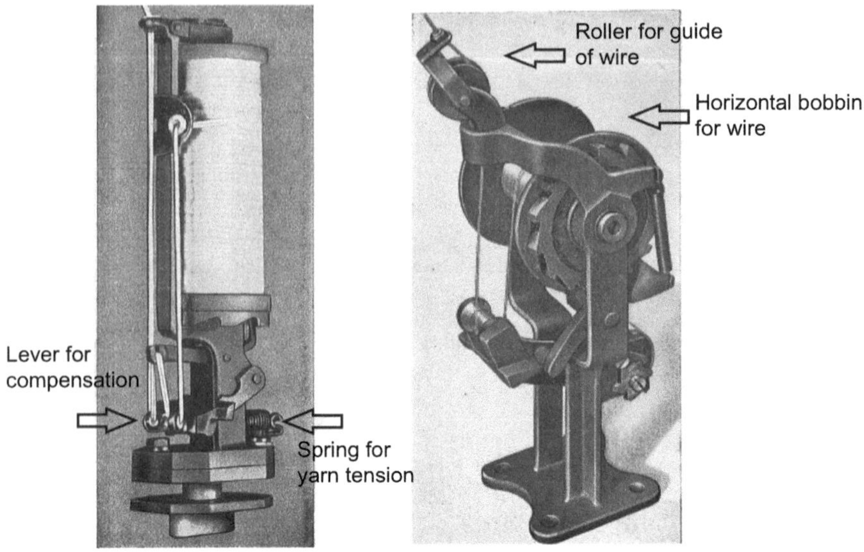

Figure 1.15 Historical industrial carrier: (a) for textile yarns and (b) for wire.

and width, but less than three dimensions), or alternatively 3D braiding (due to the braided product having thickness in all three directions). However, according to other classifications, they may also be named 3D, 4D, etc. braiding, where D means not "dimension" but "diagonal." The term employed here for the general case of thicker braid production will be form braiding. Form braiding means that the machine has more than two tracks so that the cross-section of the product takes some (non-flat and non-circular) form. An example of four-track form braiding for a structure with four "diagonals" and a square cross-section is given in Fig. 1.16

With the new generation of braiding machines, which are computer-controlled devices (these will be specified later), the track is not constant. The carrier can change its motion and can travel along a set of connected curves, and the selection of the next curve can change dynamically during the braiding process. Because in this case the braids do not have a constant cross-section, they are called 3D braids, and the process itself 3D braiding (Büsgen, 1993).

1.2.8 Terminological issues

The machines which follow this principle of braiding, like the maypole dance, are called "maypole" braiding machines, and they are currently the most widely used braiding machines in the industry. They are also known as "rotary" braiding machines because the gears, which move the carriers, rotate. This term will not be employed here, however, as another type of braiding machine, known as "fast" or "rapid" or "Wardwell" (Wardwell, 2013) or "Wardwellian" machines, are also referred to as "rotary braiding machines" by some companies and sources (Douglass, 1964).

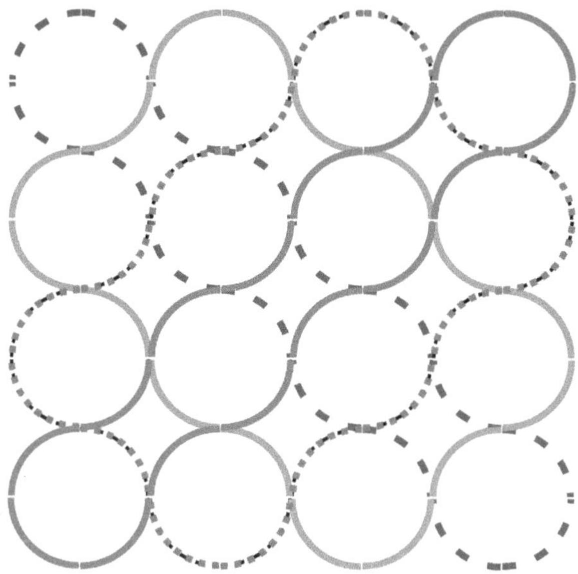

Figure 1.16 Tracks for production of braids with square cross-section.

1.2.9 Rapid or Wardwellian braiding

If during maypole braiding half of the dancers were to stay in their places and the another half moved in a circle, while the stationary dancers moved their yarns up and down so that the moving dancers could go under or over their yarns, then tubular braid would be produced. This type of braid is manufactured using machines known as rapid or Wardwellian braiding machines, and the principle is demonstrated in Fig. 1.17. Half of the bobbins have fixed positions on one ring, with a yarn guiding lever swinging up and down. It swings and moves the yam between two rotating bobbins (carriers) from the other group. In this way, the rotating carriers follow their (steady) motion. Because the motion is steady, these machines can work significantly faster than the classical maypole braiding machines, where each carrier has to be stabilized under dynamic conditions. The disadvantage of such machines is their limited variability – changing the pattern requires replacement of large machinery parts (cam profiles) – and because of this they are used mainly for overbraiding of cables and ropes for high pressure and other tubular products with wire.

1.3 Spiral braiding

A modification of the maypole braiding machines leads to the group of braiding machines which work on the principle of spiral braiding. These machines use carriers similar to the maypole braiding machines, have the same horn gears, but have a

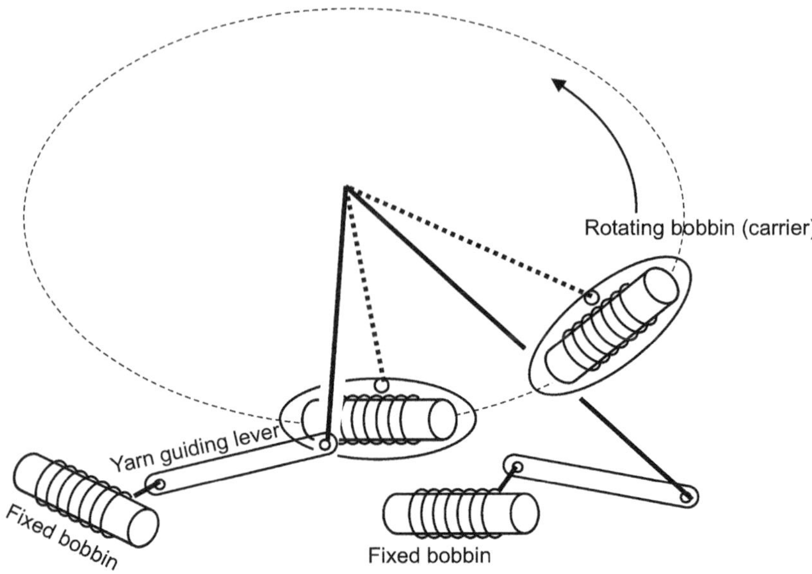

Figure 1.17 Principle of Wardwellian (rapid) braiding.

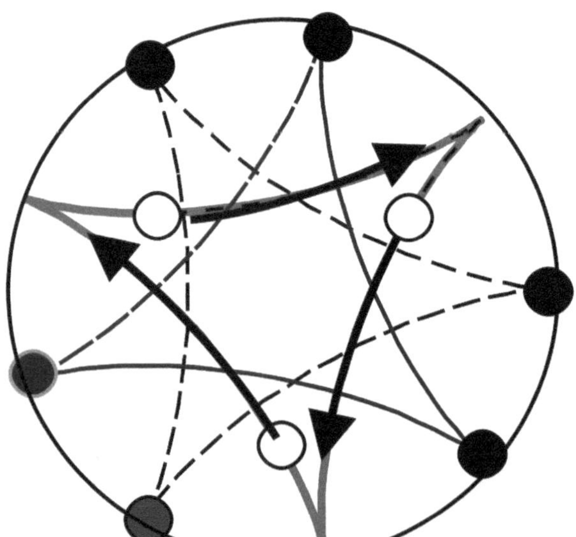

Figure 1.18 Carrier motion during spiral braiding.

different construction and a different track, so that a different type of interlacement between the yarns is performed.

Close observation of tubular braiding shows that each yarn goes clockwise or counter-clockwise (Fig. 1.18). This is true also for spiral braiding, but with one difference – during the interlacements, only one of the yarns changes its place (on the Fig. 1.18 at the current step, only the yarns/carriers with white circles). The other yarns (filled black circles) are waiting their turn. These waiting yarns build vertical yarn pieces in the braid and remain at 0° (parallel) to the product axis. Only the yarn pieces interlacing the parallel yarns are at an angle. This different orientation leads to the completely different mechanical properties of these products. More details about the spiral braiding can be found in Chapter 11.

1.4 Lace braiding

If the dancers in the maypole dance, or respectively the carriers, can stop at some place and are able to switch to another track, then significantly more interlacement types between the yarns would be possible. The designer in this case can decide at each step which two yarns to interlace (Figs 1.19–1.22). Braiding of this type is known as "pillow laces braiding," and is used for producing laces for underwear, ladies accessories, home textiles, etc.

Pillow lace braiding can be performed on a machine under industrial conditions (Krenzler, 1930; Bucksch, 2013). This technique is used today not only for fashion products but also in medical applications for elastic bandages, some safety applications, smart textiles for the integration of wire, etc. The basic principle of such a machine is demonstrated in Fig. 1.23.

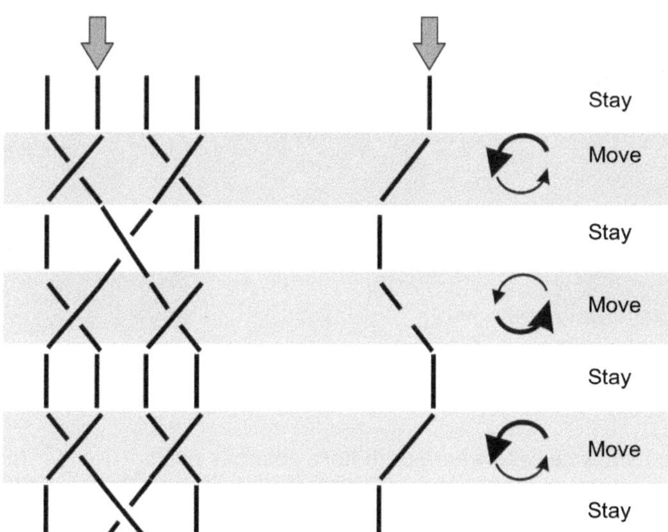

Figure 1.19 Interlacement of pillow laces. The marked yarn stays half the time, and it moves once to the left, once to the right, twisting with another yarn, which is not possible with maypole braiding.

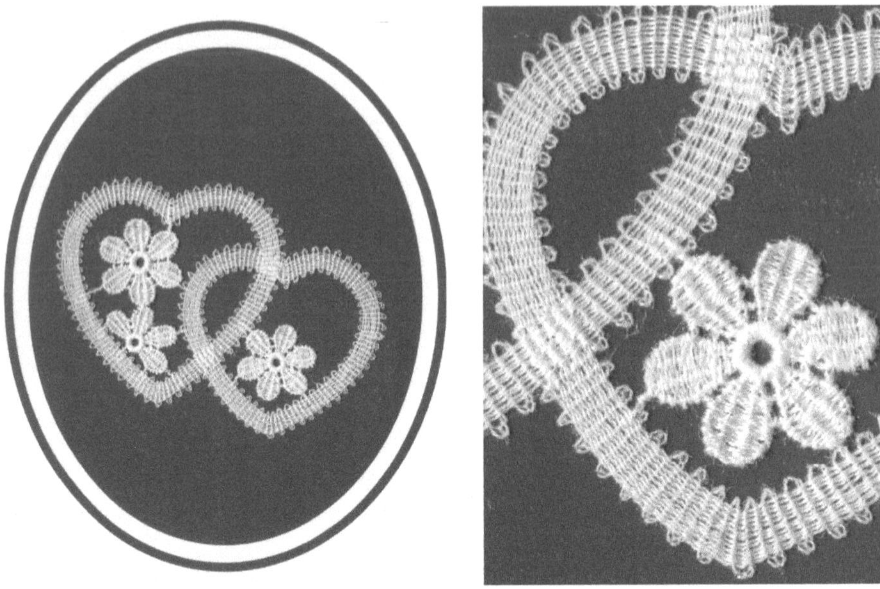

Figure 1.20 Handmade pillow laces made in the city of Bruges, Belgium. The pillow lace technique allows interlacements between the yarns from three types – braiding (not presented in the figure), weaving and twisting.

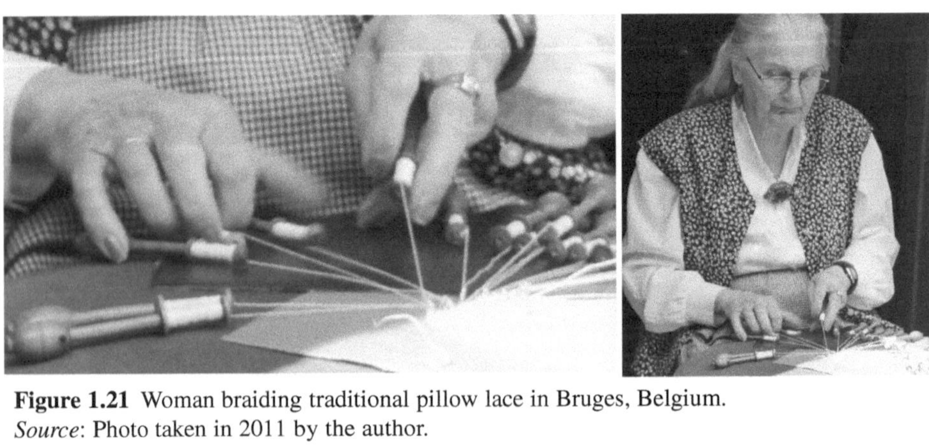

Figure 1.21 Woman braiding traditional pillow lace in Bruges, Belgium.
Source: Photo taken in 2011 by the author.

Figure 1.23 shows a case where each horn gear has two sections; if the horn gear rotates, it will take both the carriers A and B during the rotation. At the same time, the horn gear with the carrier C and D can (but should not) rotate independently. After the rotating horn gears have rotated through 180° they stop and their carriers can be taken and rotated from other horn gears, or again from the same horn gears. In the case of Fig. 1.23, the carriers A and D change their places during the next rotational step.

Figure 1.22 Wooden carriers.

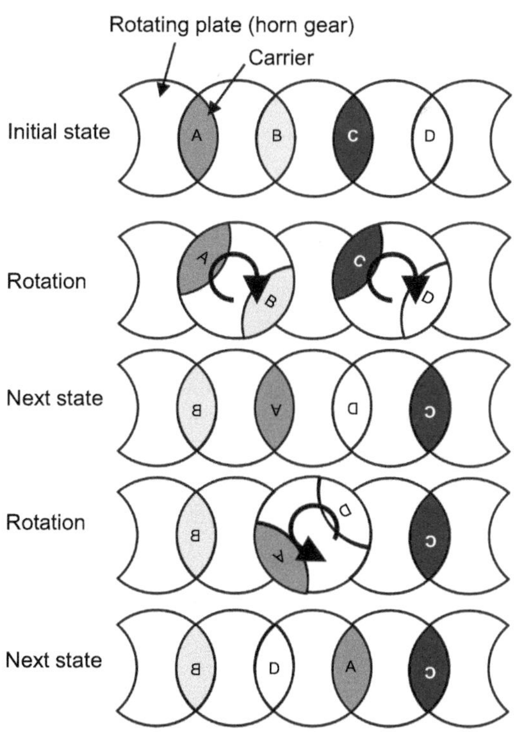

Figure 1.23 Principle of motion of the carriers in a pillow lace machine.

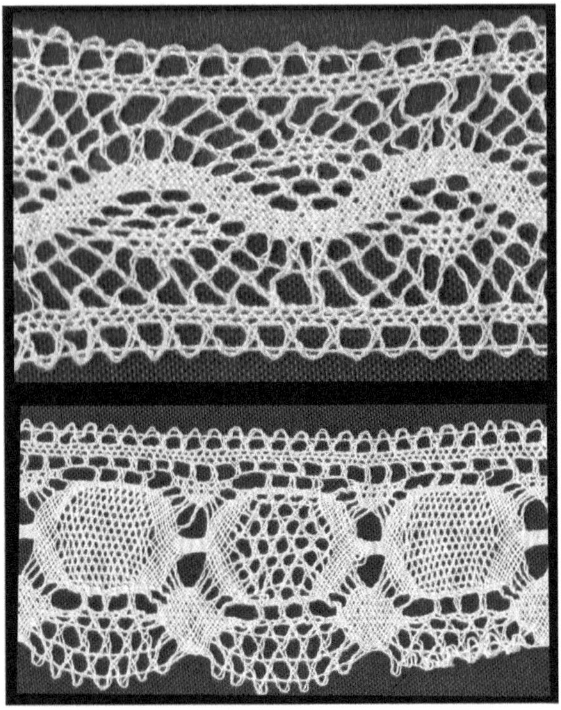

Figure 1.24 Machine-produced pillow laces.

Usually such machines have jacquard control of the horn gears, and each of these can receive the command to rotate or not, independently of the other horn gears. The rotation is possible only if both the neighbouring horn gears of the selected one are fixed during that time. Examples of machine-produced laces are presented in Fig. 1.24.

There are some extensions of this principle to a larger set of carriers, arranged in a rectangular (Jung, 1993) or hexagonal mesh with each of the horn gears having four or six slots, respectively (Mungalov and Bogdanovich, 2002; Schreiber *et al.*, 2009). Putting two carriers on the plate between the horn gears so that their plate can rotate as well extends the patterning possibilities of this technology.

The extended patterning possibilities have a drawback, however, in that each motion and stop causes acceleration, which leads to inertial forces, high loads on the machinery parts, and noise. Because of these dynamic issues, such machines cannot run fast, but their advantage is the flexibility and the variety of the yarn paths which can be realized.

1.5 Bobbinet machines

In all the previously mentioned cases, the bobbins (or, with machine braiding, their holders, the carriers) move along paths which consist of several (nearly) circular arcs.

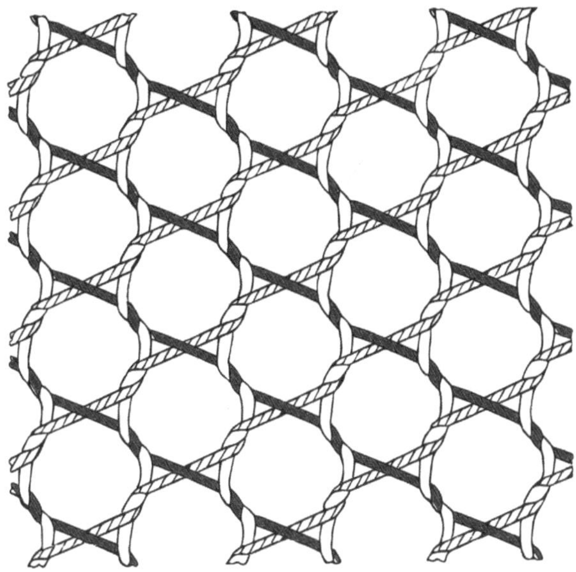

Figure 1.25 Bobbinet structure.
Source: Courtesy of David Eilbeck, G2 Innovation Group.

In some cases, yarn interlacement becomes more efficient if the bobbins are moving linearly. Such machines exist and were invented in 1807 by John Heathcoat to produce the finest bobbinet tulle (net lace). They are still working today in the company Swisstulle (Swisstulle, 2013). In this case, one system of yarns remains vertical (called warp yarns, the white yarns in Fig. 1.25) and one or more yarn systems interlace (called weft yarns). The bobbins are in small flat "carriers" or "shuttles," which swing between the warp yarns, and then each group moves as a packet to the left or to the right at each step (Fig. 1.26).

This technique allows for the production of very fine nets of large width.

1.6 Cartesian braiding

Moving carriers around a rectangular mesh, the so-called two-step and four-step braiding processes involve the placement of the carriers over a special table into rows and columns. Motion is performed by moving an entire row or column in a given sequence, usually alternately. This process is reported to be used mainly for composite profiles with a 3D cross-section (Zhang *et al.*, 2007). The machine for four-step braiding is shown in Fig. 1.27 (Florentine, 1982) One cycle of interlacing of the yarns with the four-step braiding machine is done in a four-fold sequence of motion, hence the name. In this machine, all the yarns are in carriers.

During two-step braiding, a set of many (heavy) inlay yarns is used, around which a lesser number of carriers are moving and interlacing (Fig. 1.28) (McConnel and Popper,

Figure 1.26 Bobbinet machine.
Source: Photo taken in 2013 from Swisstulle, Switzerland, by the author.

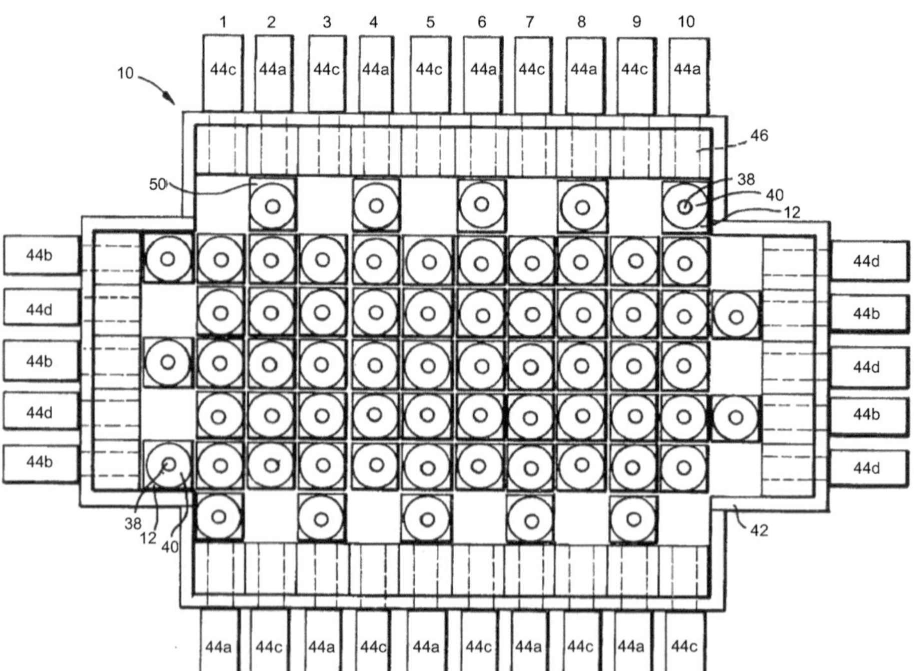

Figure 1.27 Machine for Cartesian four-step braiding (Florentine, 1982).

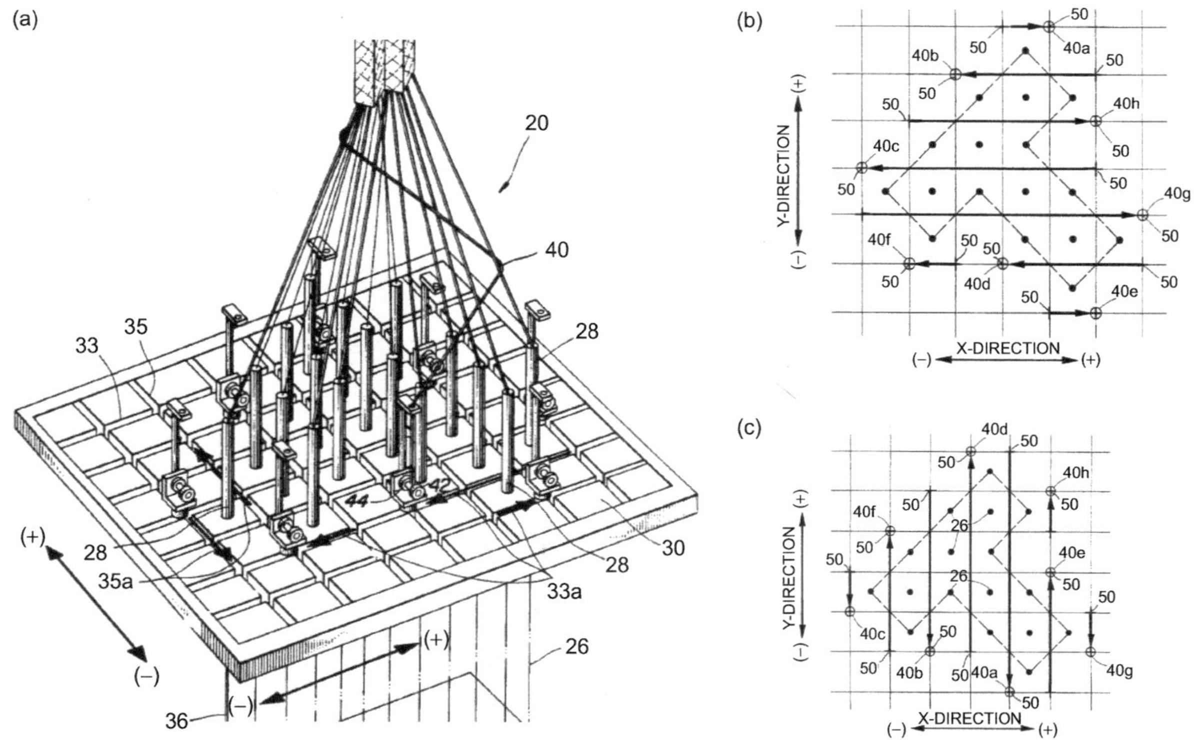

Figure 1.28 Machine for Cartesian two-step braiding (McConnel and Popper, 1988). (a) View of the machine; (b) carrier motion step in X-direction and (c) carrier motion in Y-direction.

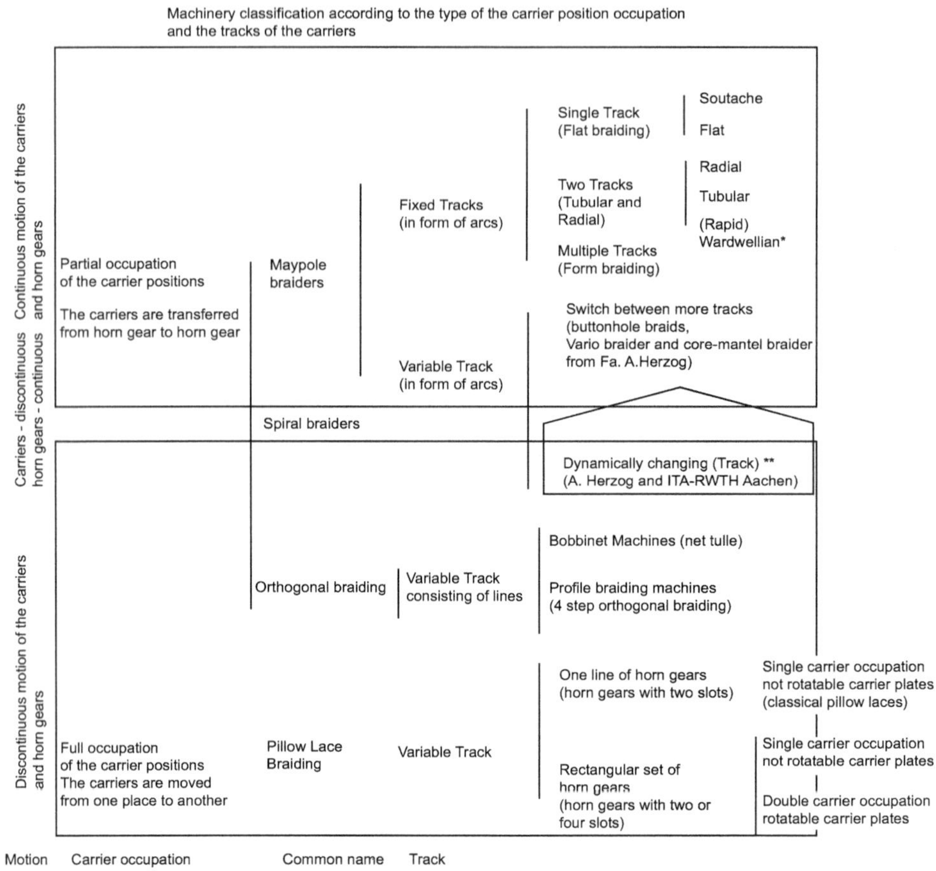

* Rapid or Wardwellian braiding do not use carrier transfer and horn gears.
** The machines with the dynamically changing track now (Year 2013) are still working on two steps (carrier motion, swich motion), but in case the electromechanical devices are able to switch fast enough, these machines will belong to the type of machines with continuous carrier and horn gear motion.

Figure 1.29 Machine classifications according to the motion type of the carriers and horn gears.

1988). The carriers are usually placed on the periphery of the profile and, during the motion, they cross completely and compact the structure (Fig. 1.28b and 1.28c). Some more examples of such prototypes of new machines can be found in (Tong *et al.*, 2002).

1.7 Machine and product classification

There are several different classifications of the braided processes, machines and products.

Figure 1.29 summarizes those introduced in this chapter, depending on the type of the motion of the horn gears and the carriers, and the track type.

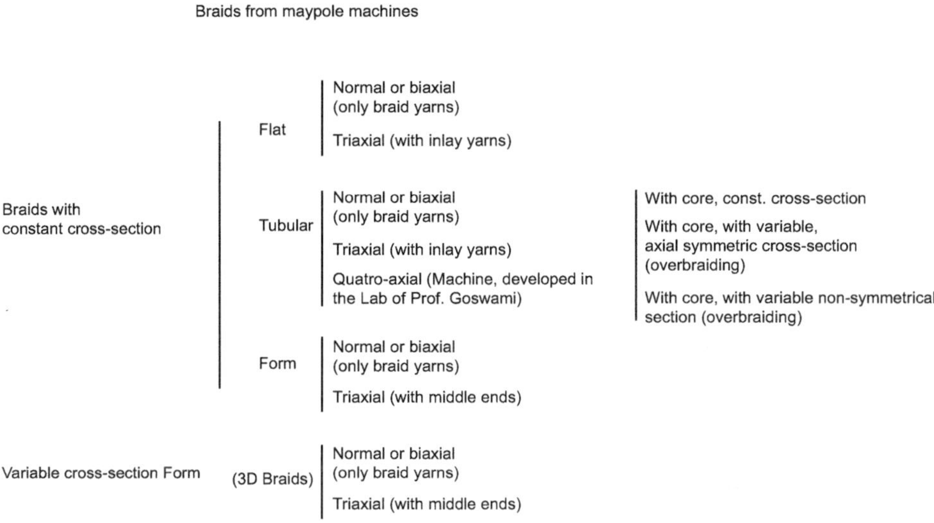

Figure 1.30 Braided products of maypole braiding machines.

Figure 1.30 shows the more common maypole braided product types. In the future, it is expected that with the increasing speed of electromechanical drives, maypole braiding machines with switches (3D braiding) will be able to work continuously. New machines are currently under development, for example the quatro-axial braiding machine developed by Like Bao and Professor Goswami (Bao and Goswami, 2008) at Clemson University, which is able to braid classical tubular braids and additionally insert a weft yarn at less than 90° to the product axis. There are additionally some product names being developed which, as well as being used for marketing, give further rights protection for certain terms and structures, for example UD-Braids (unidirectional braids), which present classical tubular braiding where the carriers in one direction work with very thin filaments, and other carriers with heavy tows from carbon or glass fibres. Topologically and structurally speaking, such structures are classical braids, but because of their special properties (non-crimp) they are introduced and used with their special names in order to differentiate them from common products. To investigate such special details is not the goal of this book, which aims purely to outline the technological basics for the braiding experts.

1.8 Sources of further information and advice

There are several introductory books and chapters available, where the basics of the main ideas and the working principles of braiding are described. Douglass (1964) provides a very good overview of the state of the machines from 1964, covering most aspects of modern braiding. Some aspects regarding the materials, and some basic ideas about the calculation of the braided products and methods for computing the properties of braided composites, can be found in the book (Ko *et al.*, 1989). A short introduction to maypole braiding and its application for composites can be found in (Cherif, 2011; Potluri, 2012) and other chapters. Good information about the development of two-step and four-step braiding from the orthogonal system is given in (Tong *et al.*, 2002).

A free online review of braiding machines and technology can be found in the *Journal of Engineered Fibers and Fabrics* (Branscomb *et al.*, 2013).

There is also an important range of German language braiding literature, dating back to 1914 with the first edition of Professor Leperhoff's book, which collected and presented the knowledge in this area for more than 40 years across three editions (Lepperhoff, 1953). More practically orientated is the book of Goseberg and team (Goseberg *et al.*, 1981). Professor Harald Engels's book in two parts is currently still the basic source for students (Engels, 1994a, b), giving a good and detailed explanation of the machines and methods for the graphical development of the braided structure. For teaching purposes, social events and just for fun, a very good book with colourful pictures and detailed instructions for hand braiding of 250 different samples is (Owen, 1995).

References

August Herzog Maschinenfabrik, Approx. (1940). Hochleistungs-garnflechter für die Kabelindustrie. *In: Oldenburg*, 21.

Bao, L. and Goswami, B.C. (2008). Braiding machine with weft insertion yarns under 90 degree: Video, *Presented During the 1st World Conference on 3D Fabrics and Their Applications,* Manchester.

Beard, L. and Beard, A.B. (1893). *How to Amuse Yourself and Others: The Amerian Girls Handy Book.* New York: Charles Scribner's Sons.

Branscomb, D., Beale, D. and Broughton, R. (2013). New directions in braiding. *Journal of Engineered Fibers and Fabrics* [online], **8** (2), 11–24. Available from: http://www.jeffjournal.org.

Bucksch, M. (2013). *Klöppelspitzenmaschinen – Produkt Katalog* [online]. Wuppertal, Krenzler Flecht- & Klöppelspitzenmaschinen GmbH & Co. KG. Available from: http://www.krenzler-gmbh.de/.

Büsgen, A. (1993). *Neue Verfahren zur Herstellung von dreidimensionalen Textilien für den Einsatz in Faserverbundwerkstoffen: Dissertation.*

Cherif, C. (2011). *Textile Werkstoffe für den Leichtbau: Techniken – Verfahren – Materialien – Eigenschaften.* Berlin: Springer.

Douglass, W.A. (1964). *Braiding and Braiding Machinery.* Eindhoven: N.V. Uitgevesmaatschappij Centrex.

Engels, H. (1994a). *Handbuch der Schmaltextilien. Die Flechttechnologie. Teil I: Maschinen und Verfahren zur Erzeugung konventioneller Geflechte.* Mönchengladbach.

Engels, H. (1994b). *Handbuch der Schmaltextilien. Die Flechttechnologie. Teil II: Maschinen und Verfahren zur Erzeugung von Flechtprodukten mit speziellen physikalischen und chemischen Anforderungen.* Mönchengladbach.

Florentine, R.A. (1982). *Apparatus for Weaving a Three Dimensional Article.* USA Patent US 4312261, 01.26.1982.

Goseberg, F., Brabender, K. and Moeller, P. (1981). *Textiltechnik Maschinengeflechte: Ausbildungsmittel Unterrichtshilfen.* Frankfurt am Mai: Arbeitgeberkreis Gesamttextil.

Hearle, J. and Chen, X. (2013). *TexEng Software LTD* [online]. Available from: http://www.texeng.co.uk/ (Accessed 2013).

Jung, J. (1993). *Flechtmaschine für die Erstellung von dreidimensionalen Geflechten.*

Ko, F.K., Pastore, C.M. and Head, A.A. (1989). *Handbook of Industrial Braiding*, Atkins & Peirce.

Krenzler, E. (1930). *Lace Braiding Machine.* USA Patent US01776361, application 1929, issued 23.09.1930

Lepperhoff, B. (1953). *Die Flechterei.* 3rd edn. Eugen G. Leuze Verlag Saulgau/WTTBG.

Lomov, S.V. (2012). *Wisetex* [online]. Available from: http://www.mtm.kuleuven.be/Onderzoek/Composites/software/wisetex.

McConnel, R. and Popper, P. (1988). *Complex Shaped Braided Structures* Patent Nr. 4719837.

Mungalov, D. and Bogdanovich, A. E. (2002). USA US Patent 6 439 096.

Owen, R. (1995). *The Big Book of Sling and Rope Braids: Patterns for Over 250 Braids.* London: Cassell.

Potluri, P. (2012). Braiding. *In:* L. Nicolais and S.M. Lee, eds. *Wiley Encyclopedia Of Composites.* 2nd ed. Hoboken, NY: Wiley, 138–152.

Schneider, M. (2000). *Konstruktion von dreidimensional geflochtenen Verstärkungstextilien für Faserverbundwerkstoffe.* Aachen: Shaker Verlag.

Schreiber, F., Ko, F.K. *et al.* (2009). Novel three-dimensional braiding approach and its products. *In: ICCM-17: 17th International Conference on Composite Materials.* London: IOM Communications.

Swisstulle (2013). *History of Bobbinet* [online]. Available from: http://www.swisstulle.co.uk/ (Accessed 2013).

Tong, L., Mouritz, A.P. and Bannister, M.K. (2002). *3D Fibre Reinforced Polymer Composites.* 1st edn. Boston: Elsevier.

Wardwell (2013)[online], Wardwell Braiding & Spirka Schnellflechter GmbH. Available from: http://www.wardwell.com/history.asp (Accessed 2013).

Zhang, W., Ding, X. and Li, Y. (2007). Microstructure of 3D braided preform for composites with complex rectangular cross-section. *Journal of Composite Materials* [online], **41** (25), 2975–2983. Available from: http://jcm.sagepub.com/cgi/content/abstract/41/25/2975.

Part One

Patterning and design of braided structures manufactured using maypole braiding technology

Patterning of braided products

2

2.1 Introduction

Industry requires a broad understanding of braiding process, as discussed in (Lepperhoff, 1953; Goseberg *et al.*, 1981; Engels, 1994). This chapter explains the basics of the most important relations between carrier arrangement, horn gear slots and the braiding structure and gives some background to the methods used in Chapters 3 and 4.

2.2 Horn gears

The arrangement of a braiding machine is usually represented by a drawing of the horn gears and the carriers, and the former may be drawn in a number of different ways (Fig. 2.1). In books it is often represented as a circle, with the slots for the carriers marked with radial lines. In Fig. 2.1a the carriers are drawn on the radius, in Fig. 2.1b they are inside the circular segment, and in Fig. 2.1c inside the segment but almost touching the radius. The notation used in Fig. 2.1c has a particular advantage as, in this case for example, the direction of the motion of the carrier is already marked, since the line (which presents the radius) is pulling the carrier. In this book variant 2.1d will be used, because it represents the horn gear form most realistically. A simple Matlab function for drawing horn gears with different numbers of slots is presented in Chapter 15, Section 15.2.7.

In order to simplify the images of the different machine configurations, the arrows with the rotational direction of the horn gears can be replaced by a mark in the middle of the horn gear, as presented in Fig. 2.2. If the circle has a dot in the middle, the vector is coming towards the reader, who therefore sees only the peak of the arrow. If the vector is going away from the reader into the paper plane, then the reader sees the back end of the arrow, represented by a cross sign. The direction of the rotation of the horn gear is then defined by the right-hand rule (Fig. 2.2): when the reader wraps the right hand around the arrow so that the thumb points in the arrow direction, then the fingers will show the direction of the rotation of the horn gear.

Braiding Technology for Textiles. http://dx.doi.org/10.1533/9780857099211.1.29

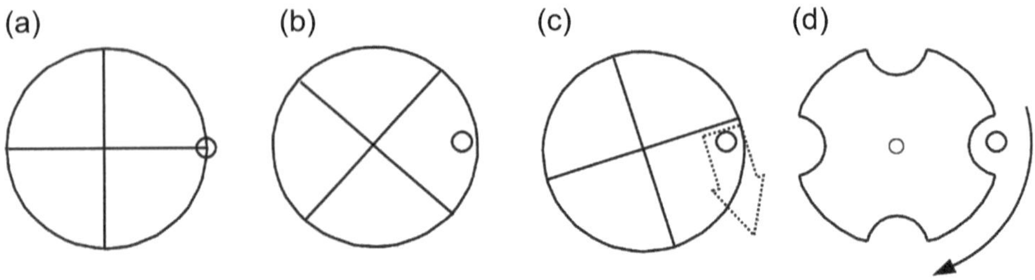

Figure 2.1 Different ways of drawing a horn gear and the carrier position in the documentation: (a) carrier placed on the edges of the arc segments; (b) carrier placed in the middle of the segment; (c) carrier placed on one side of the radius, as if the radius line pushes the carrier; and (d) horn gear drawn similar to the real part.

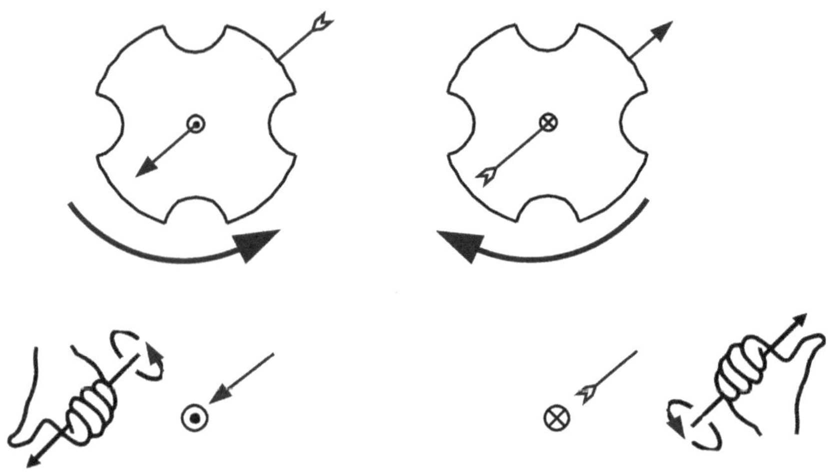

Figure 2.2 Centre marks for notation of the rotation direction of the carriers.

2.3 Carrier and bobbin arrangement

2.3.1 Carrier arrangement

The horn gears move the carriers along their track(s), and Fig. 2.3 shows a segment of a braiding machine with three horn gears and two tracks, one of which (track one) is marked with a thick line and the second with a dashed line. The arrangement of the carriers along the track is determined from the number of full positions (with carriers) and empty positions (horn gear slots without a carrier). Every slot of each horn gear can receive a carrier but with the maypole braiding machines, only *half* the positions can be filled with carriers, the other half being empty in order to be able to receive the carriers from the neighbouring horn gears.

For every track, the arrangement is notated; for example, the arrangement in Fig. 2.3 is "1 Full – 1 Empty," written short 1F 1E:

1F 1E = 1 Full – 1 Empty

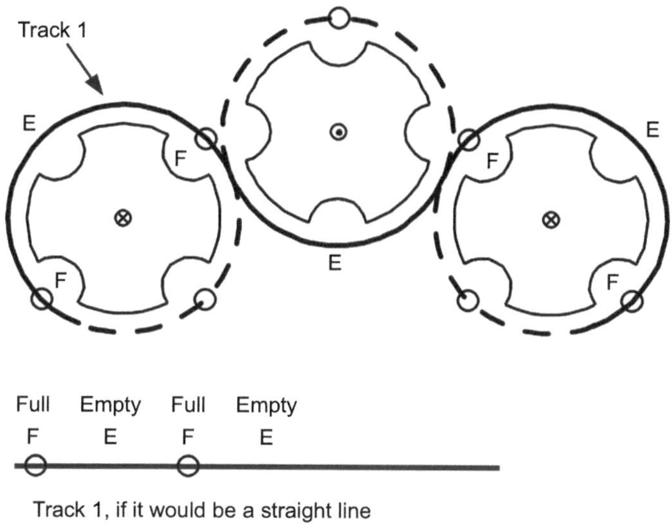

Figure 2.3 Track notation.

Flat braiding machines usually have one track, and tubular braiding machines usually have two, where the carriers have the same arrangement. In both cases (flat and tubular braiding machines) it is sufficient to notate the arrangement only once. If some of the tracks have a different carrier arrangement, then the data for all of the tracks is required and have to be listed separately.

Where the horn gears are in the position of transferring a carrier from one to another, as in Fig. 2.4, difficulty can arise in determining to which track the carrier belongs.

In fact, in Fig. 2.4 the carrier belongs to track 2. Identification of the carrier affiliations on the machine is easy, as they can be identified after small rotations of the horn gears in the working direction. In the case of Fig. 2.4, track 1 has to have an "Empty" position, because there is only room in the transition area for one carrier. If both of the tracks have a carrier, the machine will not be able to work, as the two carriers will collide shortly before the crossing point of the tracks.

Figures 2.5 and 2.6 demonstrate the track and carrier arrangements on a flat and a tubular braiding machine, respectively. In both cases a 1F 1E arrangement is selected.

Normally, the information about the arrangement (in this case 1F 1E) is enough to specify the carrier arrangement.

2.3.2 Repeat in the carrier arrangement

"Repeat" is the shortest repeatable sequence in one large set of objects. For the carriers for a specific track, one repeat consists of a set of "Full" and "Empty" places. The sum of the numbers of "Full" and "Empty" places in one repeat gives the length

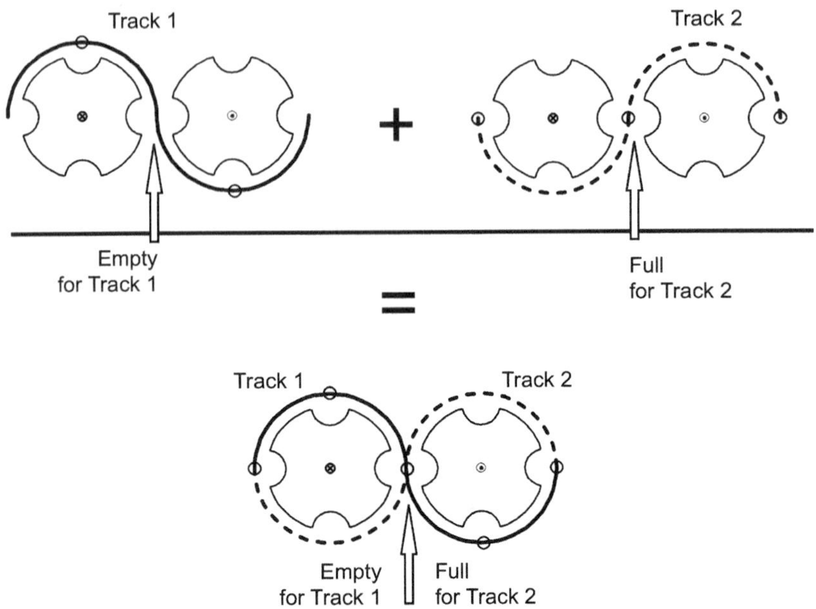

Figure 2.4 Common slot of two tracks.

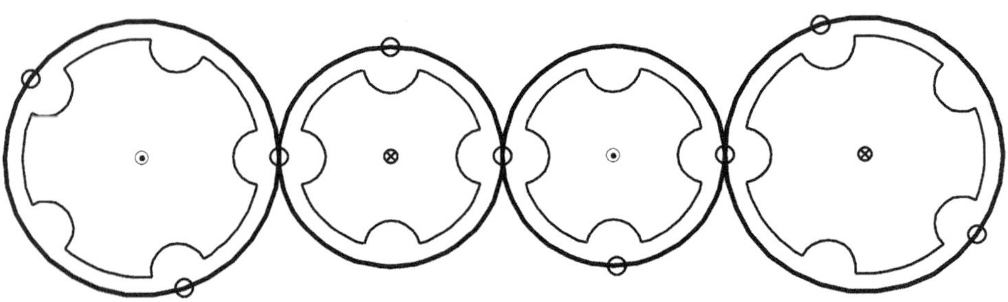

Figure 2.5 Track and carriers on a flat braiding machine with arrangement 1 Full 1 Empty (1F 1E).

of the repeat. The most commonly used carrier arrangements and the corresponding repeats are given in Table 2.1

The repeat length also defines the required number of horn gear slots which have to be available and the minimum number of slots for one horn gear. For instance, for horn gears with four slots, only the structures with repeat two and four can be produced.

2.3.3 Colour arrangement

For the production of coloured braids for shoe laces and decorative applications, the colour arrangement of the bobbins has to be specified as well as the carrier arrangement.

Table 2.1 Common used carrier arrangements and the corresponding repeats

Carrier arrangement	Repeat length
1 Full: 1 Empty	1 + 1 = 2
1 Full: 2 Empty	1 + 2 = 3
1 Full: 3 Empty	1 + 3 = 4
2 Full: 2 Empty	2 + 2 = 4
3 Full: 3 Empty	3 + 3 = 6

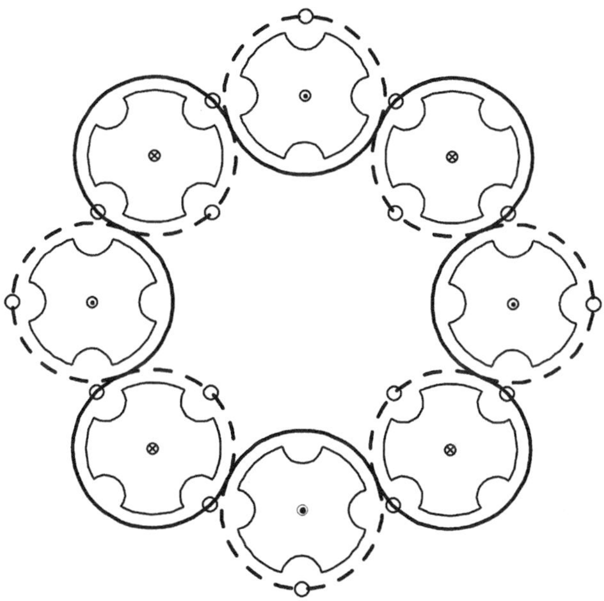

Figure 2.6 Tracks and carriers on a tubular braiding machine with arrangement 1 Full 1 Empty (1F 1E).

The colour arrangement represents the list of the colours of the yarns on the carriers in one track.

For instance, the colour arrangement of the machine shown in Fig. 2.7 can be:

3 black
3 grey
3 white

which means that if somebody follows the carrier in the track, the first three carriers will have black yarn, the next three grey yarn, and the last three white yarn (Fig. 2.7).

For tubular braids, the colour arrangement of both tracks has to be specified (if different), and for the form braiding machines, all tracks must be colour specified.

Figure 2.7 Carriers and empty positions at arrangement 1F 1E, where the carriers are with colour arrangement of the bobbins three black, three grey, three white.

2.4 Structural and pattern representation of maypole braids

The pattern design of the braided structures requires two types of drawing – one for the identification of the structure (interlacing type) and one for the colour design. Figure 2.8 presents the principle of these drawings.

The structural drawing presents in a simplified way the orientation of the visible yarn pieces. They are drawn on a paper prepared with squares, where each square corresponds to one interlacement between two yarns. For a given square cell, only small pieces of the yarn are shown. The orientation of the diagonal "S" – from the left upper corner to the bottom right corner "\" corresponds to a yarn piece, which was moved from the carrier during the motion from left to right, being on the outer side of the track and vice versa.

Each row of the sheet corresponds to the building of one row of interlacing points between the yarns which happens during the rotation of the horn gears, corresponding to the central angle of one slot.

The structural drawing can be used, for instance, to analyse the type of interlacement of the yarns if the machine configuration and the carrier arrangement are also given. It is not suitable for colour design of the braids, as it is too sparse.

For the colour design, again a square mesh is used but rotated 45° so that the edges of the square cells are parallel to the yarn direction. This presentation is absolutely identical to the pattern design of woven structures, the only difference being that the picture is rotated (Kyosev, 2012). Non-orthogonal meshes can also be used to represent a more realistic structure where the braiding angle is other than 45°. This requires the use of either the prepared paper with the different angles or the software presented in Chapter 13 (Computer assisted design (CAD) software for the design of braided structures).

With braiding machines for flat, tubular and form braids, the horn gears do not change their rotation direction. For a single horn gear, all the carriers from one track are moving in the same way if carried from this horn gear and their yarns build interlacements in the same direction. As a result, the so-called "ridges" are built along the product axes. Each ridge consists of parallel yarn pieces of the same length.

The next horn gear along is rotating in the opposite direction, so the neighbouring ridge comprises yarn pieces which are crossing the yarn pieces of the first one (Fig. 2.9). Therefore, the number of ridges is equal to the number of horn gears for one tubular or flat braided structure. For tubular braids, only the front half is visible and, in most cases, only this part is visualized.

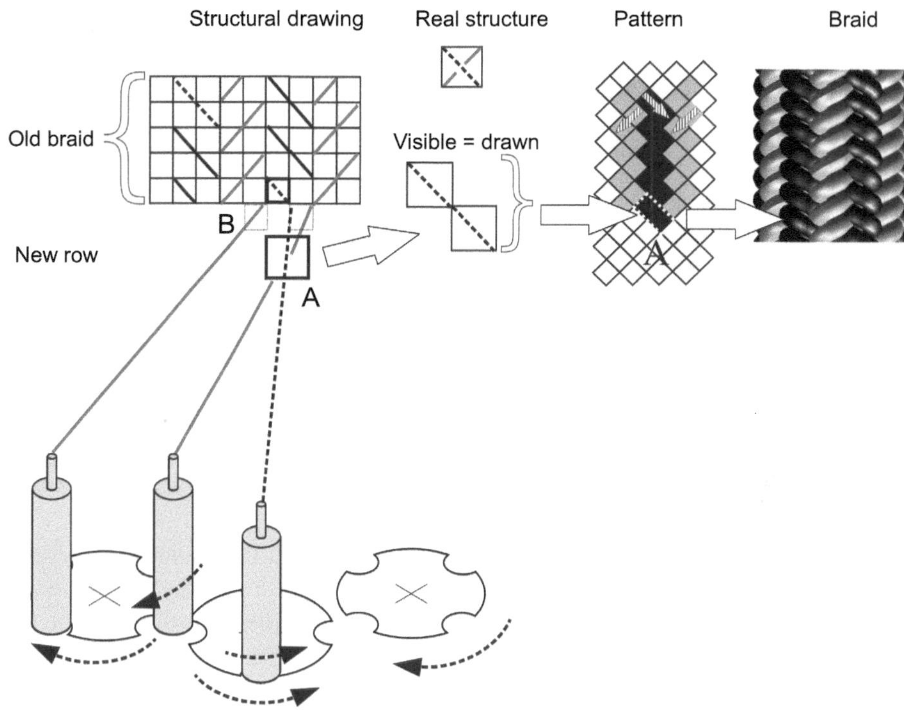

Figure 2.8 Basic principle for the development of the structural drawing and the colour pattern drawing of braided structures.

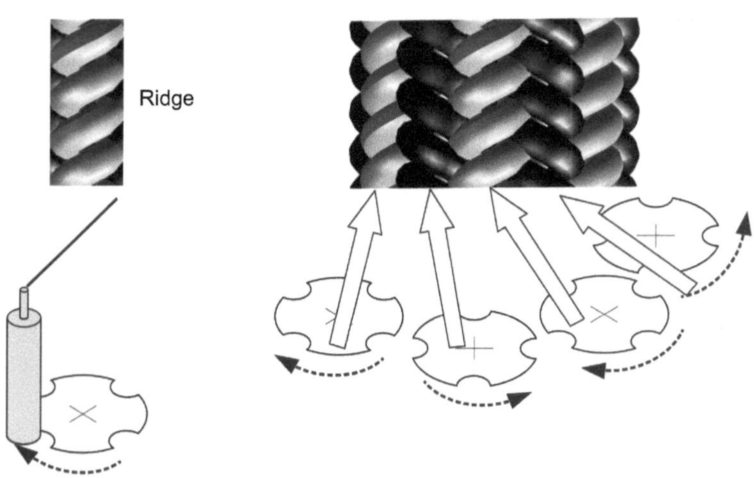

Figure 2.9 Visualization of the principle – each ridge corresponds to one horn gear. On the right-hand side of the picture, only some of the horn gears are presented.

2.5 Braiding pattern basics

2.5.1 Notation systems

There exists no worldwide unified system for the notation of braids. Probably the most common are the English and Germans ones.

The "English" system consists of a few specified names for the popular structures – regular, basked and diamond. Some analogies to weaving are used as well for the plain and twill structures. One more clear system from the engineering point of view describes structures in a form following "one over and under one," "one over and under two," etc. (Ko *et al.*, 1989)

The "German" system is very well developed and based on two numbers – the first describing the intersection type and the second the number of yarns per group. In this system the words "Flechtigkeit" and "1-, 2- and 3-flechtige" structures are used, which describe how many yarns build the braid. The word "flechtig" means *floating length* for the given interlacement. Using the floating length simplifies the analysis and design of the machines and the braids.

The system "one over one and under one" has one main disadvantage: in several cases, such as "two over two and under two" (Fig. 2.10c), it can lead to misunderstanding and does not conform to the analytical methods for determination of the structure type. "Two over two and under two" actually designates a structure with floating length *one* for a group of two yarns (Fig. 2.11), but the floating length "one" is never mentioned in the structure name.

The most common braiding patterns and their commonly used names are presented in Fig. 2.10.

A structure where each strand passes alternately over and under one of the other strands (Fig. 2.10a) is called a diamond braid or basked stitch (Douglass, 1964; Ko *et al.*, 1989). The floating length here is 1 (every strand is going over and then under one other strand). A braided structure where one strand passes alternately over and under two other strands is called a "regular braid" and has a floating length of 2 (Fig. 2.10b).

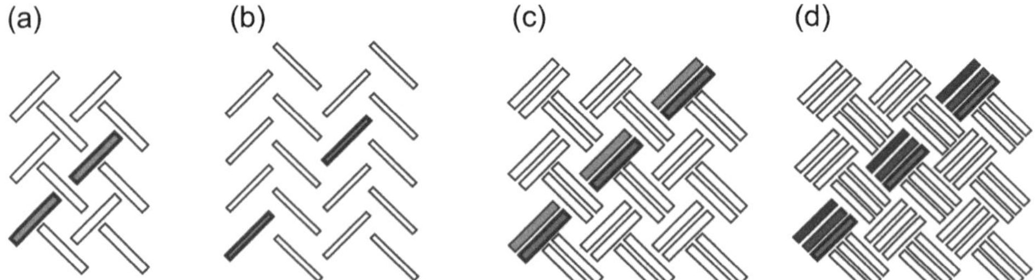

Figure 2.10 Common pattern types and their popular names. (a) Plain, diamond, basked braid, 1:1 Stitch, one over and under one; (b) regular braid, one over two, twill; (c) two over two; (d) Hercules Braid, three over three.

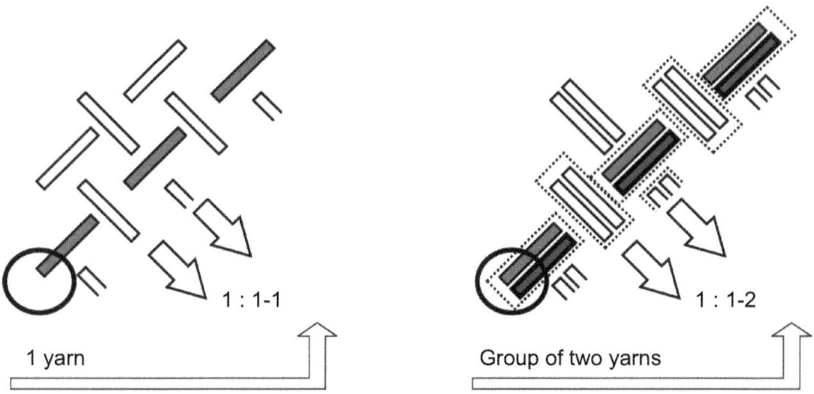

Figure 2.11 Examples for the pattern notation rule for 1:1-1 and 1:1-2 structures.

For these two cases, only one yarn is building interlacements with the other yarns.

For cases where two or more yarns interlace as **a group** over another group of yarns, the number of yarns in the group must also be specified. The structure in Fig. 2.10c, for example, has a floating length of one group, where the group consists of two yarns. Figure 2.10d also has a floating length of one group, where one group in this case consists of three yarns.

These two numbers (the floating length and the number of yarns in one group) completely specify the braiding type, but in order that these two numbers are not confused, analogical to the weave notation (DIN 61 101) will be used to denote them (Fig. 2.11):

- the interlacement type is defined using the **floating length** of one yarn or of a group of yarns by "over X under X," where the two numbers are separated with a colon – X:X
- the number of yarns in a group (strand), separated by a hyphen "-" has to be given. These yarns can be wound on a single bobbin as a ply yarn, or the machine can be prepared so that each one is carried by a separated carrier, which does not influence the pattern type.

In this way, the structures in Fig. 2.10 are denoted as follows:

(a) 1:1-1; (b) 2:2-1; (c)1:1-2; (d) 1:1-3.

This new system conforms to all earlier systems, since it uses the interlacing notation of the English system (over: under), and also fully represents the German system.

The pattern notation rule is:

X: X – Y

Over X groups: Under X groups – As a group of Y yarns

2.5.2 Basic pattern

The following figures present the basic pattern in the braiding structures denoted according to the rule defined above. Figures 2.12 and 2.13 demonstrate braids with floating lengths from one to six, all with one yarn per group. Structures with larger

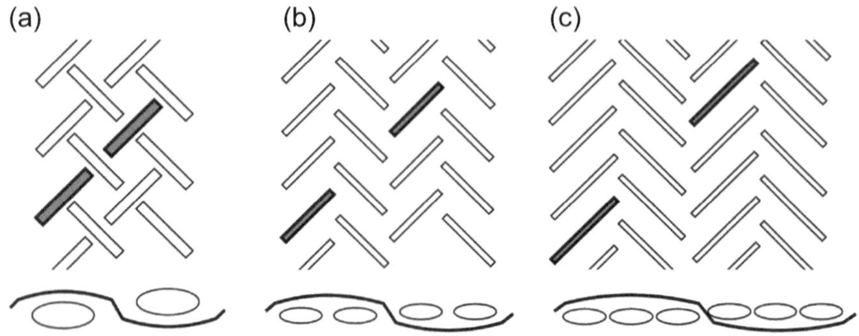

Figure 2.12 Basic pattern with one yarn per group: (a) with floating length 1, structure 1:1-1; (b) with floating length 2, structure 2:2-1; (c) with floating length 3, structure 3:3-1.

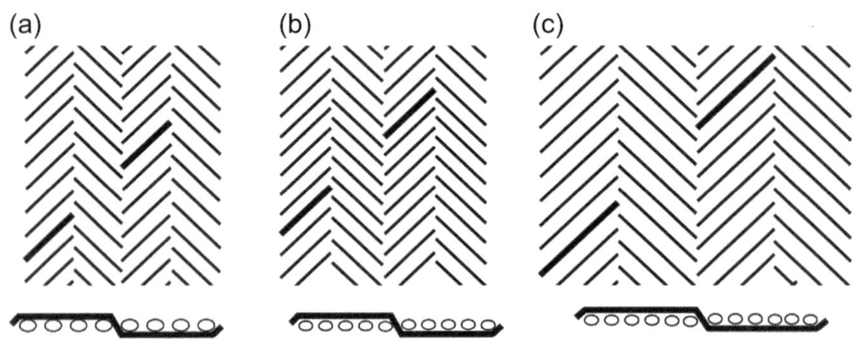

Figure 2.13 Pattern with higher floating lengths: (a) 4:4-1; (b) 5:5-1; (c) 6:6-1.

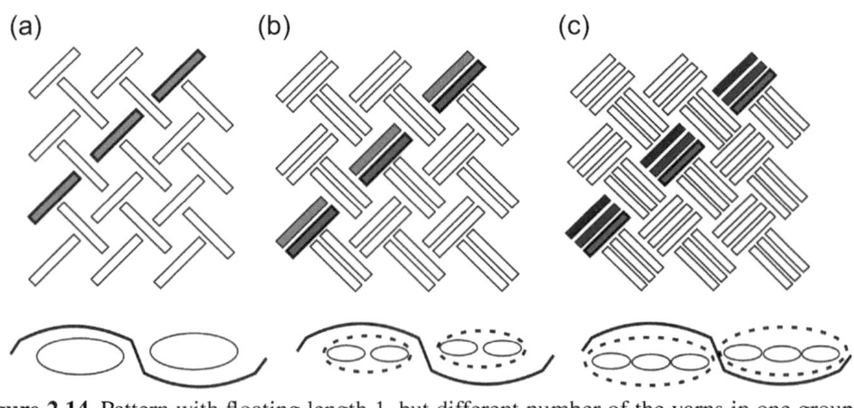

Figure 2.14 Pattern with floating length 1, but different number of the yarns in one group: (a) 1:1-1 as basic structure; (b) 1:1-2; (c) 1:1-3.

floating lengths are not popular, but are used in some special cases of technical, fashion or other types of products.

Figure 2.14 shows the 1:1-1 pattern, which uses two or three yarns in a group, each following the same path and having no intersection points.

2.6 Pattern type and carrier arrangement

2.6.1 Floating length and carrier arrangement

Braiding types such as regular (2:2-1), basked (1:1-2), etc. are valid for the entire braided structure if all horn gears on the machine have the same numbers of slots.

For the production of fancy braids, a combination of horn gears with different numbers of slots can be used, and in this case the braiding type is defined **per ridge**. The integer part of the ratio of the number of slots per horn gear and the repeat of the carrier arrangement determines the floating length in the structure (Lepperhoff, 1953; Goseberg *et al.*, 1981):

$$\text{Floating\_length} = \left[\frac{\text{Sections\_per\_horn\_gear}}{\text{Repeat\_in\_the\_arrangement}} \right] \qquad [2.1]$$

Table 2.2 gives the floating lengths for the most common cases of carrier arrangements and horn gear sizes.

2.6.2 Common carrier arrangements

Figure 2.15 demonstrates braiding of a "regular braid," with a floating length of

4 slots/(1F+ 1E) = 4/2 = 2.

Table 2.2 Floating length depending on the horn gear and carrier arrangement

Arrangement	1F:1E	1F:2E	1F:3E	2F:2E	3F:3E
Repeat	2	3	4	4	6
Slots per horn gear					
2	1	–	–	–	–
3	1	1	–	–	–
4	2	1	1	1	–
5	2	1	1	1	–
6	3	2	1	1	1
7	3	2	1	1	1
8	4	2	2	2	1
9	4	3	2	2	1
10	5	3	2	2	1
11	5	3	2	2	1
12	6	4	3	3	2
13	6	4	3	3	2
14	7	4	3	3	2

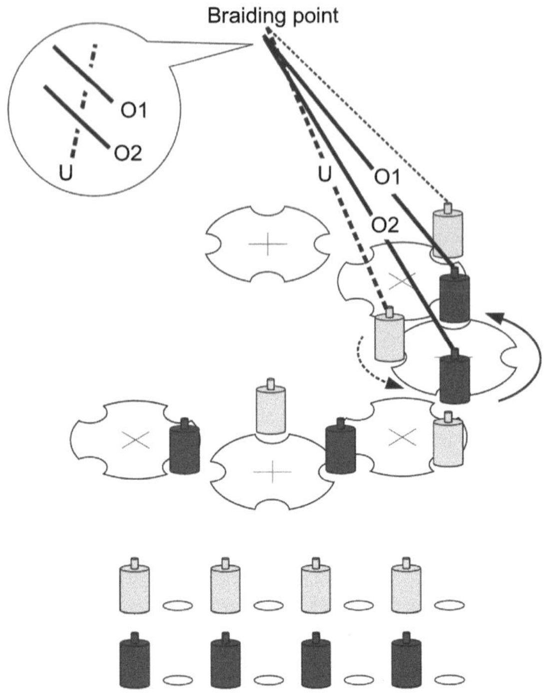

Figure 2.15 Carriers, arranged "1 Full 1 Empty" on a braiding machine with horn gears with four slots. Since the yarn U (moved from its carrier) is going under two and then over two other yarns, the floating length is 2. (Structure type 2:2-1, "regular braid").

The yarn denoted by a dashed line and marked with "U" is currently moving under the yarns O1 and O2. During the horn gear motion, the carrier places yarn U under the yarns O1 and O2. After that, yarn U will go over the yarns of the next two carriers.

During the motion of the carrier U around the horn gear, two carriers will move on the other side of the horn gear. Yarn U will build a float under these two yarns. After that, the same yarn will float over two other yarns. This pattern is of type 2:2-1, and is very commonly used.

The other commonly used braid 1:1-2, known as the diamond braid type, has two carriers moving, one after another, with the arrangement 2F: 2E. The yarns will cross the yarns of two other carriers moving around the same horn gear (Fig. 2.16). The two yarns U1 and U2 are carried on separate carriers in the figure, but it is also possible that both yarns are wound as a one-ply yarn on one bobbin, leaving the other position empty. Because both of these yarns move as one group and have the same orientation, they are treated as **one group** of two yarns, which floats over and under another group of yarns. The floating length here is

4 slots/(2F+ 2E) = 4/4 = 1.

As there are two yarns in a group the pattern is notated 1:1-2.

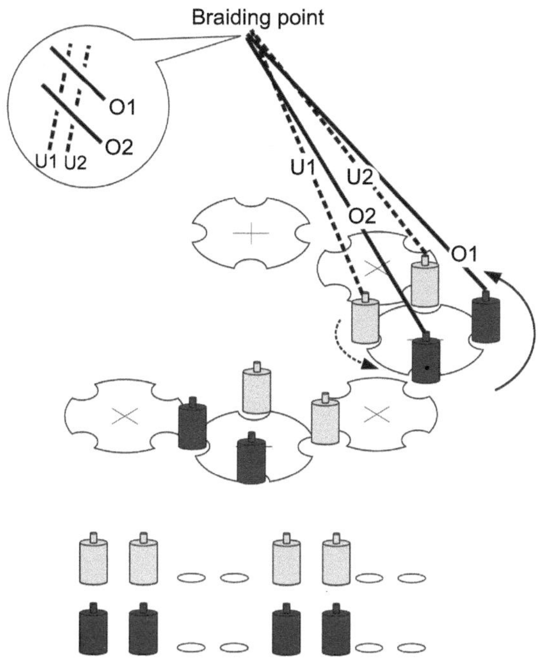

Figure 2.16 Carriers, arranged "2 Full 2 Empty" on a tubular braiding machine with horn gears with four slots, for production of 1:1-2.

If in Fig. 2.16 every second carrier were removed, a 1 Full 3 Empty arrangement would result (see Fig. 2.17). In that case a 1:1-1 structure would be produced and the floating length would be

4 slots/(1F+ 3E) = 4/4 = 1.

This arrangement of the carriers is termed "half occupation," and the structure can be produced by horn gears with two slots (See Table 2.2). However, in reality it is not practical to build such machines, as they would have no flexibility and only braids 1:1-1 could be produced. Using horn gears with four slots, on the other hand, allows for the production of three different types of braid (2:2-1, 1:1-2 and 1:1-1).

Machines with horn gears with higher numbers of slots allow for the production of patterns with larger floats, especially advantageous for glass and carbon fibre preforms for composites. The larger floats in such structures lead to lower crimp of the yarns. Table 2.3 demonstrates the configurations for machines with horn gears having six slots.

With horn gears having six slots, however, a repeat pattern 2F 2E = 4 is not possible, because it will lead to a collision with the carriers in the next repeat.

Table 2.3 shows that a machine with six-slot horn gears can replace one with four slots and allow for two additional patterns; 3:3-1 and 1:1-3. The only disadvantage

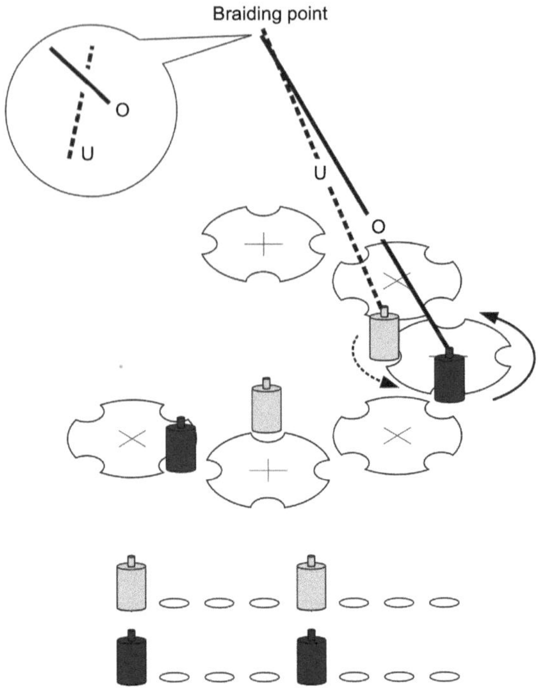

Figure 2.17 Carriers, arranged "1 Full 3 Empty" on a tubular braiding machine with horn gears with four slots ("half occupation," 1:1-1 braid).

Table 2.3 **Main carrier arrangements for horn gear with six slots**

Arrangement	Repeat	Floating length	Structure	Occupation
1F 1E	2	6/2 = 3	3:3-1	Full
1F 2E	3	6/3 = 2	2:2-1	Partial
1F 5E	6	6/6 = 1	1:1-1	Partial
2F 4E	6	6/6 = 1	1:1-2	Partial
3F 3E	6	6/6 = 1	1:1-3	Full

of having a six-slot horn gear is the requirement for a larger machine size, which has both economic and technological implications.

The case of a 1 Full 1 Empty carrier arrangement is demonstrated in Fig. 2.18, with the yarns of three carriers crossing one yarn from the opposite system before changing their position to build a 3:3-1 structure.

The common configurations for machines with eight-slot horn gears are given in Table 2.4. With such a machine, as well as the patterns from a four-slot machine, the following additional patterns are possible; 4:4-1, 2:2- 2, 1:1-3 and 1:1-4.

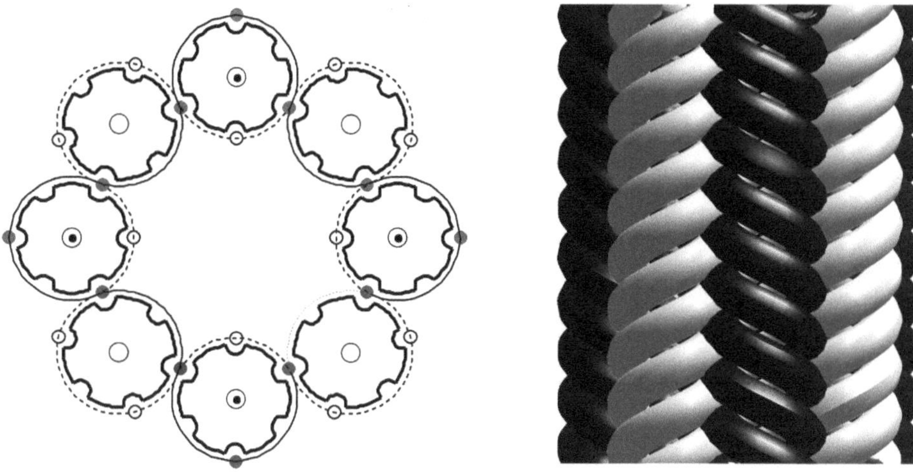

Figure 2.18 Carrier arrangement "1 Full 1 Empty" on a tubular braiding machine with six-slot horn gears and corresponding tubular braid with pattern type 3:3-1 (Kyosev, 2011).

Table 2.4 Possible regular carrier arrangements for eight-slot horn gear machines

Arrangement	Repeat	Floating length	Structure	Occupation
1F 1E	2	8/2 = 4	4:4-1	Full
1F 3E	4	8/4 = 2	2:2-1	Partial
2F 2E	4	8/4 = 2	2:2-2	Full
1F 7E	8	8/8 = 1	1:1-1	Partial
2F 6E	8	8/8 = 1	1:1-2	Partial
3F 5E	8	8/8 = 1	1:1-3	Partial
4F 4E	8	8/4 = 1	1:1-4	Full

2.6.3 Arrangements for flat (fancy) braids

Figure 2.19 shows two 13-slot horn gears with a 1 Full 1 Empty carrier arrangement. As with all flat braiding machines, this machine has one track.

The current floating length for this machine is

$$13 \text{ slots per horn gear}/(1F + 1E) = 13/2 = 6{,}5 = 6,$$

with a building pattern 6:6-1. Each yarn floats over six other yarns, then changes direction and floats under six yarns.

Fancy braids are produced on machines with different horn gears, so that often different floating lengths and interlacings per ridge appear. In this case, the above rule applies per horn gear, i.e. per ridge. As an example, Fig. 2.20 shows a machine with an arrangement of five different areas – two side areas where the carriers are

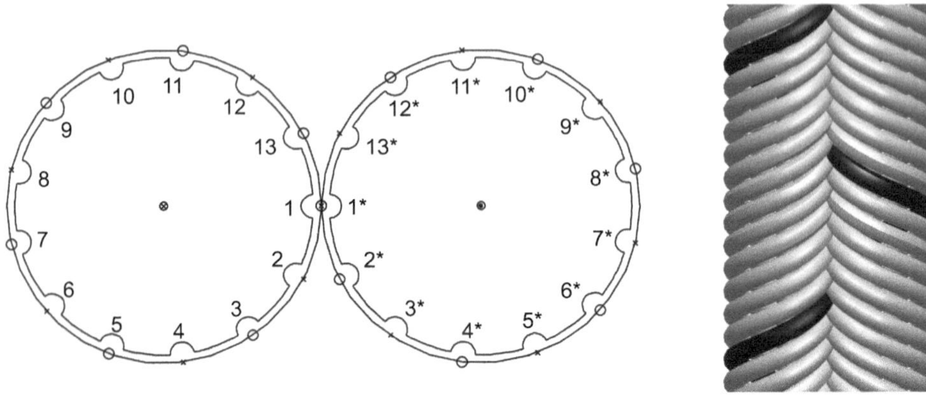

Figure 2.19 Carrier arrangement "1 Full 1 Empty" on a flat braiding machine with 13-slot horn gears and corresponding braided structure (Kyosev, 2011).

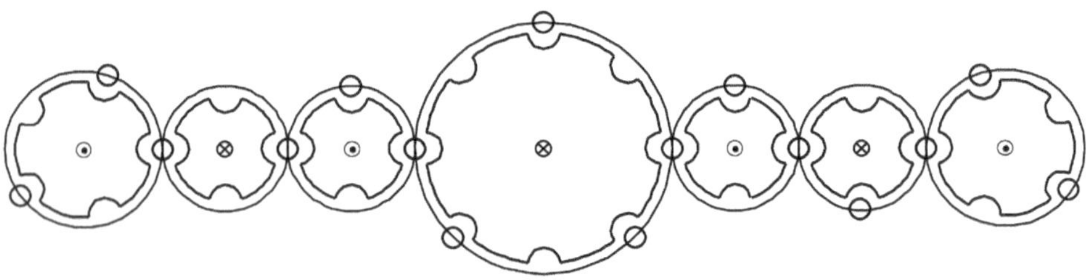

Figure 2.20 Carrier arrangement "1 Full 1 Empty" on a flat braiding machine with 13-slot horn gears and corresponding braided structure (Kyosev, 2011).

changing direction, two areas with two ridges produced from four-slot horn gears, and one middle area produced from the larger eight-slot horn gear.

The floating lengths are:

5 slots/(1F + 1E) = 5/2 = 2,5 = 2 floats => structure 2:2-1

4 slots/(1F + 1E) = 4/2 = 2 = 2 floats => structure 2:2-1

8 slots/(1F + 1E) = 8/2 = 4 = 4 floats => structure 4:4-1

2.6.4 Not-full carrier occupations and the pattern

From any (collision free) carrier arrangement of the braiding machine, arrangements exist whereby one or more carrier can be excluded, either by removal from the machine or, in the case of a small test series, having its yarn fixed to the carrier body. Simply breaking the yarn and keeping the carrier moving is very dangerous, because if the machine continues working, the free yarn end will wind over the nearest carriers, which can cause serious damage.

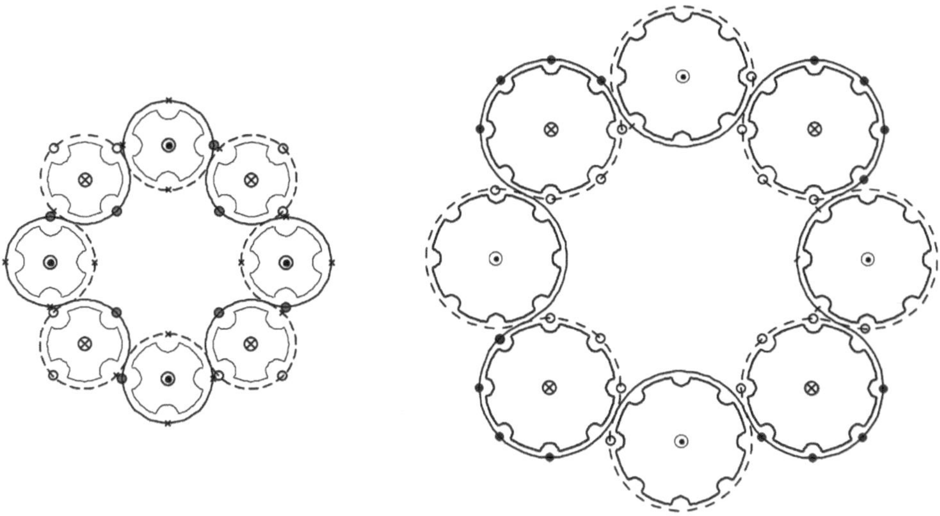

Figure 2.21 Arrangements for production of 1:1-4 structure: (a) using two-ply yarns per carrier; (b) using single yarn per carrier. The scale of both pictures is not exactly calculated, but is kept identical in order that the differences in the dimensions of both the machines are visible.

The interlacement between the yarns changes in areas where the yarn of the excluded carrier would have been. At that place, the floating lengths of the other yarns are longer, and special patterns can be created in this manner.

2.7 Quality and length issues of using ply yarns

Patterns with two, three or more yarns per group (such as 1:1-2, 1:1-3, 1:1-4, 2:2-2 etc) can be produced in two different ways (Fig. 2.21 demonstrates the case 1:1-4):

- using the carrier arrangement for the basic pattern (i.e. 1:1-1) and winding multiple yarns on one bobbin (ply yarn)
- using a single yarn per bobbin, with the normal carrier arrangement.

In the first case, significantly smaller machines can be used, although winding multiple yarns onto a single bobbin reduces the length of the braided product (without knots) and increases the changing frequency of the bobbins (negatively impacting machine productivity). Further, if one of the yarns breaks during the winding process and is missing on some part of the bobbin this may not be detected during braiding, and also the unwinding process of multiple yarns is connected with a higher risk of defects.

The last, but most visible, problem is that the position of the single yarns in the braid cannot be guaranteed as these can twist in places instead of remaining parallel. This is a significant defect, which is usually not acceptable to customers. As well as changing the braid's appearance, it can also lead to variations in the tensile properties of the final product.

This problem is solved by the second case, which uses a single yarn per bobbin. Here the interlacing is clear, because each yarn is guided to the correct place with well-controlled tension from its carrier. The disadvantage of using a single carrier per yarn is that machines with more or larger horn gears are required, although this allows for a larger yarn length per carrier.

References

Douglass, W. A. (1964): Braiding and braiding machinery. Eindhoven: N.V. Uitgevesmaatschappij Centrex.

Engels, H. (1994): Handbuch der Schmaltextilien. Die Flechttechnologie. Teil I: Maschinen und Verfahren zur Erzeugung konventioneller Geflechte. Mönchengladbach.

Goseberg, F., Brabender, K. and Moeller, P (1981): Textiltechnik Maschinengeflechte. Ausbildungsmittel Unterrichtshilfen. Frankfurt am Mai: Arbeitgeberkreis Gesamttextil.

Ko, F., Pastore, Ch. and Head, A. (1989): *Handbook of Industrial Braiding*: Atkins & Peirce.

Kyosev, Y. (2011): TexMind Braider. Version 0.9. Mönchengladbach: TexMind UG. Available online at www.texmind.com.

Kyosev, Y. (2012): Simulation of wound packages, woven, braided and knitted structures. In D. Veit (Ed.): *Simulation in Textile Technology*: Woodhead Publishing Limited, pp. 266–309.

Lepperhoff, B. (1953): *Die Flechterei*. 3rd edn.: Eugen G. Leuze Verlag Saulgau/WTTBG.

Structural design of flat and tubular braids

3

3.1 Introduction

This chapter extends the basics of braid patterning presented in Chapter 2, and covers the creation of drawings of braided structures for evaluation prior to application, and the possible required purchase of a new machine.

The method presented is used mainly for flat and tubular braided structures. Flat structures have only a front and a back, and the interlacement of the yarns can be represented in a 2D drawing. Tubular braids have outer and inner sides where for the outer side only half is visible and they can also be represented in 2D.

Pattern drawings are in common use and are described in (Lepperhoff, 1953; Goseberg *et al.*, 1981; Engels, 1994) where the rules for creation of flat and tubular braids are slightly different. In this chapter, a unified approach for both types of braid is proposed.

3.2 Flat braids

3.2.1 Basic steps for flat braids with three yarns (corn rows)

The pattern drawing for these braids consists of small lines which represent the visible parts of the structure. For representation of the structure, a square mesh is required (Fig. 3.1). Each row of this mesh corresponds to the part of the braided structure built during the rotation of the horn gears to angle, corresponding to the central angle of one slot. For three-slot horn gears (Fig. 3.1), this corresponds to a rotation angle of 120° and, for the classical four-slot horn gears, to a rotation angle of 90°.

During this motion, the yarn of carrier A (Fig. 3.1) crosses the yarn of carrier B. These two yarns can be represented as diagonals on the corresponding cell, and on the drawing only the upper yarn is drawn because only this is visible. The bottom yarn (B) remains under yarn A and is not visible to the observer.

The track and horn gears determine the type of interlacement between the yarns. The line connecting the horn gear centres splits the drawing into two parts. If during one step of the motion of one carrier the yarn remains above the centre line of the drawing, nothing is presented in the corresponding cell; if the motion is below the line, a diagonal line in the corresponding cell is drawn in the same direction as the

Braiding Technology for Textiles. http://dx.doi.org/10.1533/9780857099211.1.47

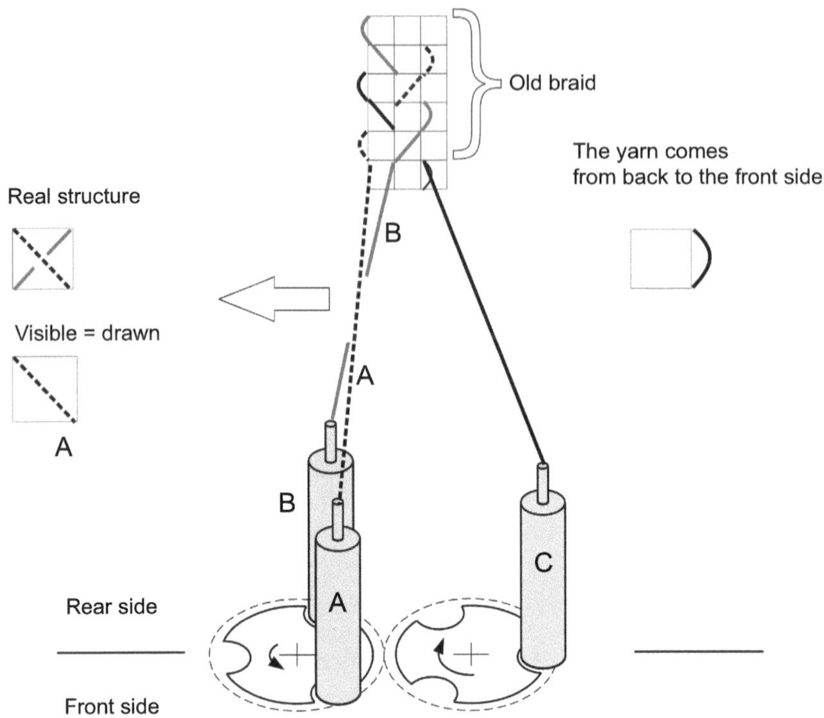

Figure 3.1 Basic ideas for creation of pattern drawing of flat braids.

carrier motion. In Fig. 3.2 the grey circle represents a carrier which will move during the next 120° rotation of the horn gear above the line, so its yarn is invisible to the observer and nothing is drawn. At the same time the carrier, marked with a dashed circle, moves under the line and its yarn is visible. Because the carrier moves from left to right and the take-off device takes the braid from down to up, the diagonal line is drawn from the upper left corner to the bottom right corner of the mesh.

If the horn gears have an odd number of slots then there are motions during which the carriers change their visibility. In Fig. 3.2 the black-filled circle represents such a carrier. The yarn of this carrier is in the invisible part of the drawing at the beginning of the motion step, and at the end of the step is on the visible side. During this motion, an arc from the bottom to the front side of the braid is built, and on the drawing it is also represented by an arc.

The process can be summarized as follows:

1. Draw a line connecting the horn gear centres and mark the half-plane below this line as the front (visible) side.
2. Prepare a mesh with square cells where, for each horn gear with *N* slots, (*N*/2) (the integer part after division by 2) columns are required since only half of the horn gear is visible.
3. Each row represents the motion of the carriers during one step (=1 slot), so the number of rows in one repeat is equal to the number of all (full and empty) carrier positions in the track. Fig. 3.2 demonstrates this principle, and the full repeat is given in Fig. 3.3.

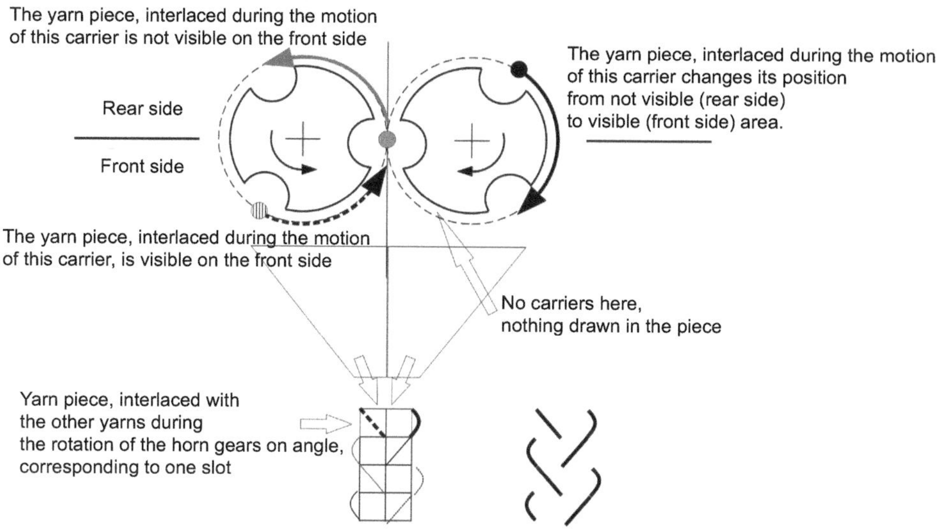

Figure 3.2 Rules for the creation of pattern drawing of flat braids.

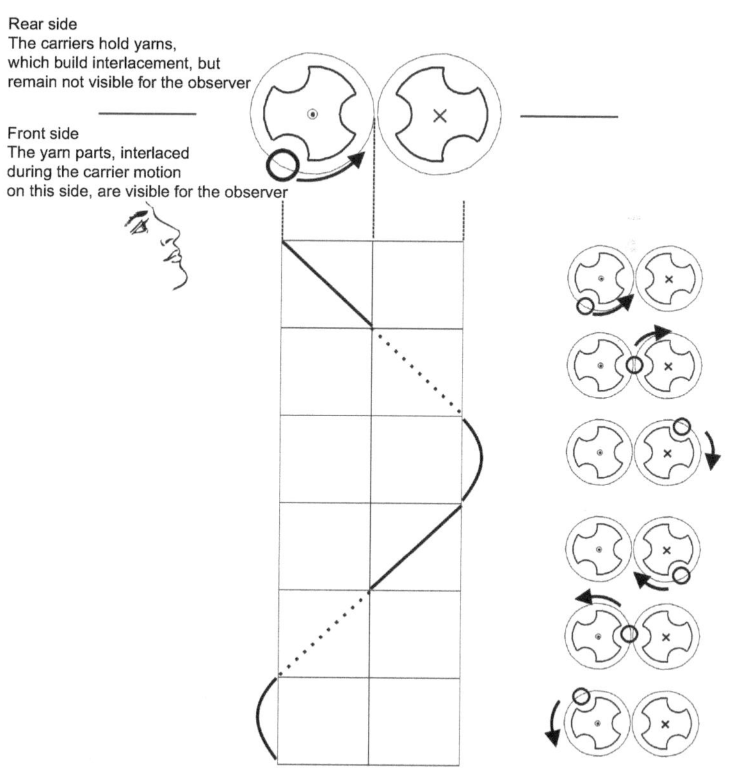

Figure 3.3 Single steps during the creation of the path of first yarn on the pattern drawing of flat braid.

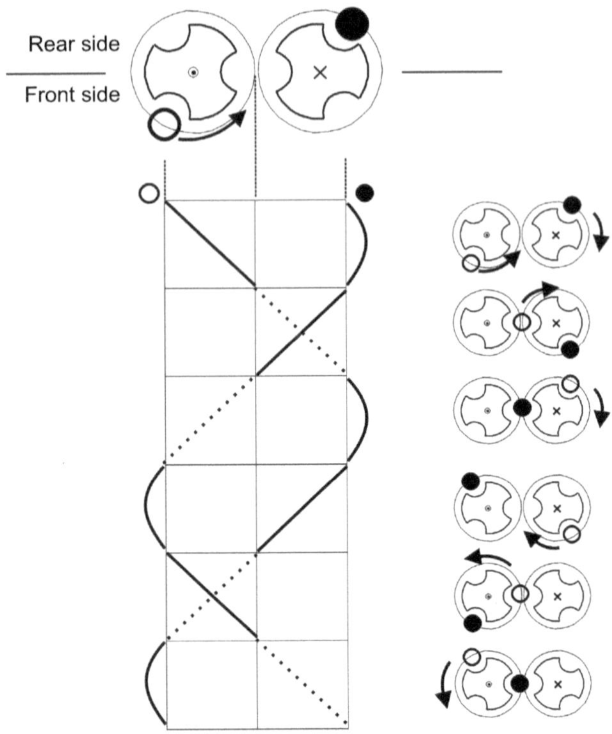

Figure 3.4 Adding the path of the second yarn of the flat braid.

4. For each carrier find which cell corresponds to its current position and draw the line for each step:
   a. If the carrier is in the visible half-plane and will remain in the visible area during the step, draw a diagonal line in the corresponding cell in the same direction as the carrier moves (Fig. 3.3, first row).
   b. If the carrier moves in the invisible part of the drawing, do not draw anything but follow its motion (Fig. 3.3 second row).
   c. If the carrier moves from the invisible to the visible area, draw an arc (Fig. 3.3, third row).

This sequence is shown step by step in Figs 3.3–3.6. Figure 3.3 represents the drawing of the path of the first yarn, Fig. 3.4 adds the path of the second yarn and Fig. 3.5 adds the path of the third yarn. On these figures the invisible yarn parts are marked with dotted lines to illustrate the yarn path. Usually these lines are not drawn, so the final pattern drawing is as represented in Fig. 3.6.

Removal of the mesh allows for a cleaner appearance of the drawing, as shown in the middle part of Fig. 3.6. The 3D simulation on the right hand side shows how the idealized structure looks.

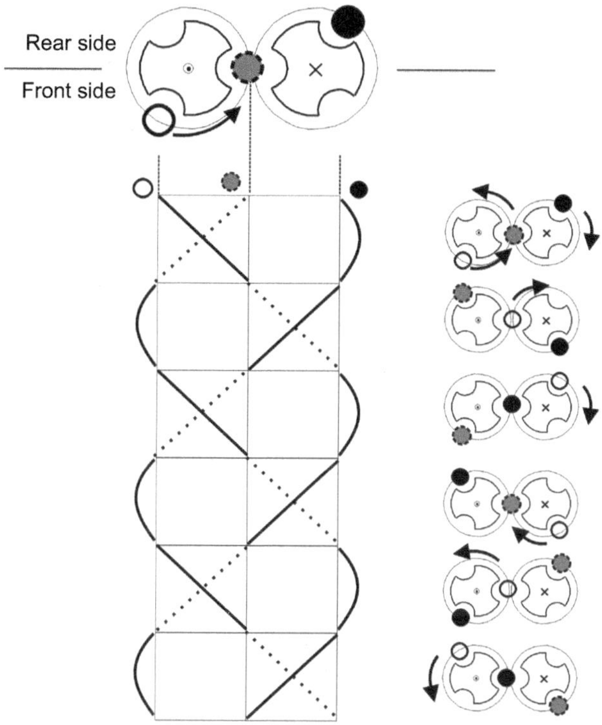

Figure 3.5 Adding the path of the third yarn of the flat braid.

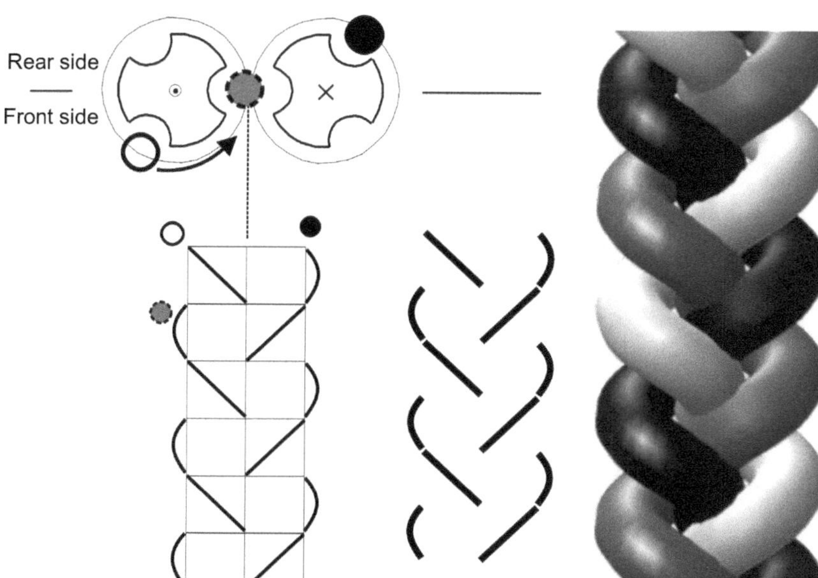

Figure 3.6 Final pattern drawing, drawing without the square mesh and 3D view (Kyosev, 2012).

3.2.2 Classical flat braids with an uneven number of yarns

The previous methodology can be applied for the creation of pattern drawings of any flat braided structure. The carriers on flat braiding machines with a low number of horn gears are arranged on one line. Machines with more than four horn gears are also available. In such machines, the gears are usually again arranged in a circle, so that the yarn length differences between the braiding point and the carrier during the braiding process are minimal. Such machines have to be presented as well (deformed), so the horn gears are placed on a horizontal line for the creation of the pattern drawing.

Figure 3.7 represents a braid built of nine yarns on a machine having four horn gears. For the end horn gears with five slots, (5/2) = (2.5) = 2 columns for each horn gear are required, and for the middle horn gears with four slots each (4/2) = 2 columns are also required, making eight columns. The track consists of 18 positions (2 * 5 + 2 * 4 = 18) and each carrier has to go through all of these positions for a single repeat, requiring 18 rows on the drawing.

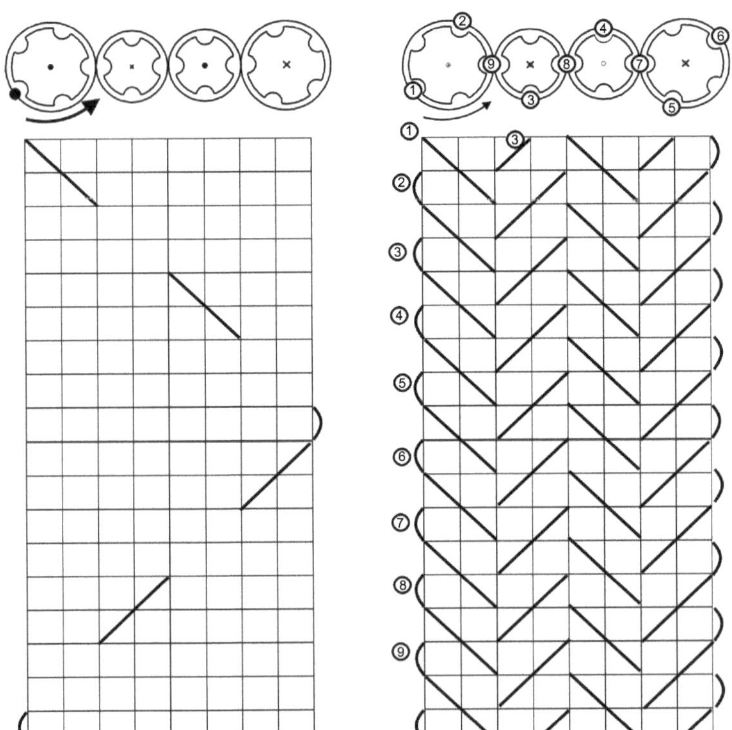

Figure 3.7 Configuration and pattern drawing of one nine-yarn flat braid. On the left-hand side the path of the first yarn is presented and on the right hand side the entire drawing.

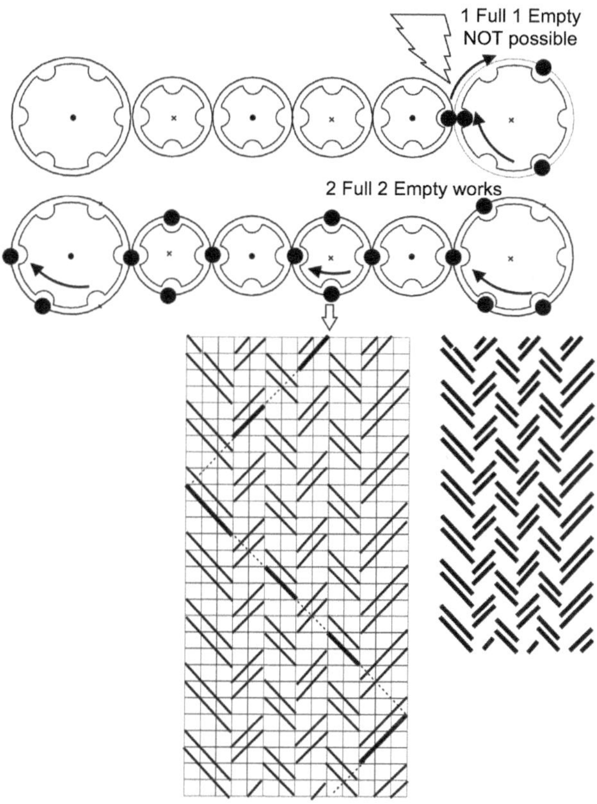

Figure 3.8 Configuration and pattern drawing of a 14-yarn flat braid.

On the left-hand side of the Fig. 3.7, the creation of the yarn path for the first carrier is represented, and on the right-hand side is the final picture. In this manner, the pattern drawings of all normal flat braids can be created.

3.2.3 Flat braids with an even number of yarns

Figure 3.8 shows the special case of an even number of carriers. Such machines have even numbers of slots at the end horn gears and an odd number of four-slot horn gears in between. In this case the carrier arrangement "1 Full 1 Empty" cannot work because after the fitting motion around the end horn gear, the first carrier will attempt to go into a place which is already occupied by another carrier (top of Fig. 3.8). Only the "2 Full 2 Empty" configuration can be used on such machines, so the braiding type will be 1:1-2 (over one: under one – as a group of two yarns). The floating length of the entire braid is equal to one.

For the four-slot horn gears: four slots: (2 Full + 2 Empty) = 4/4 =1;

For the six-slot horn gears: six slots: (2 Full + 2 Empty) = 6/4 =1.5 => 1.

The difference in the creation of the drawing consists in the representation of the yarn paths of the end horn gears, each of which requires three columns, so in all 3 + 2 + 2 + 2 + 2 + 3 = 14 columns are required. The track consists of 28 slots, so for the repeat 28 rows are required. Because the end horn gears have an even number of slots, no arc elements are in this drawing (see the bottom part of Fig. 3.8). The drawing does not show all the yarn pieces which can be visible if the structure is not a dense braid. Theoretically, for such more open structures this drawing can be extended as presented in Fig. 3.20. For dense structures braided with a lower take-off velocity, the empty places on the drawing will be filled.

3.2.4 Flat braids with different floating zones

For fancy braids used mainly in the clothing industry, braiding machines with different numbers of horn gears are used so that zones with a different floating length can be created. The creation of a pattern drawing for such structures is very valuable because the designs change very often and usually new machines must be purchased.

Figure 3.9 shows an example of such a machine, where in the middle of a classical flat braiding machine a horn gear with eight slots has been placed. In this case, a middle area with a floating length of 4 will be built and long floatings will appear. Having an odd number of horn gears leads to asymmetric sample, which could tend to roll or deform to one side. For stability, a second such horn gear has

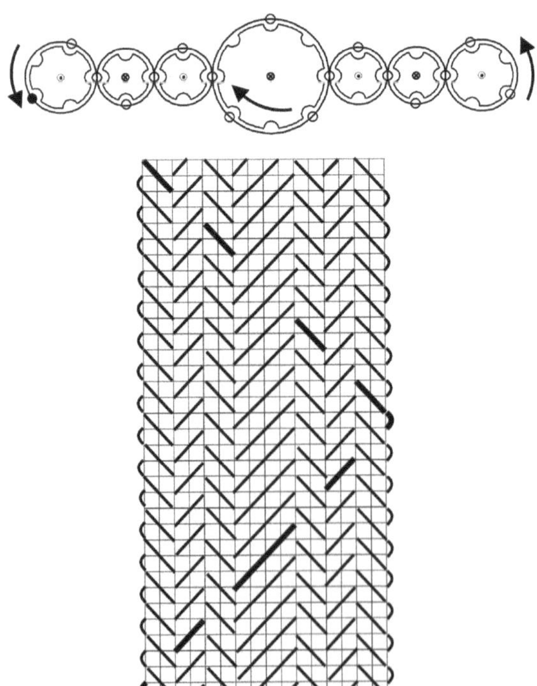

Figure 3.9 Configuration and pattern drawing of a flat fancy braid with 17 yarns.

Figure 3.10 Configuration and pattern drawing of a flat fancy braid with 21 yarns.

to be included as shown in Fig. 3.10. For both of the drawings, four columns per eight-slot horn gear are required; otherwise the process is the same as that for flat braided structures.

3.2.5 Flat braids on two-horn gears with a higher number of slots

There are a large number of machine variants based on two-horn gears where different configurations are used. For instance, in one structure, similar to that represented in Section 3.2.1, braid can also be produced on horn gears with six slots each and a carrier occupation of 2 Full – 2 Empty. The resulting structure has again a floating length of 1:

6 slots: (2 Full + 2 Empty) = 6: 4 = 1.5 => 1

with a group of two yarns. Such a machine is used, for instance, for candlewick production.

The pattern drawings of all samples with more than one yarn in a group are not so dense, so that empty areas are available, although these areas do not remain empty in the real structure. Figure 3.11 shows a machine where six columns and 12 rows are required for one repeat. Figure 3.12 presents the machine and carrier configuration and the pattern drawing of flat braid with 5 yarns.

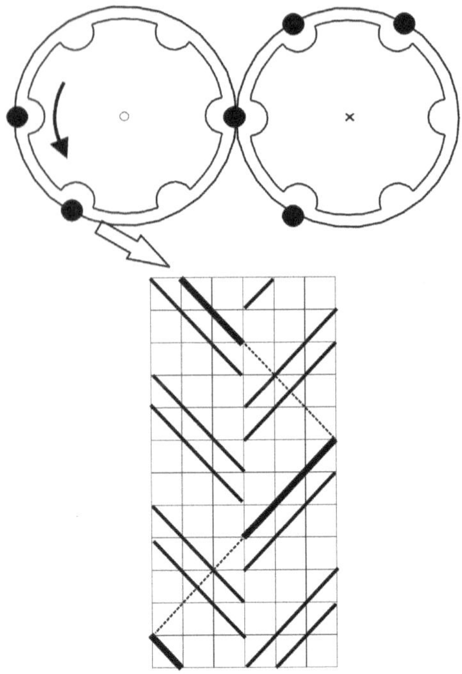

Figure 3.11 Configuration and pattern drawing of a six-yarn flat braid in 1:1-2 structure.

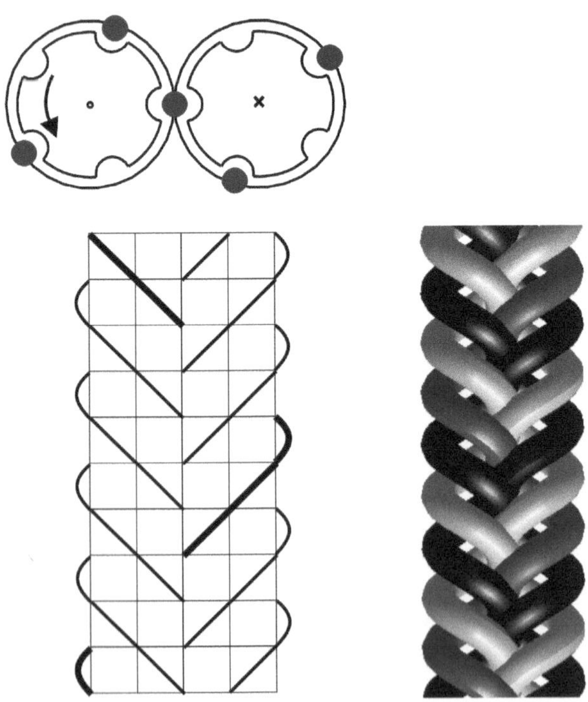

Figure 3.12 Configuration and pattern drawing of a five-yarn flat braid in 2:2-1 structure.

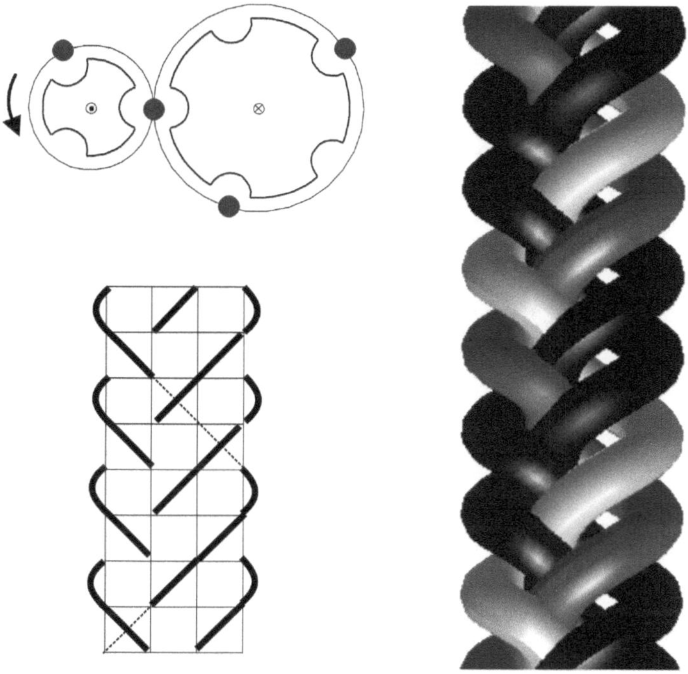

Figure 3.13 Configuration and pattern drawing of a four-yarn flat braid in 1:1/2:2-1 structure (one ridge with floating length of 1, one ridge of floating length of 2).

On such a machine, asymmetric braids are also possible and are mainly for the fashion industry. Figure 3.13 represents a machine with one three-slot horn gear and one five-slot horn gear with four carriers in a 1 Full 1 Empty occupation, together with the corresponding pattern drawing. Figure 3.14 shows the case of horn gears with five and seven slots respectively, and the braid is based on six yarns. The five-slot horn gear builds a ridge with a floating length of 2:

5 slots: (1 Full + 1 Empty) = 5:2 = 2.5 => 2

and the seven-slot horn gear builds a ridge with floating length of 3:

7 slots: (1 Full + 1 Empty) = 7:2 = 3.5 => 3.

3.3 Application examples

Figures 3.15–3.19 represent examples of flat braided structures for clothing accessories (reproduced with the kind permission of Mr Ezio Rossini of "Manifattura C. Rossini & C. srl," Italy, http://www.manifatturarossini.it/). Two flat braids with a larger number of carriers and different kinds of yarn are shown in Fig. 3.15. In Fig. 3.16, colour effects are also included. The first sample (left and middle photos) is

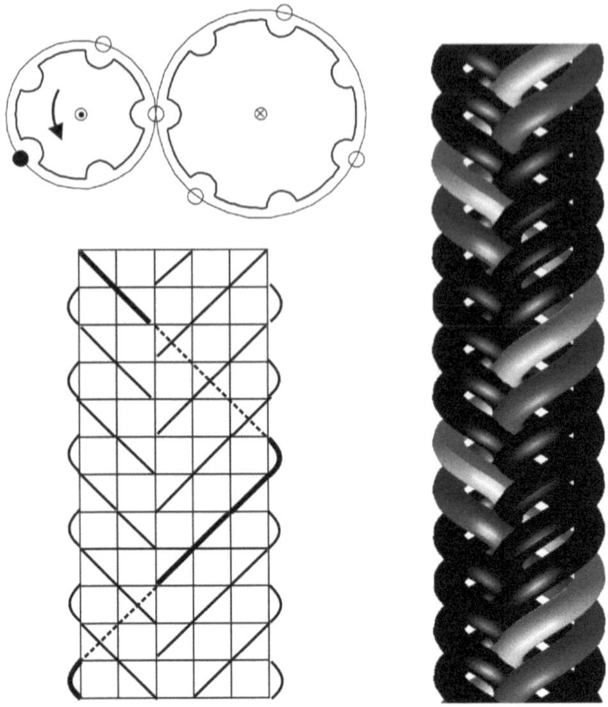

Figure 3.14 Configuration and pattern drawing of a six-yarn flat braid in 2:2/3:§-1 structure (one ridge with floating length of 2, one ridge of floating length of 3).

Figure 3.15 Examples of flat braided structures with different yarn material.
Source: Courtesy of Manifattura C. Rossini & C. srl, http://www.manifatturarossini.it/

Figure 3.16 Examples of flat braided structures with colour effect.
Source: Courtesy of Manifattura C. Rossini & C. srl, http://www.manifatturarossini.it/

Figure 3.17 Examples of flat braided structures using braid as primary material or thick monofilaments.
Source: Courtesy of Manifattura C. Rossini & C. srl, http://www.manifatturarossini.it/

Figure 3.18 Examples of flat braided structures combining different materials or different floating lengths.
Source: Courtesy of Manifattura C. Rossini & C. srl, http://www.manifatturarossini.it/

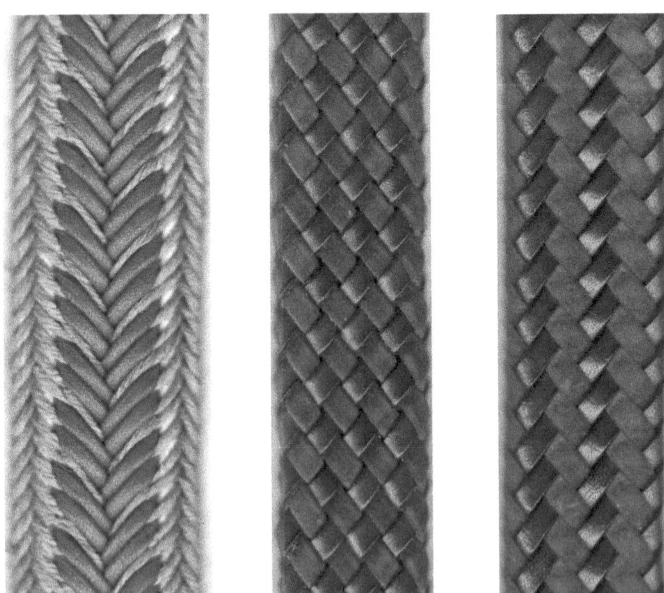

Figure 3.19 Examples of flat braided combining different materials (left sample) or using leather.
Source: Courtesy of Manifattura C. Rossini & C. srl, http://www.manifatturarossini.it/

braided from braided yarns, and is thus double braided. The sample on the right-hand side combines four yarns per group and colour effects. Note that there are at least three possible machine configurations for the production of such a sample with structure type 1:1-4. One possibility is to use a braiding machine with a floating length of 1 (four-slot horn gears with a 1 Full 3 Empty configuration) and four yarns on each bobbin. The second way is to use two yarns per carrier and an arrangement of 2 Full 2 Empty, again on four-slot horn gears. The third way is to use a special machine with eight-slot horn gears, where a 4 Full 4 Empty carrier arrangement is used and every carrier holds only one yarn. Such a configuration could guarantee ideal yarn guidance and longer braiding time in the absence of bobbin changes, but the machine is significantly larger. Figures 3.18 and 3.19 demonstrate the combination of different materials and floating lengths.

3.4 Tubular braids

The tubular braids are used for ropes and other tubular products. They are seldom used in the fashion industry, so the main tasks during the structural design are to clar-

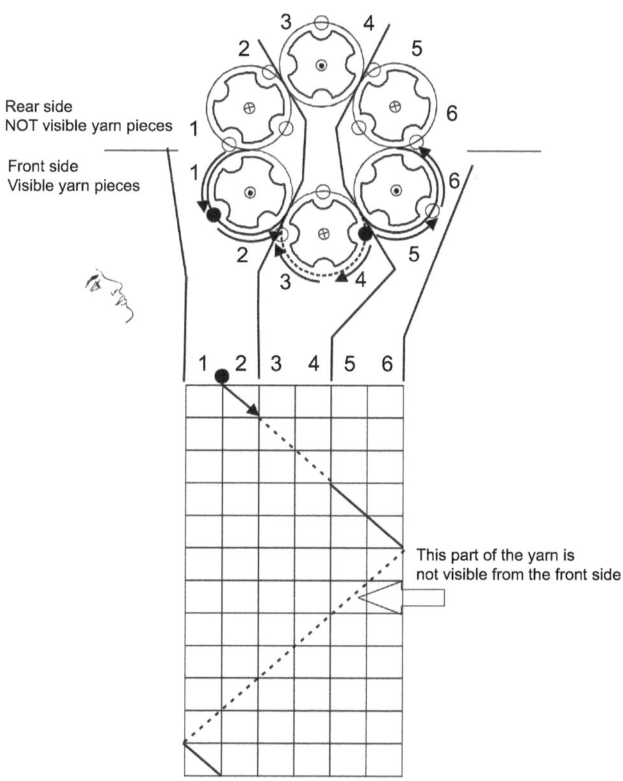

Figure 3.20 Tubular braiding machine with its visible areas and the path of one yarn.

ify the interlacement of the yarns and to understand which possibilities are available on a given machine.

3.4.1 Basic steps for tubular braids

For the structural design of tubular braids the same steps can to be used as for flat braided fabrics, with some adjustments. With tubular braids, the observer can see only one fourth of the sample surface, as the rear and the complete inner surface of the rope remain invisible. Depending on whether or not the rear side of the rope must be drawn there are some differences in the drawing. Figure 3.20 demonstrates this concept.

For tubular braids, yarn parts which are interlaced during the motion of the carriers within the circle on which all horn gear centres are placed are not visible. These parts build the inner region of the rope or tube. Additionally, depending on the position of the observer, only half of the braid is visible. In Fig. 3.20, the visible areas are below the lines which split the machine onto two symmetrical parts.

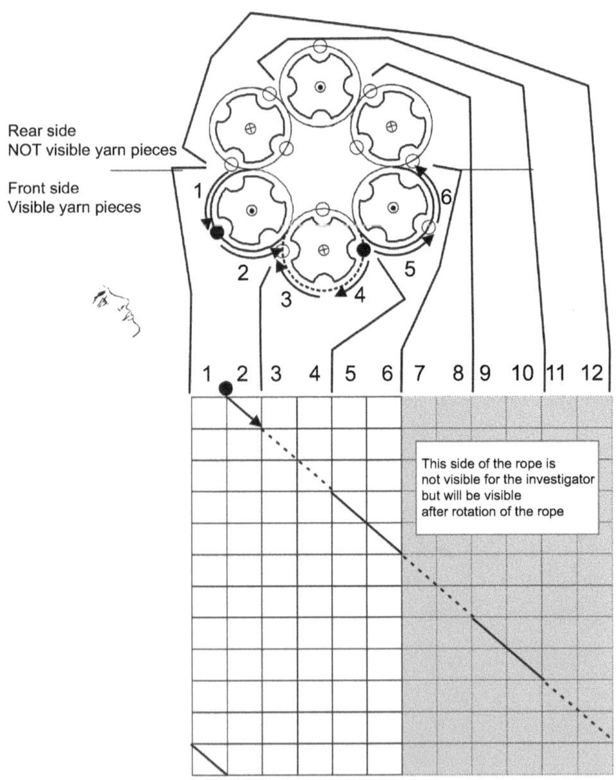

Figure 3.21 Alternative representation of the visible areas of a tubular braiding machine and the path of one yarn.

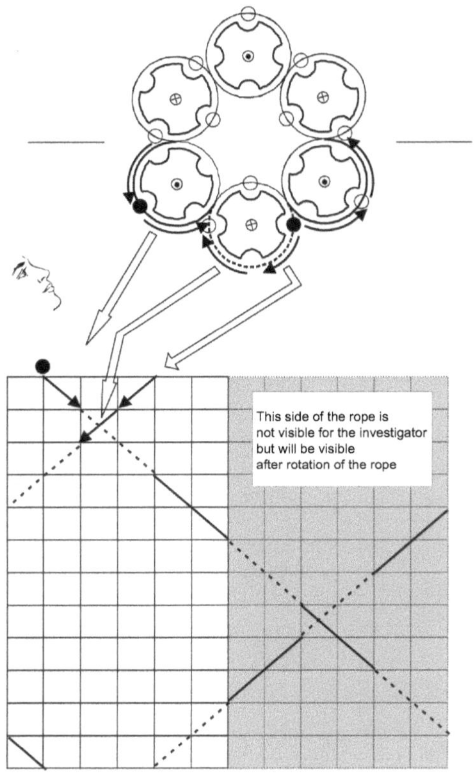

Figure 3.22 The path of second yarn added to the Fig. 3.21.

Each horn gear requires $N/2$ columns, where N is the number of slots. For the example shown in Fig. 3.20, there are three horn gears in the visible area, each of which has four slots, so that $3 \times 2 = 6$ columns are required for the drawing. The machine consists of two tracks, each with 12 slots, so 12 rotations of the horn gears are required for one full repeat, corresponding to 12 rows on the drawing. The carriers which are in the invisible area can be tracked on the mesh if desired. During one repeat, less than the half of the yarn length is visible. One half of the yarn remains on the rear side (the dashed line on the bottom part of the mesh in Fig. 3.20) during the motion of the carrier over the upper half of the machine.

An alternative representation is detailed in Fig. 3.21. Here the braid is virtually cut and uncoiled, so that the rear side is also represented. Figure 3.22 presents additionally the yarn of one second carrier. It can be seen, that the right-hand side of the picture can be prepared (extended) on request at any time. Because of this and for simplicity, in the following examples only the visible part will be used. In Fig. 3.23 the path of the next one yarn is given and in Fig. 3.24 the entire drawing is displayed. Usually, after some experience, the numbering of the yarns with the carriers is not very essential, so the final pattern drawing for the 12 yarn ropes in 2:2-1 structure (floating length of 2) is demonstrated. Figure 3.25 shows the final presentation of the

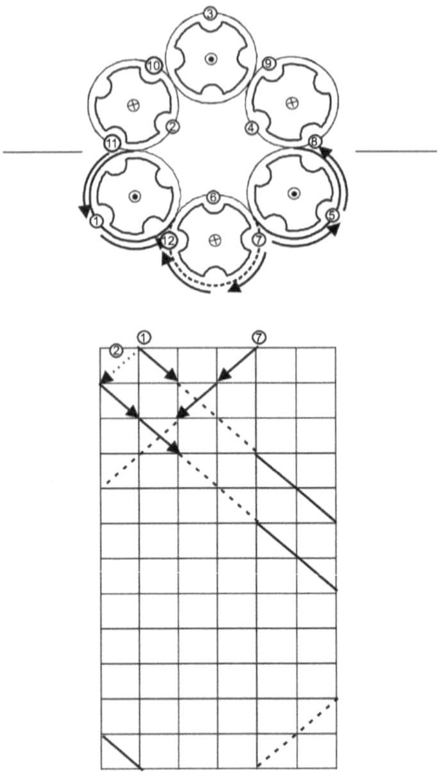

Figure 3.23 Third yarn path added to the drawing, using only the really visible area.

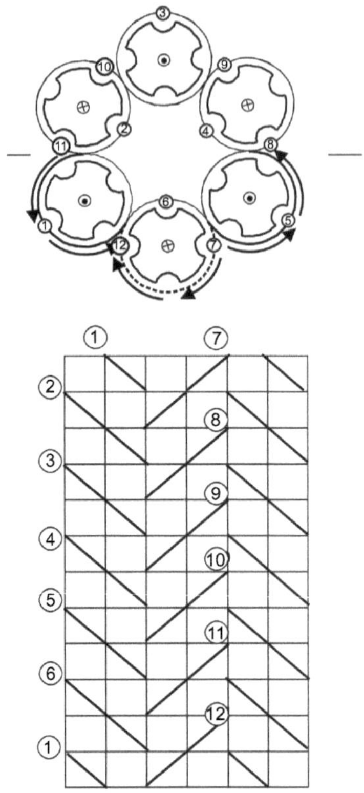

Figure 3.24 Complete structure drawing of the 12-yarn tubular braid.

machine configuration, once with mesh, with removed mesh and as 3D visualization. In the same way the pattern drawings of all tubular braids with a floating length of 2 can be developed.

3.4.2 Other common tubular braids

Figure 3.26 shows structures with floating length of one, but grouped of two yarns. The bottom left drawing has been prepared according to the above methodology and appears to show more empty places in the mesh. Normally, the structure looks like the variant shown in the bottom middle picture of Fig. 3.26 where, in the mesh cells, where no upper yarn is presented, the bottom one is drawn with a dotted line. This dotted line indicates that the yarn is there and if the braid is produced very loosely at that place the bottom yarns will be visible, as shown on the bottom right-hand side. The reality is actually exactly between the both pictures, because the two yarns following one after another build interlacement like panama weaves. However, this cannot be presented in the same way with the square mesh, so the reader has to decide

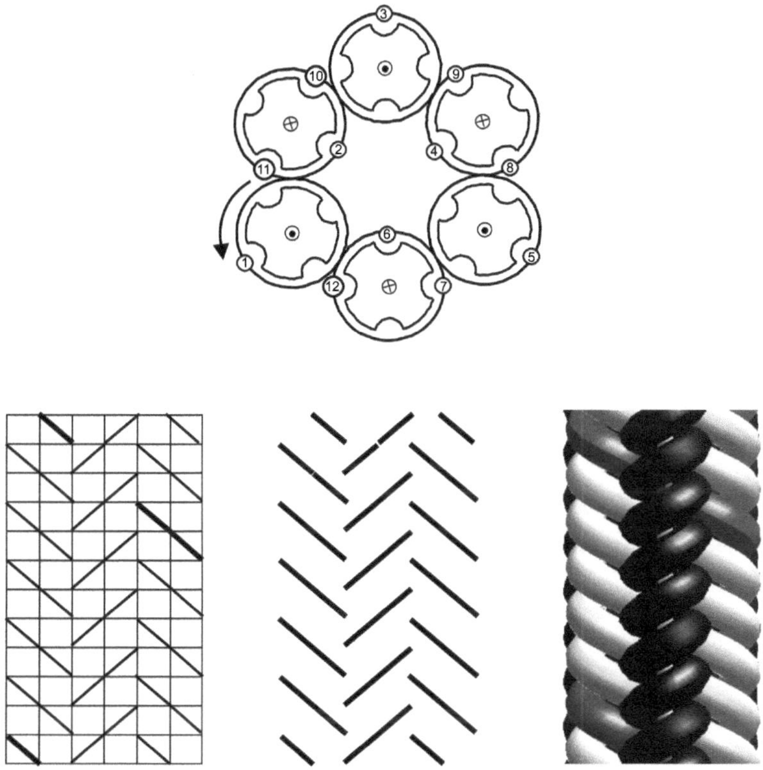

Figure 3.25 Final presentation of the machine configuration, structural drawing with and without mesh and 3D simulation of the 12-yarn tubular braid with 2:2-1 structure.

alone which method of representation would be preferable to use. The first makes the picture clearer, and thus is preferred here. Figure 3.27 represents the configuration for producing a tubular braid on the same machine with six horn gears, but with a floating length of one (1:1-1). In this case a carrier arrangement of 1 Full – 3 Empty must be used.

3.4.3 Tubular braids for six-slot horn gears

Six-slot horn gears are not very commonly used, as they require larger machines which work slowly compared to four-slot horn gear machines. Some uses do exist, however, primarily with composites of glass and carbon fibres. The availability of six slots allows the production of samples with a floating length of 3, so that less crimp in the structure is available. The development of the pattern drawing for such machines follows the same rules as for all other machines. Figure 3.28 represents a machine with six horn gears and 18 carriers arranged in a 1 Full 1 Empty configuration. The floating length in this case is as follows: six slots: (1 Full +

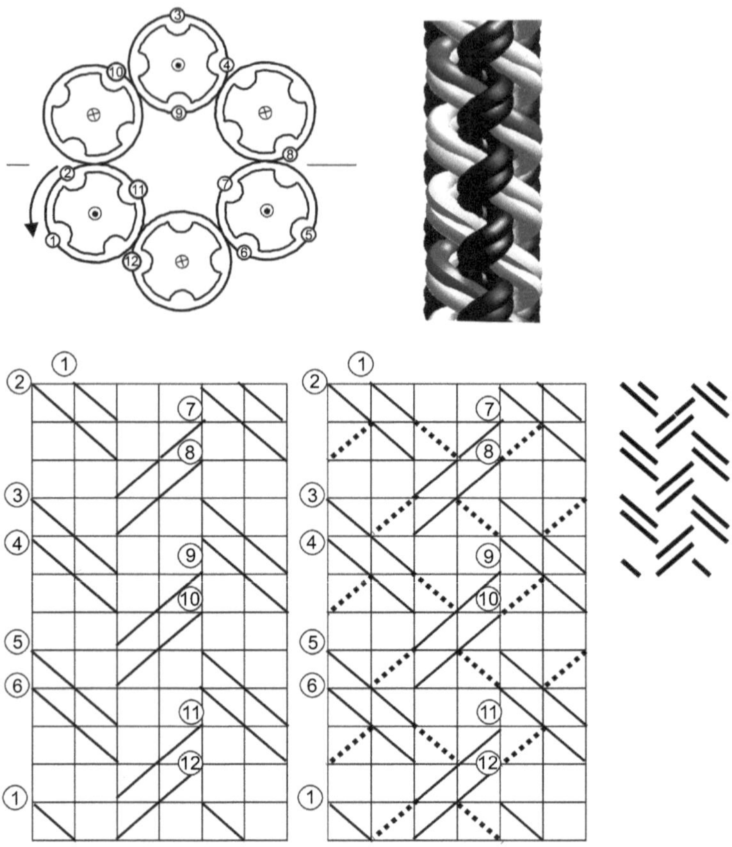

Figure 3.26 Machine configuration and pattern drawing of 12-yarn tubular braid with 1:1-2 structure.

1 Empty) = 6:2 = 3. For each horn gear on the visible side, three columns must be used so that, in all, nine columns and 18 rows are required for the repeat.

The representation of the yarn path of carrier one (1) and carrier B from the opposite track can be drawn with no problem. A problem does occur, however, in determining the current position of carrier A. If, starting from the place of carrier 1, the slots from the bottom side of the horn gear are counted, the yarn has to be placed in column 4. If the slots are counted along the track where carrier B is, then only six slots are there, so the yarn has to be in column 2. The correct position is in fact column 3. Drawing the path of the carriers, starting from their interlacement point, can avert such problems. The yarns of carriers 1 and B cross in the upper left cell of the mesh. After that, following the carrier arrangement "1 Full 1 Empty" in both of the tracks one row is skipped (for the empty position) and one is filled

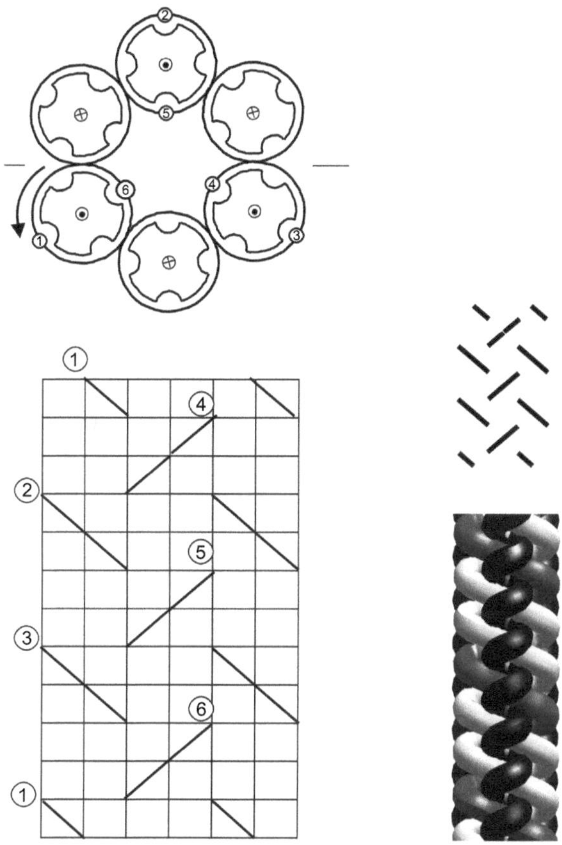

Figure 3.27 Machine configuration and pattern drawing of six-yarn tubular braid with 1:1-1 structure. The machine uses 1 Full 3 Empty carrier arrangement.

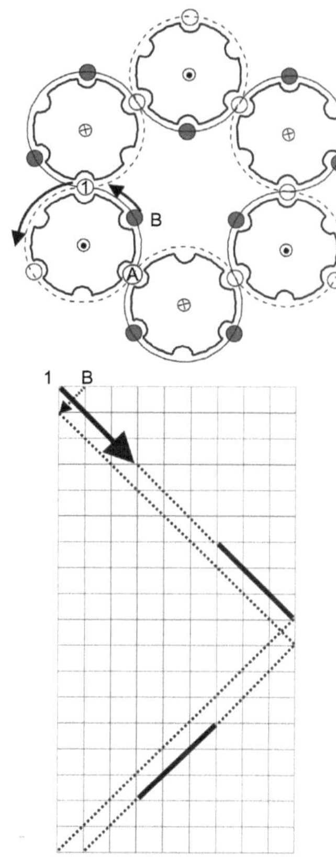

Figure 3.28 Machine configuration and yarn paths of two yarns of a tubular machine for producing 18-yarn braid with floating length of 3 (3:3-1).

with yarn (for the full position). In this manner the drawing shown in Fig. 3.29 is created. The same issue exists for smaller tubular braids, an example of which is shown in Fig. 3.30 (left), contrary to the braids with more horn gears (right) where the standard procedure works.

3.5 Advanced patterning

The drawings discussed in this chapter cover only the most commonly used basic structures. Further possibilities are available, however, if some of the carriers are removed (Goseberg *et al.*, 1981; Lepperhoff, 1953; Engels, 1994).

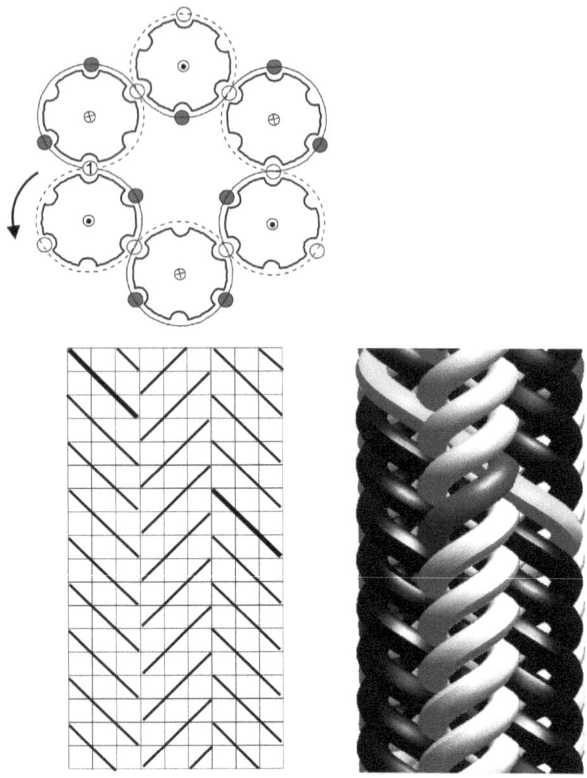

Figure 3.29 Machine configuration, pattern drawing and simulated structure of a tubular machine for producing 18-yarn braid with floating length of 3 (3:3-1).

Figure 3.30 Machine configurations and pattern drawings for the smallest tubular braid with floating length of 3 (using 12 yarns) and larger tubular structure with the same structure.

References

Engels, H. (1994). *Handbuch der Schmaltextilien. Die Flechttechnologie. Teil I: Maschinen und Verfahren zur Erzeugung konventioneller Geflechte*. Mönchengladbach.

Goseberg, F., Brabender, K. and Moeller, P. (1981). *Textiltechnik Maschinengeflechte: Ausbildungsmittel Unterrichtshilfen*. Frankfurt am Mai: Arbeitgeberkreis Gesamttextil.

Kyosev, Y.K. (2012). *TexMind Braider* [online]. Mönchengladbach. Available from: www.texmind.com.

Lepperhoff, B. (1953). *Die Flechterei*. 3rd edn.: Eugen G. Leuze Verlag Saulgau/WTTBG.

Colour design of tubular braids

4

4.1 Introduction

The classical and historical applications of braided products are in the clothing industry. Shoe laces, belts and decorative laces have to match the rest of the clothing, and so for these applications the colour design is as important as the structural design. For the colour design of the braids, a stylized presentation of the structure is used to enable the designer to concentrate on the colour arrangement. Since the yarns in flat and tubular braids have different paths, some differences in design are required, and these groups are discussed separately. This chapter concentrates on tubular braids. Tubular braids include all kinds of braided and overbraided ropes, electric cables, decorative laces and so on. In this chapter, products mainly employing effects based on the combination of two colours are described, making the most effective use of a book printed in black and white.

4.2 Colour patterning basics

For colour patterning, a square mesh is usually used (Fig. 4.1). The mesh is rotated to 45° so that the cells have the same orientation as the yarns in the braids. As in weaving, each cell is filled with the colour of the yarn which is visible to the observer.

As described in the Chapter 2 (Patterning of braided products), the yarns of the carriers which move in the same direction around one-horn gear build one ridge in the braid. The length of the visible yarn piece in this ridge is equal to the floating length for this ridge. The path of each yarn is followed in the mesh, and so many cells are alternately filled or skipped, depending on the floating length for the pattern. For instance, the floating length of regular braids is two, so every two cells have to be filled with the yarn colour, and after that two cells have to be kept empty, because the yarn is underneath another two yarns and so is not visible, with this pattern repeating again and again (Fig. 4.2).

For structures where one group of yarns consists of more than one yarn, the representation is identical (Fig. 4.3). In this case a number of "columns" have to be filled, according to the number of the yarns in the group.

Braiding Technology for Textiles. http://dx.doi.org/10.1533/9780857099211.1.71

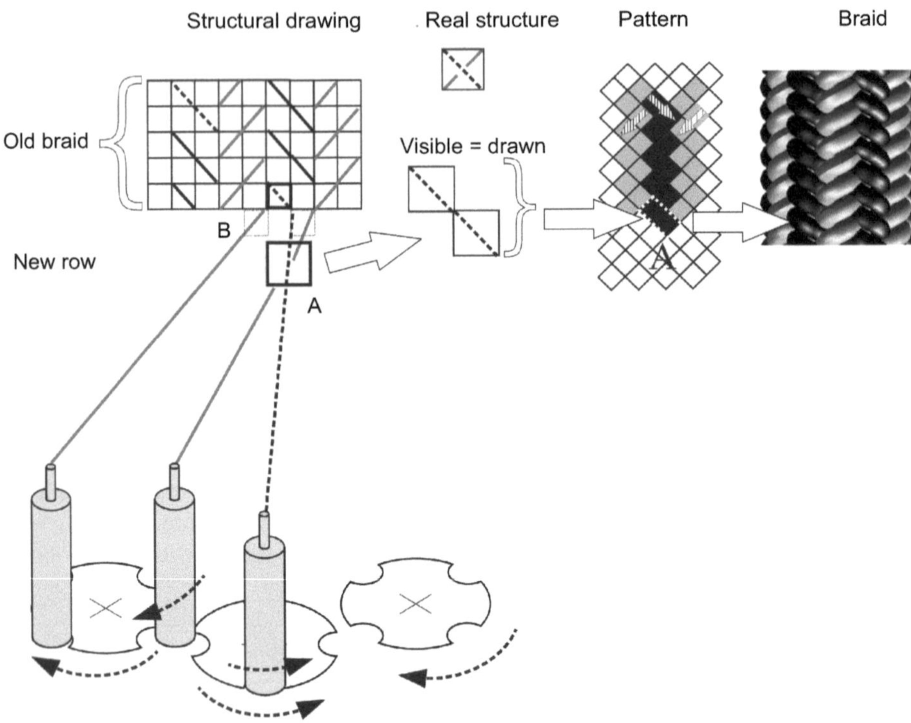

Figure 4.1 Connection between the structural presentation, mesh for the colour design and the structure.

To simplify the colour design process, a separated mesh per structure type can be prepared in advance. Examples of such meshes are given in Fig. 4.4, and are available for download in larger sizes from the author's home page www.kyosev.com.

4.3 General remarks about the colour design of tubular braids

As explained in Chapter 2, each horn gear builds a separate ridge in the braided structure. Machines for tubular braids always consist of an odd number of horn gears, so an odd number of ridges will be visible. With any tubular structure, the customer can see only the front half of the braid (Fig. 4.5), so only this part is developed and painted during the colour design process. One yarn carrier, moving in a clockwise direction, follows the simplified path shown as a helix in Fig. 4.5. Only half of this path will show visible yarn on the pattern (Fig. 4.5c); the other half of the yarn will be visible on the other (back) side of the braid. After one full

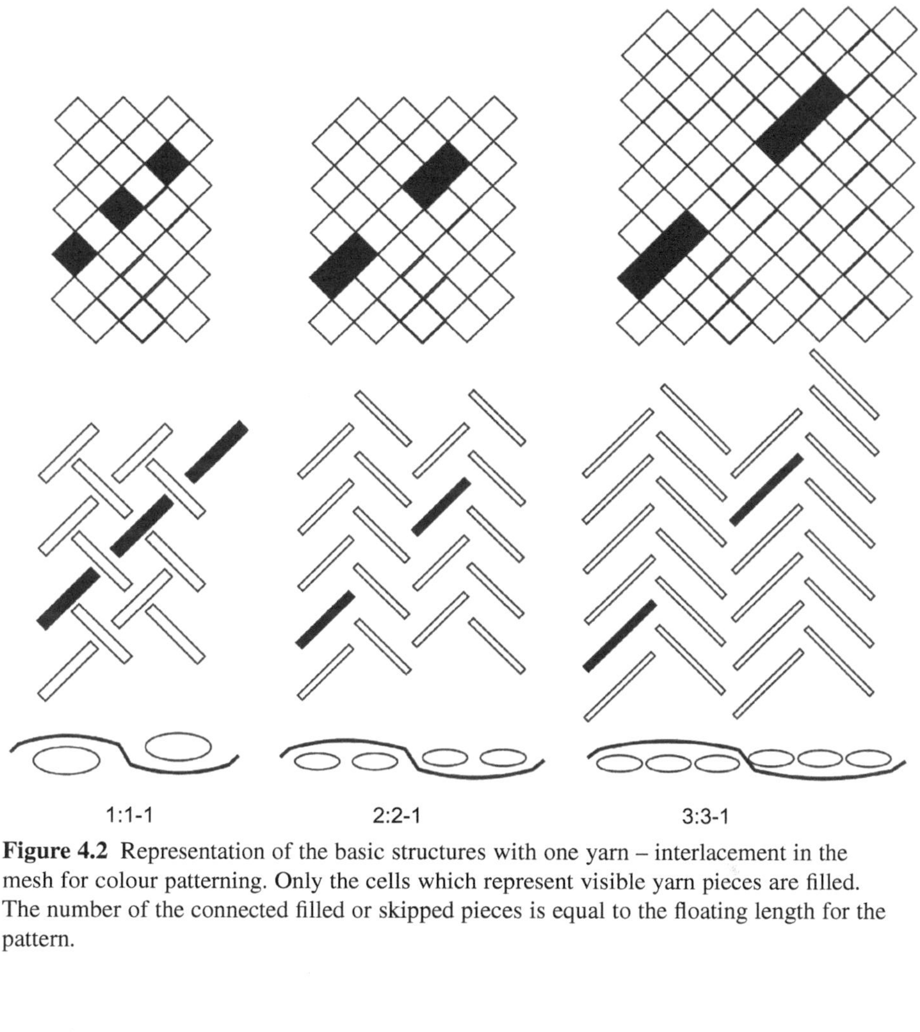

Figure 4.2 Representation of the basic structures with one yarn – interlacement in the mesh for colour patterning. Only the cells which represent visible yarn pieces are filled. The number of the connected filled or skipped pieces is equal to the floating length for the pattern.

revolution of the carrier, the yarn will be back in the same position. The length of the braid between the two places where one yarn begins the same pattern is called the "repeat" of the braid (Fig. 4.6). Usually, only one repeat has to be completely drawn.

4.4 Pattern development sequence for tubular braids

4.4.1 Development steps

Using the above basic information, the colour pattern of a tubular braid can be developed in the following sequence.

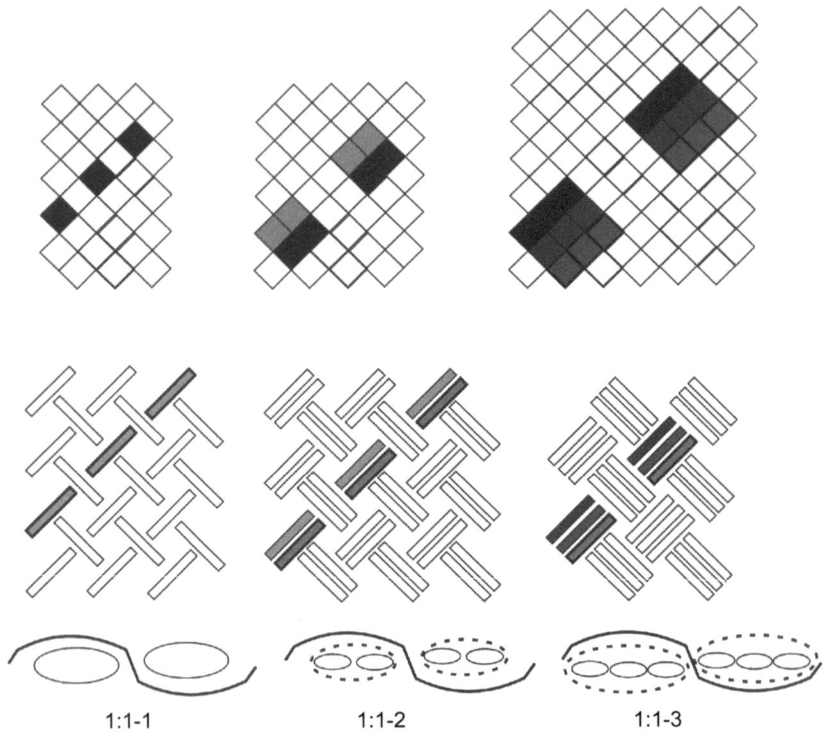

Figure 4.3 Representation of the basic structures with one yarn (1:1-1) for comparison and with two or three yarns in a group. For each yarn a separate channel is used.

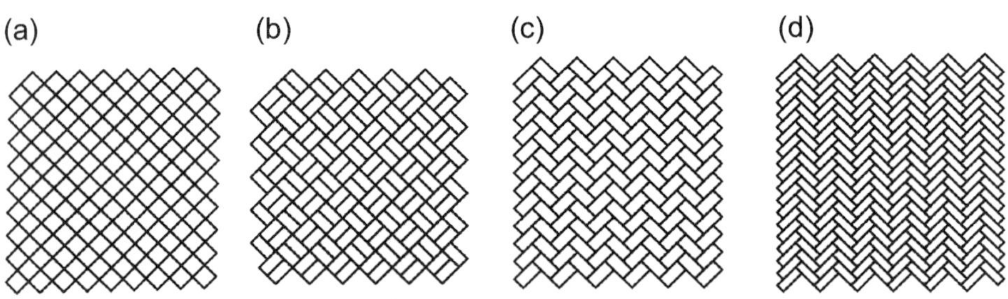

Figure 4.4 Prepared meshes for simplification of the colour patterning of tubular and flat braids for (a) 1:1-1; (b) 1:1-2; (c) 2:2-1 and (d) 3:3-1 structures.

1. Determine number of ridges to be drawn
   The number of horn gears of the machine is equal to the number of the ridges of the tubular braid (rope). Since only half of the ridges are visible, a mesh with

$$\text{Drawn ridges} = \frac{\text{Number of horn gears}}{2} \quad [4.1]$$

   has to be prepared.

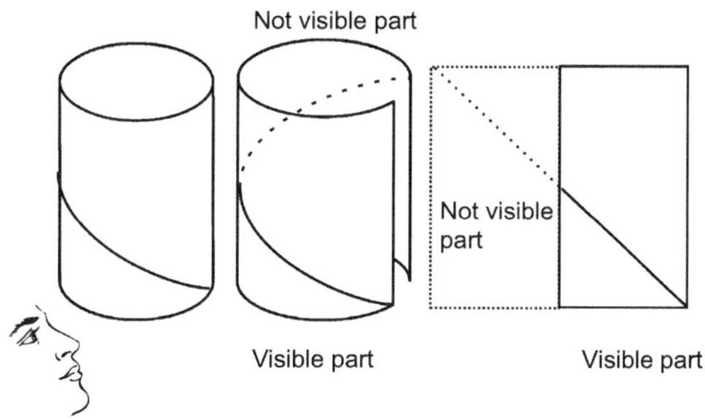

Figure 4.5 Visible half of the braided product, the circumference of the surface, and its evolved presentation.

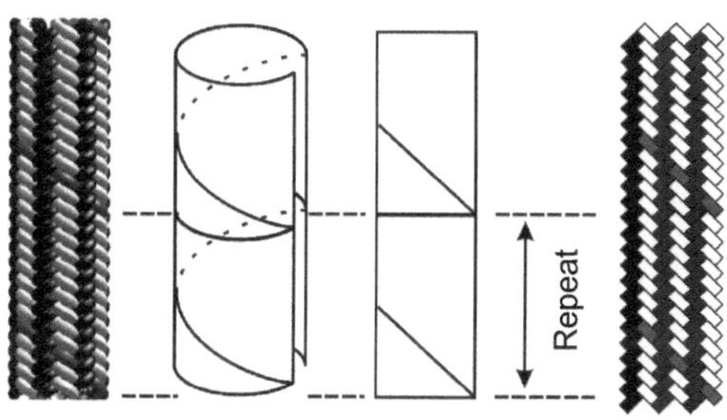

Figure 4.6 Tubular braid with its visible evolved part, its flat presentation and the mesh for the colour design.

2. Determine the floating length
   The floating length (the pattern type) is determined by the number of slots per horn gear and the carrier arrangement (see Chapter 2):

$$\text{Floating length} = \frac{\text{Slots per horn gear}}{\text{Carrier arrangement repeat}} \quad [4.2]$$

3. Determine the number of yarns in one group
   The number of full (occupied) positions of the carriers, which follows one after the other without empty places, determines the number of the yarns in one group. If these yarns are ply yarns, the number of yarns in the group is determined by the following equation:

$$\text{Yarns in group} = (\text{no. of full positions without interruption}) \times (\text{strands in ply yarns}) \quad [4.3]$$

For instance, for the carrier arrangement "2 Full, 2 Empty," each group of yarns consists of two yarns, and if each of the yarns is double ply, then the group is 2 × 2 = 4 yarns.

4. Determine the repeat of the sample
The repeat of the sample is determined by the number of the carriers moving in one direction.
Usually (for the regular pattern discussed in this chapter) they are half of the total number of carriers on the tubular machine

$$\text{Repeat} = \frac{\text{Total number of carriers}}{2} \quad [4.4]$$

5. Set the colour arrangement
The colour arrangement is determined by the sequence of the colours of the bobbins in one track. For instance:

all carriers in clockwise direction: black
all carriers in counter-clockwise direction: white

6. Select or prepare a suitable mesh and mark the visible area
The suitable mesh consists of the determined number of ridges with the floating, where so many cells are parallel, calculated as the number of yarns in one group. The visible area should have at least as many lines (rows) as the repeat. The length of each line (number of cells) is equal to:

$$\text{Visible cells} = \text{Ridges} \times \text{Floating length} \times \text{yarns in one group} \quad [4.5]$$

7. Fill the cells along the path of each yarn for the carriers of one of the tracks
For a floating length of F, F cells have to be filled and then F cells have to be kept empty, with this pattern repeating, representing the yarn. During the filling, one line corresponds to each yarn, as shown in Fig. 4.7. If one carrier has two-ply yarn, then two lines have to be used.
8. Fill the cells for the yarns in the carriers of the second track.

4.4.2 Example

A pattern for a machine with 16-horn gears and 32 carriers, set as one full and one empty, has to be created (Fig. 4.8). The horn gears have four slots each. The colour carrier arrangement for the clockwise moving carriers is all black; for the counter-clockwise moving carriers, it is seven white and one grey (red).

1. Determine the number of ridges to be drawn
The visible ridges are 4:

$$\text{Drawn\_Ridges} = \frac{\text{Number\_Horn\_Gears}}{2} = \frac{16}{2} = 8 \quad [4.6]$$

2. Determine the floating length
The floating length is:

$$\begin{aligned}\text{Floating\_length} &= \left[\frac{\text{Slots\_per\_Horn\_Gear}}{\text{Carrier\_arrangement\_repeat}}\right] \\ &= \left[\frac{4}{1\text{Full} + 1\text{Empty}}\right] = \left[\frac{4}{2}\right] = 2\end{aligned} \quad [4.7]$$

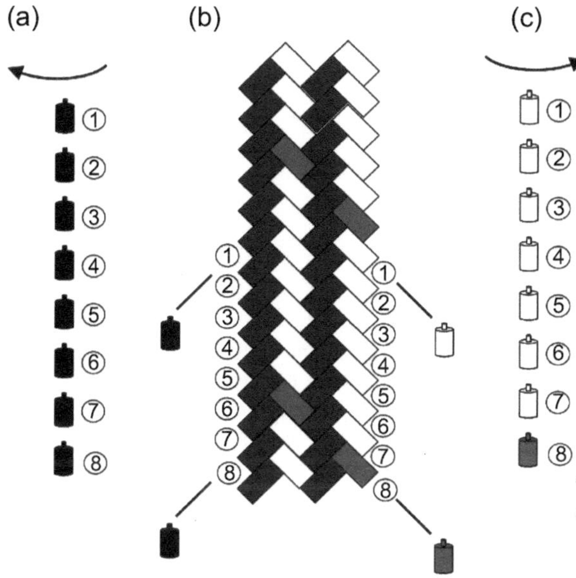

Figure 4.7 Colour design of a tubular braid with 16 carriers on 2:2-1 pattern: (a) colour arrangement of the carriers, moving clockwise; (b) developed picture of the pattern; (c) colour arrangement of the carriers, moving counter-clockwise.

Figure 4.8 Arrangement of the braiding machine with 16 carriers for the pattern.

3. Determine the yarns in one group

Here a single yarn per carrier is used. For a 1 Full 1 Empty arrangement, the group consists of one yarn.

4. Determine the repeat of the sample

The repeat of the sample is:

$$\text{Repeat} = \frac{16}{2} = 8 \qquad [4.8]$$

5. Set the colour arrangement

Clockwise moving carriers: all black;

Counter-clockwise moving carriers: seven white and one grey (red)

6. Select or prepare a suitable mesh and mark the visible area

From the floating length 2, it follows that along the yarn path every two cells can be merged together, and paper of the type in Fig. 4.4c (2:2-1) can be used. At least eight rows are required for the repeat, where each is 4× 2 × 1 = 8 cells long.

7. Fill the cells along the path of each yarn for the carriers of the first track. The filled cells for the first yarn of this track are shown in Fig. 4.9

Fill so many cells as the floating lenth is, after that keep the same number of cells empty, as the yarn is not visible and repeat this until the end of the repeat. During filling, each line corresponds to the yarn of one carrier (Fig. 4.10)

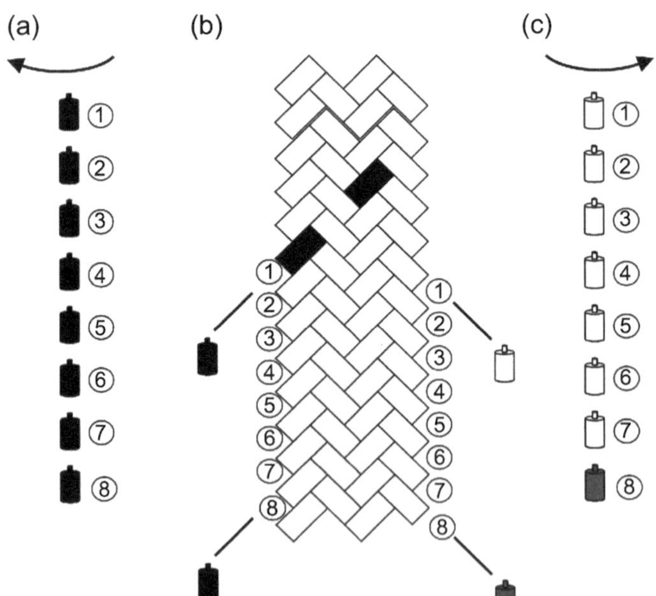

Figure 4.9 Mesh, prepared for the drawing of the colour design of for 2:2-1 structure of tubular machine with 24 carriers, moving on 12 four-slots-horn gears (b) and the colour arrangement of the yarns of the both tracks (a) and (c).

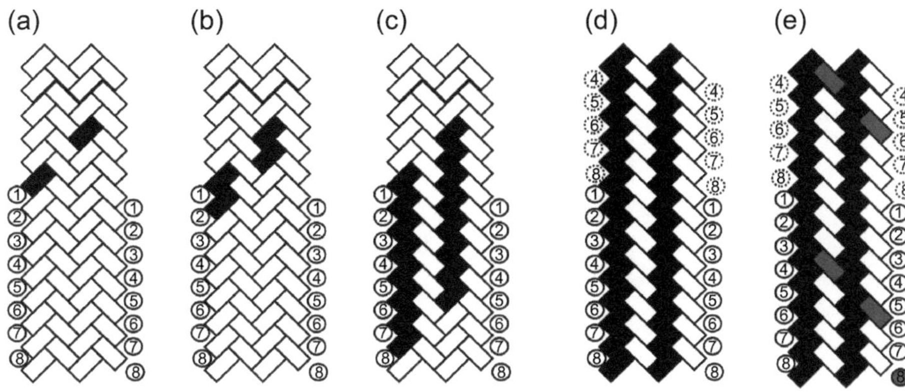

Figure 4.10 Pattern development steps: (a) filled the visible yarn pieces for the yarn of the first carrier; (b) yarn of the second carrier added; (c) the yarn of all carriers of the one track are filled – this is the repeat of the braid; (d) the repeat is extended over the visible picture; (e) the yarn of the carrier number 8 is as well marked. The other seven yarns are white, thus in this case no action is required.

8. Fill the cells for the yarns in the carriers of the second track

In the same way, the yarns of the carriers from the second track have to be visualized by filling the corresponding cells. In the example here, only carrier 8 is not white, so all the other cells can remain white.

4.5 Common patterns for tubular structures

4.5.1 Common patterns for tubular structures with a floating length of 2

Patterns with a floating length of 2 (2:2-1) are commonly used, because all possible (though not all available) slots on the horn gears are occupied, and the machine is used effectively.

If all carriers in one track have yarns of one colour and the carriers from the opposite track have yarns of another colour, then vertical stripes are created (Fig. 4.11a and 4.11b), as already demonstrated in the previous example.

Horizontal stripes are created if the colours are alternated in both the tracks "1 black, 1 white," as demonstrated in Fig. 4.11c and 4.11d. These two figures demonstrate that it does not matter when the counting of the colour starts. Figure 4.11 shows a colour effect with half horizontal stripes, where the colour arrangement is, for the clockwise moving carriers, "1 black, 1 white," and for the counter-clockwise moving carriers, "all white."

For machines with a large number of carriers, broken lines in one direction, or crossing lines, are commonly used, as shown in Fig. 4.12. In both cases two carriers,

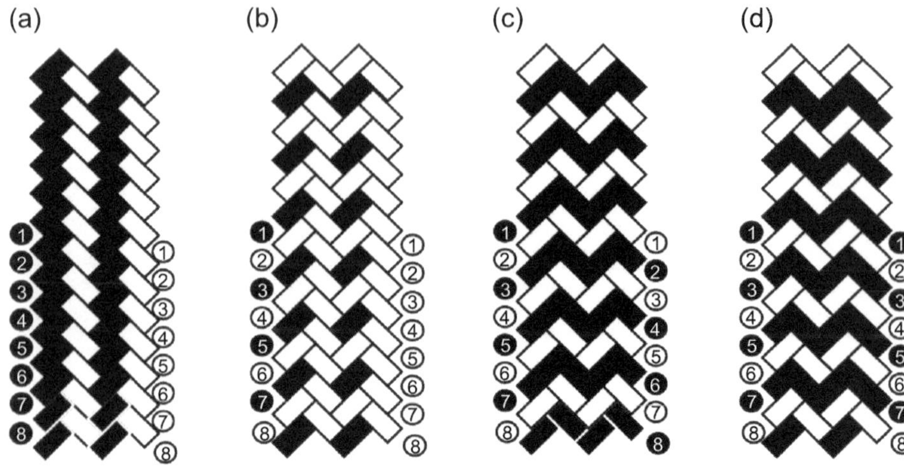

Figure 4.11 Colour pattern for 2:2-1 tubular structures: (a) vertical stripes – clockwise "all black," counter-clockwise "all white"; (b) broken vertical (or horizontal) lines – clockwise "one black one white", counter-clockwise "all white"; (c) horizontal lines – clockwise "one black one white", counter-clockwise "one white one black"; (d) same as (c), but the carrier counting starts one carrier later and in both direction "one white one black".

one after the other, are used with different colours in order for the stripes to be more clearly visible. The arrangement for Fig. 4.12a is:

Clockwise carriers: ten white, two black, two white, two black
Counter-clockwise carriers: all white.

and for Fig. 4.12b, two sets of black yarns in the counter-clockwise direction are added, so that the arrangement in both directions is:

ten white, two black, two white, two black.

These two examples are for a machine with 16-horn gears (each with four slots), having a total of 32 carriers.

4.5.2 Common pattern for tubular structures with a floating length of 1

The structures of type 1:1-1 are not very commonly used, because the machines are not used effectively – the carrier arrangement in such a case is "1 Full, 3 Empty." At such half occupation, twice as many horn gears are required as would be the case with 2:2-1 or 1:1-2 structures for the production of rope of the same diameter. Nevertheless, there are customers and cases which require such a pattern. The colour patterning of 1:1-2 structures is identical to that of 1:1-1, so it makes sense to analyse and understand this patterning as well. Figure 4.13 shows some of the basic variations for colour design of 1:1-1 tubular structures with 12 carriers.

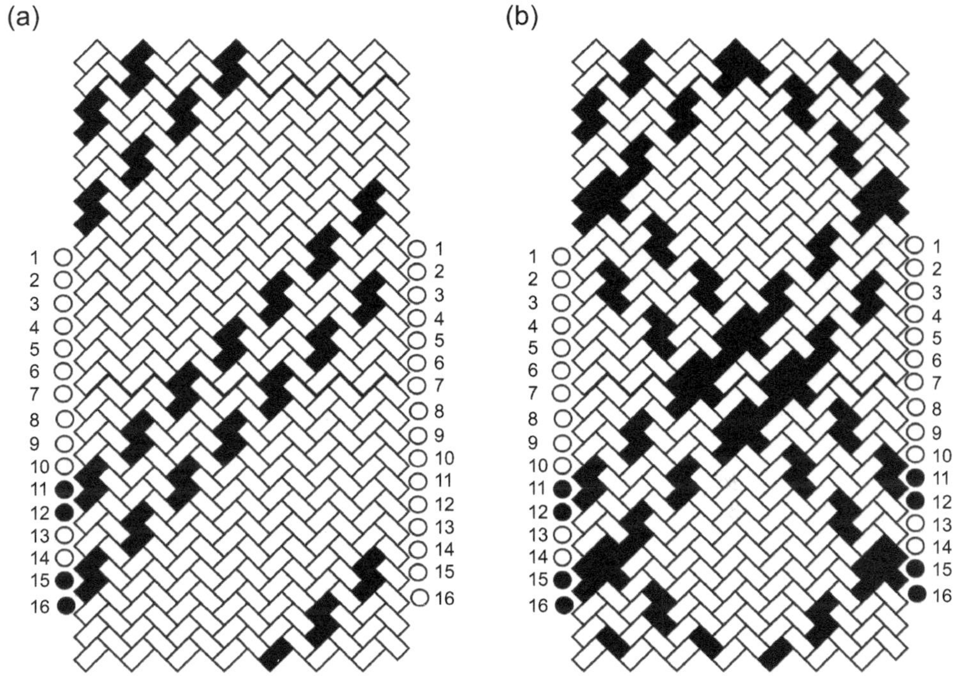

Figure 4.12 Colour pattern for 2:2-1 tubular structures with more carriers, here 32 carriers: (a) two broken diagonal lines in one direction – arrangement clockwise "10 white, 2 black, 2 white, 2 black," counter-clockwise "all white"; (b) crossing diagonal lines – in both directions "10 white, 2 black, 2 white, 2 black" (Following Goseberg *et al.*, 1981).

One typical effect is the checkmate effect (Fig. 4.13a), realized by separating the colours into the different tracks – one direction black, the other white.

Other typical effects for 1:1 structures are "twist" effects. Figure 4.13b shows the visible yarns for the colour arrangement of clockwise moving carriers set to "1 white, 1 black," and counter-clockwise carriers set to "all white."

If a "1 white, 1 black" colour arrangement is used for the counter-clockwise carriers as well, both "Z" (Fig. 4.13c and 4.13e) and "S" (Fig. 4.13d and 4.13f) direction stripes can be produced, where the direction depends on the starting position for the arrangement of the second track. If the carrier with coloured yarn (black or grey) of the second (counter-clockwise) track is set, as shown in Fig. 4.14a), then an "S" stripe will be produced, whereas if the carrier with the coloured yarn is shifted one position, a "Z" stripe will be produced.

Using larger machines, some more complex patterns can be achieved, since more combinations of colour arrangements are possible. For instance, when using a machine with 24-horn gears and 48 carriers (at full occupation), at half occupation – with a "1 Full, 3 Empty" arrangement for 1:1-1 structures – 12 carriers will move in each direction. Since 12 is divisible of 2, 3, 4 and 6 there are $4 \times 4 = 16$

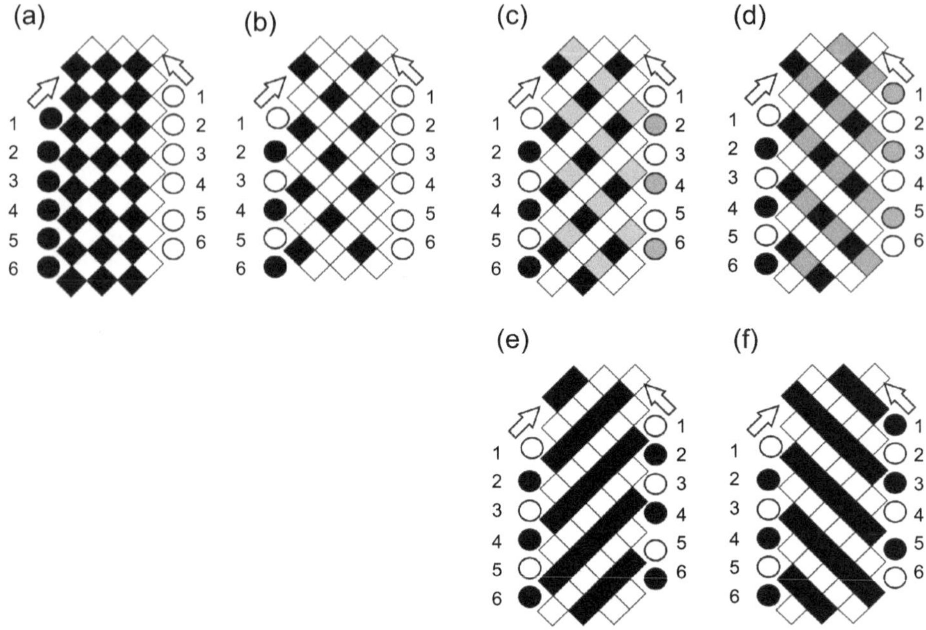

Figure 4.13 Colour pattern for 1:1-1 tubular structures with 12 carriers: (a) checkmate effect by separating the colours into the different tracks; (b) single dots, used as a preliminary step or basis for the next figures, here clockwise “1 white – 1 black,” counter-clockwise all white; (c) Z – twist effect, both directions “1 white – 1 grey”; (d) S – twist effect, both directions “1 white – 1 grey,” but the arrangement of the second one is shifted at one position; (e) is like (c) and (f) is like (d) but instead of grey a black colour is used, so that the twist effect is clear.

configurations possible. The most commonly used are presented in Table 4.1. and Figs 4.15 and 4.16.

4.5.3 Common pattern for tubular structures with a floating length of 1 and ply yarns

All the patterns, discussed in the previous section, for braiding patterns with a floating length of 1 (type 1:1-1) can be applied and used for a pattern of 1:1-2, 1:1-3, etc., using both ply yarns or using separate carriers for each yarn. The most common are 1:1-2 structures, because here the carriers of four-slot horn gear machines are fully occupied, and the diameter of the product is larger. Figures 4.17, 4.18 and 4.19 demonstrate the appearance of some of the braids. It can be seen that two yarns of the same colour and same path reinforce the effects and make them more visible. The colour arrangements for these figures are also shown in Table 4.2.

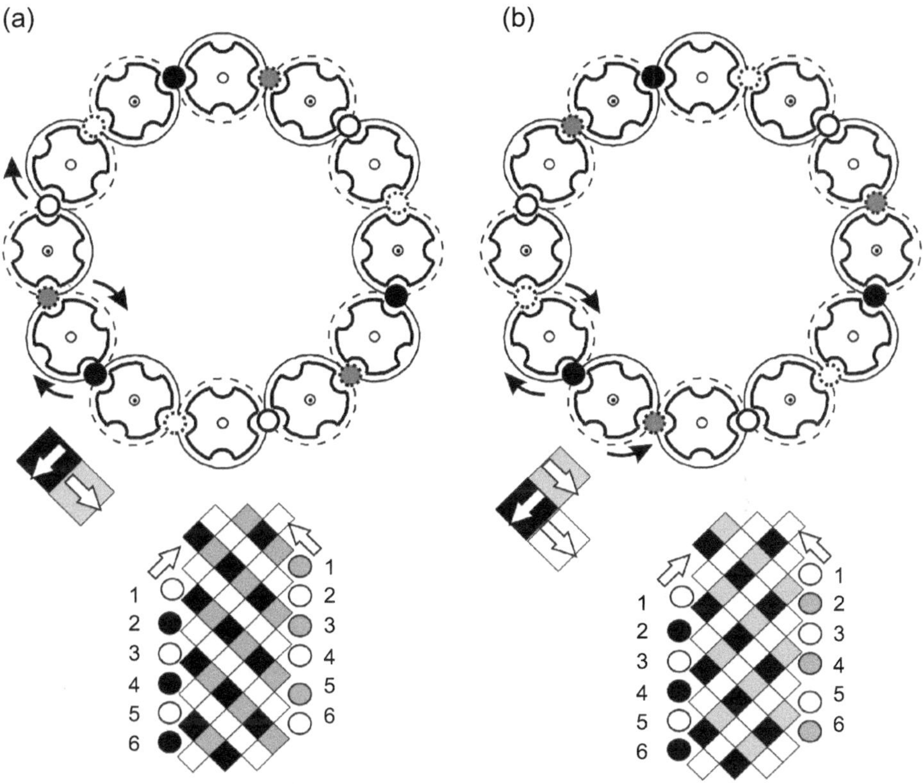

Figure 4.14 Colour arrangement and the strips direction for 1:1-1 pattern: (a) the grey yarn builds a visible piece after the black one and S strip is created; (b) because of the shift to one position colour arrangement the grey yarn builds a visible piece before the black one and Z strip is created.

4.5.4 Common pattern for tubular structures with a floating length of 3

Tubular braids with a floating length of 3 are not common. They require machines with six-slot horn gears where the carriers are arranged "1 Full, 1 Empty," but such machines are of larger size and are not standard products. Such machines are mostly used in the overbraiding of cables with metal wire or, more recently, for braiding with glass or carbon fibres, in all cases because of the lower crimp and larger floating length for the yarns or wire. The colour effect of combining two colour yarns or fibres in the case of floating length of 3 are similar to those with a floating length of 2. Figure 4.20a shows vertical stripes produced using one colour for all carriers in one direction and another colour for all carriers in the opposite direction. Horizontal stripes can be produced as well, by alternating one black and one white yarn in both directions, as shown in Fig. 4.20b, or equally by using one black then two white yarns in both directions, as presented in Fig. 4.20c.

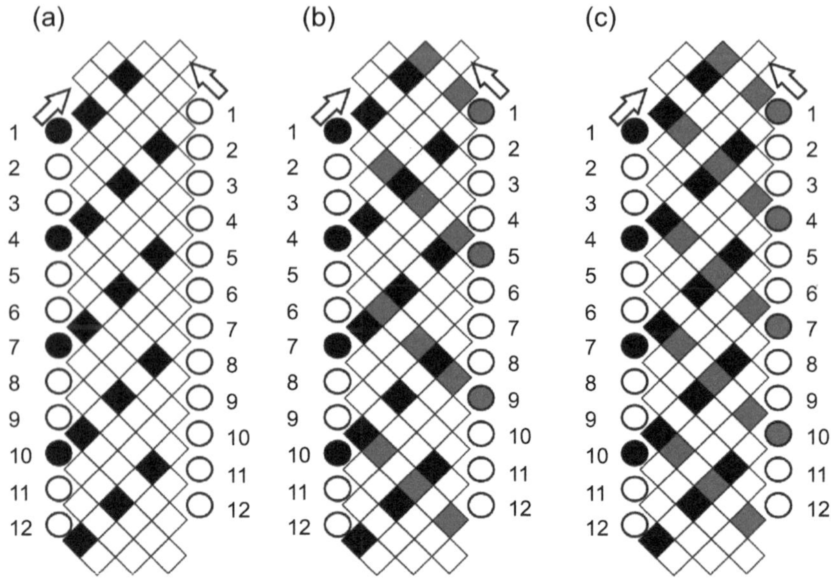

Figure 4.15 Colour arrangement and the strips direction for 1:1-1 pattern: (a) appearance, if only the clockwise moved carriers are arranged with 1 black: 2 white as a basis case; (b) like (a) but counter-clockwise 1 black: 3 white; (c) like (a) but counter-clockwise 1 black: 2 white.

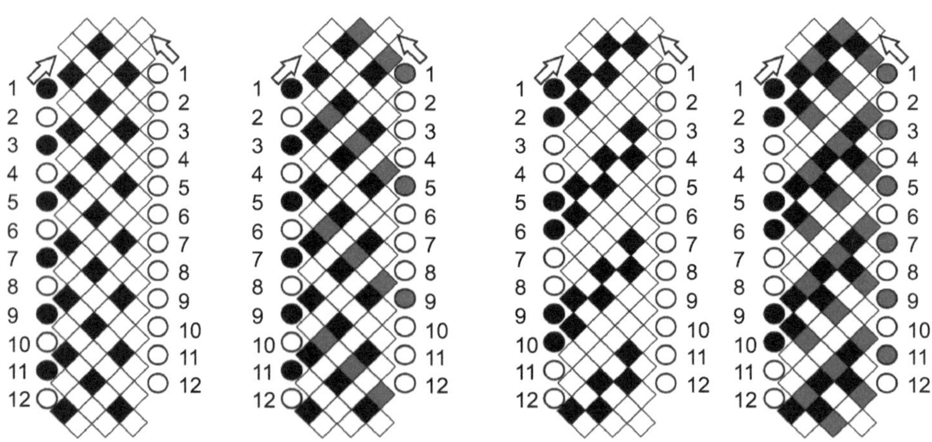

Figure 4.16 Colour arrangement for the cases 1 black: 1 white; and 2 black- 2 white in the clockwise moving carriers.

Alternating three black and three white yarns in both directions can be used to produce "V" like effects in both directions, as shown in Fig. 4.21b and 4.21c. The basic "V" pattern can still be easily recognized if only one of the tracks is arranged with two colours and the other kept all white, as shown in Fig. 4.21a.

Table 4.1 Colour arrangement combinations for machine with 24 carriers (12 carriers in one direction)

	Clockwise	Counter-Clockwise
Figure 4.15a	1 black: 2 white	All white
Figure 4.15b	1 black: 2 white	1 black: 3 white
Figure 4.15c	1 black: 2 white	1 black: 2 white
Figure 4.16a	1 black: 1 white	All white
Figure 4.16b	1 black: 1 white	1 black: 3 white
Figure 4.16c	2 black: 2 white	All white
Figure 4.16d	2 black: 2 white	1 black: 1 white

Table 4.2 Colour arrangement combinations for machine with 48 carriers (24 carriers in one direction) (1:1-2 pattern)

	Clockwise	Counter-Clockwise
Figure 4.17a	2 white: 2 black	2 white: 2 black
Figure 4.17b	2 black: 2 white	2 white: 2 black
Figure 4.18a	4 black: 4 white	2 black: 2 white
Figure 4.18b	2 black: 4 white	2 black: 4 white
Figure 4.19a	2 black: 4 white	2 black: 6 white
Figure 4.19b	2 black: 2 white	2 black: 6 white

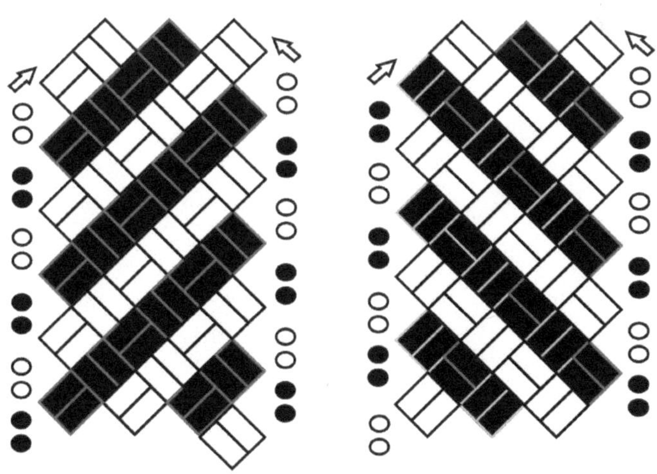

Figure 4.17 Diagonal lines at 1:1-2 structures. Identical to the explanation on Fig. 4.14, shifting the arrangement with carriers changes the crossing points of the yarns and thus the orientation of the diagonal lines as well.

Figure 4.18 Colour arrangements for pattern 1:1-2 on a machine with 48 carriers with diagonal short lines effects (following Goseberg *et al.*, 1981).

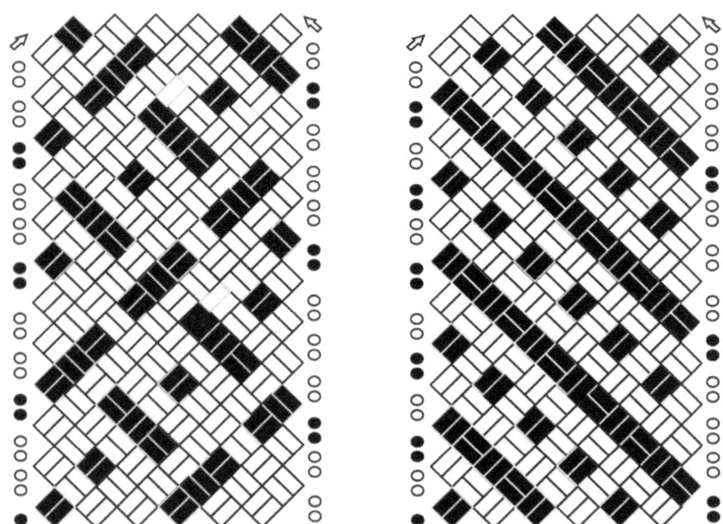

Figure 4.19 Colour arrangements for 1:1-2 pattern on a machine with 48 carriers with diagonal and points effects (following Goseberg *et al.*, 1981).

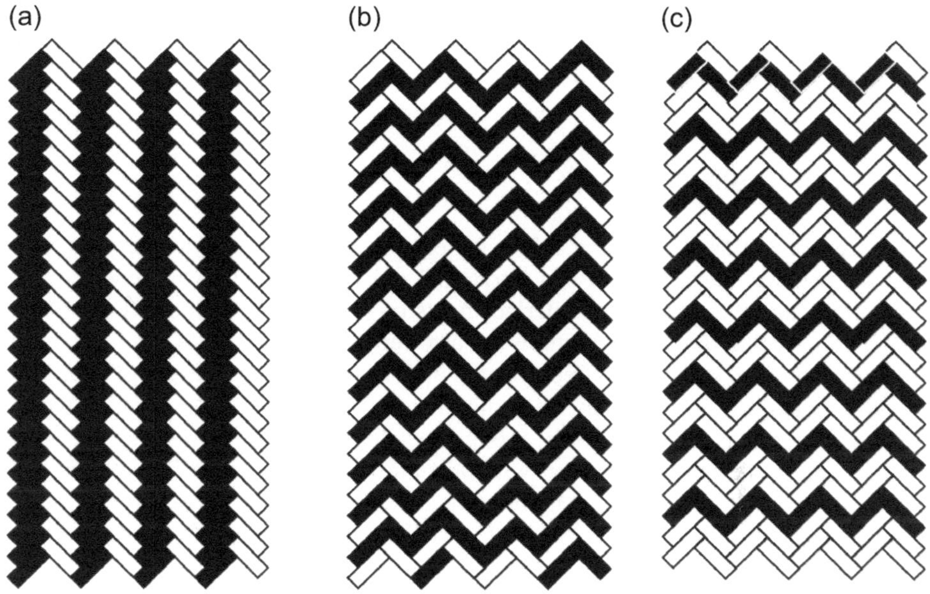

Figure 4.20 Vertical stripes for sample with floating length of 3 (a) and two kinds of horizontal stripes (b) and (c).

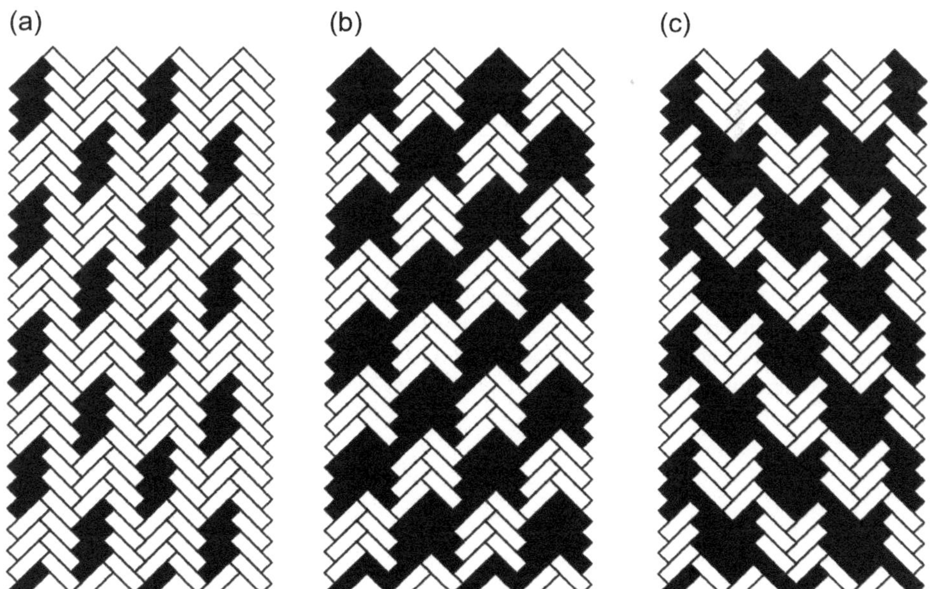

Figure 4.21 Colour effects using three black and three white yarns one direction and all white in the other direction (a) as a basic for the pattern with floating length of 3; (b) both directions three black – three white, (c) like (b) but the arrangement of the second track shifted with three carriers.

Generally, a floating length of 3 permits more combinations of colours, not only because of the larger floating length, but also because of the larger number of carriers required for producing such structures.

4.6 Conclusions

As presented in this chapter, the colour patterns for tubular braids are very limited. Mainly diagonal lines, vertical and horizontal stripes, and some checkered effects or similar, can be created using interlacements of yarns of two colours. The effects which are possible for 1:1-1 structures are also reproducible on 1:1-2 (or 1:1-3, etc.) machines and vice versa. The effects for the braids with a floating length of 2 (2:2-1) can also be applied for machines with a floating length of 3, but these permit a larger number of combinations of colours. Using specially prepared paper, where the cells are merged in the proper direction with suitable floating length, new patterns can easily be created by hand. Some modern methods for the pattern design of the fabrics using computer software are presented in Chapter 13 'Computer assisted design (CAD) software for the design of braided structures'.

4.7 Sources of further information and advice

The patterning of tubular structures is not well described in older literature, because at that time flat fabrics were more used in the clothing sector, and tubular structures were basically used only for ropes. However, a more detailed explanation with examples can be found in Engels (1994) and Goseberg *et al.* (1981).

References

Engels, H. (1994): Handbuch der Schmaltextilien. Die Flechttechnologie. Teil I: Maschinen und Verfahren zur Erzeugung konventioneller Geflechte. Mönchengladbach.

Goseberg, F., Brabender, K. and Moeller, P (1981): Textiltechnik Maschinengeflechte. Ausbildungsmittel Unterrichtshilfen. Frankfurt am Mai: Arbeitgeberkreis Gesamttextil.

Colour design of flat braids

5

5.1 Introduction

Flat braids have been used historically for decorating special and ceremonial clothing and uniforms in many cultures. Friendship bracelets are also often handmade flat braids, as are some shoe laces, although tubular braids are preferable for this purpose. Flat braids are also used as accessories for clothing and for bags and as leather belts. The colour design for all these is a very important part of the design of the braid, together with its structure. The basic principles of the development of colour patterns for tubular braids are valid for flat braids as well, with some changes connected with the specifics of flat braided products. This chapter focuses on the possible colour patterns of flat braids, with their specific properties, assuming that the reader is already familiar with colour designs for tubular braids.

5.2 Basics of colour patterning of flat braids

For the colour patterning of flat braids the same square mesh is used as for tubular braids (Fig. 4.1 of Chapter 4). The design process is speeded up and simplified if, in the yarn direction, every so many cells are merged in accordance with the floating length of the structure. The yarns of the flat braids can be represented as lying in one plane, rather than on a tube as with tubular braids. The observer of such a braid is able to see only one of the sides (Fig. 5.1a). Every yarn is visible every so many cells, depending on the floating length of the ridge, and after that the yarn moves under other yarns and so is not visible (Fig. 5.1b). Representing the visible yarns as filled or coloured cells, the colour pattern can be represented as shown in Fig. 5.1c. The yarns on the edges, which are moving from the back (invisible) side to the front (visible) side, are commonly represented by triangles (Fig. 5.1d). For flat braids, each yarn has visible parts in *both* directions, in both ridges, because every carrier turns at the edge horn gears and then moves back in the opposite direction. This is one of the most important differences in the patterning of the flat braids compared to tubular braids.

Because the carriers, moving at every horn gear, create a visible part in one direction, the number of ridges in the flat braids is equal to the number of horn gears (and not to half of it, as for tubular braids where half of the tube is not visible to the observer), as shown in Fig. 5.2. Thus, it can be assumed that more carriers on one braiding machine permit wider patterning possibilities. Accordingly, one common

Braiding Technology for Textiles. http://dx.doi.org/10.1533/9780857099211.1.89

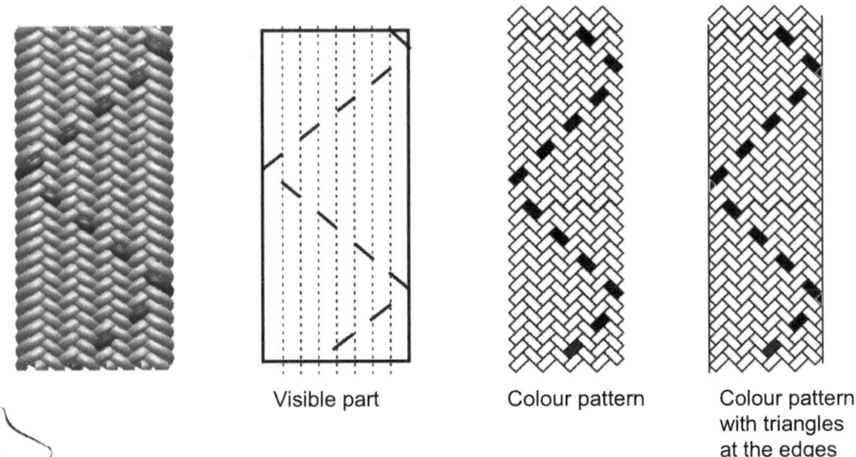

Figure 5.1 Principle of the simplified presentation of the flat braids for colour design.

binomial notation for describing braiding machines gives the maximum number of carriers which can be used on that machine and the maximum floating length. For instance, a 33/2 machine has 33 carriers with floating length of 2, and a 50/1 machine has 50 carriers and a floating length of 1. So the higher the machine number, the more the availability of patterning possibilities.

5.3 Pattern development sequence for flat braids

5.3.1 Development steps

For developing colour pattern a method identical to that used for tubular braids is used, with specific differences taken into account.

1. Determine number of ridges to be drawn
   The number of horn gears of the machine is equal to the number of ridges of the flat braid. Therefore, a mesh with

$$\text{Drawn\_Ridges} = \text{Number\_Horn\_Gears} \quad [5.1]$$

   has to be prepared.
2. Determine the floating length
   The floating length (that is, the pattern type) is determined by the number of the slots per horn gear and the carrier arrangement (see Chapter 2).

$$\text{Floating\_length} = \frac{\text{Slots\_per\_horn\_gear}}{\text{Carrier\_arrangement\_repeat}} \quad [5.2]$$

3. **Determine the yarns in one group**
   The number of the full (occupied) positions of the carriers, which follow one after the other, without any empty places, determines the number of the yarns in one group. If these yarns are ply yarns, the number of yarns in the group is determined from the equation:

$$\text{Yarns\_in\_group} = (\text{Nr\_full\_positions\_without\_interruption}) \times (\text{Strands\_in\_ply\_yarns}) \quad [5.3]$$

   For instance, for the carrier arrangement "2 Full, 2 Empty", each group of yarns consists of two yarns. If each of the yarns is a double-ply yarn then the group is $2 \times 2 = 4$ yarns.
4. **Determine the repeat of the sample**
   The repeat of the sample is determined by the number of the carriers moving in one direction, which for flat braids is equal to the number of carriers (for full occupation equal to the number of the machine).

$$\text{Repeat} = \text{Total\_Number\_carriers} \quad [5.4]$$

   In some cases, for machines with a larger number of carriers, the repeat for the colour arrangement can be equal to a divisor of the carrier repeat.
5. **Set the colour arrangement**
   The colour arrangement is determined by the sequence of the colours of the bobbins. Only the shortest repeat is given; for instance:
   two carriers black: two carriers white.
6. **Select or prepare the suitable mesh and mark the visible area**
   The suitable mesh consists of the determined number of ridges with the floating calculated where a certain number of cells are parallel, depending on the number of yarns in one group. The visible area should have at least as many lines (rows) as the repeat. The length of each line (number of cells) is equal to:

$$\text{Visible\_cells} = \text{Ridges} \cdot \text{Floating\_length} \cdot \text{Yarns\_in\_one\_group} \quad [5.5]$$

7. **Fill the cells along the path of each yarn for the carriers of one of the tracks.**
   For a structure of type X:X–Y, which means a floating length of X and Y yarns in one group, $X \times Y$ cells have to be filled, then $X \times Y$ cells have to be kept empty, with this pattern repeating for all cells, representing the yarn.
   This process is demonstrated in Fig. 5.3 for different patterns on "normal" square mesh, and on specially prepared meshes for different patterns. Figure 5.3a shows the basic mesh, suitable for all cases. On Fig. 5.3b the mesh is filled for 1:1-1 structure (floating length of 1, one yarn in the group). One cell is filled for the visible yarn part and one cell is skipped for the yarns of the opposite ridge. Filling the cells for yarn with a floating length of 3 on a normal mesh is shown in Fig. 5.3c, and on specially prepared mesh in Fig. 5.3d. Similarly, for a floating length of 2, the filling is shown on a normal mesh in Fig. 5.3e and on a specially prepared mesh in Fig. 5.3f. For structures with more yarns in one group, such as 1:1-2 (floating length of *one* group, *two* yarns in the group), every two yarns run parallel to each other. In this case $1 \times 2 = 2$ cells are filled, and two cells are skipped, as shown in Fig. 5.3g for a normal mesh and in Fig. 5.3h for a specially prepared mesh.
8. **Extend the filling in both regions, upper and lower, so that the complete interlacement is visible.**

5.3.2 Example

Create the pattern for a machine with eight horn gears and 17 carriers, 1 Full, 1 Empty (Fig. 5.2a). The horn gears have four slots each, the turning gears five slots. The given colour arrangement is 4 black, 6 white, 1 black, 6 white.

1. Determine number of ridges to be drawn
   The visible ridges are 8.

$$\text{Drawn\_Ridges} = \text{Number\_Horn\_Gears} = 8$$

2. Determine the floating length

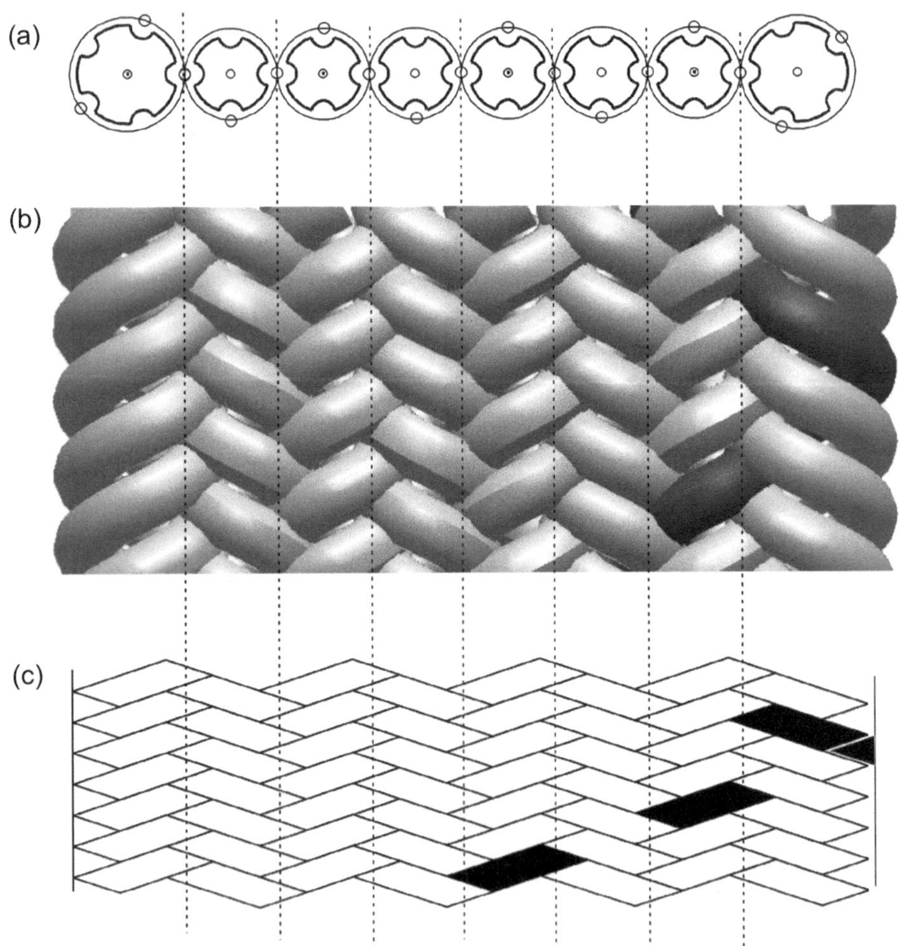

Figure 5.2 Horn gears and ridges for the flat braids.

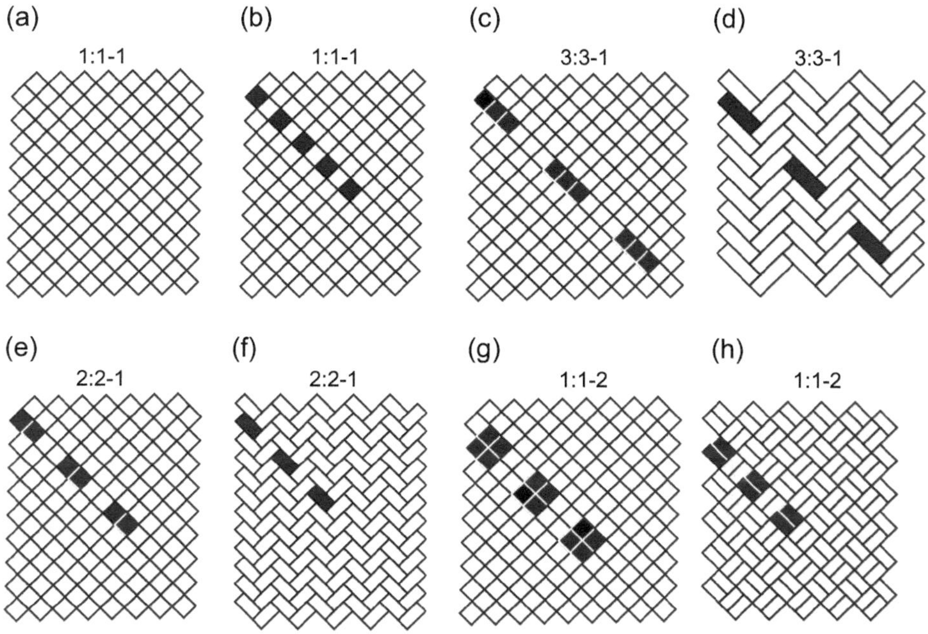

Figure 5.3 Filling the cells at different pattern and floating lengths, once for simple square mesh (b, c, e, g) and once for specially prepared mesh for these braiding types (d, f, h).

The floating length is:

$$\text{Floating\_length} = \left[\frac{\text{Slots\_per\_Horn\_Gear}}{\text{Carrier\_arrangement\_repeat}}\right] = \left[\frac{4}{1\text{Full} + 1\text{Empty}}\right] = \left[\frac{4}{2}\right] = 2 \quad [5.6]$$

The floating length at the turning edges is the same, as only the integer part of the division has to be taken into account, and so (5 slots)/(2 repeat) = [2.5] = 2.

3. Determine the yarns in one group
Here a single yarn per carrier is used. For a "1 Full, 1 Empty" arrangement the group consists of one yarn.
4. Determine the repeat of the sample
The repeat of the sample is 17 (total number of carriers).
5. Set the colour arrangement
4 black, 6 white, 1 black, 6 white.
6. Select or prepare the suitable mesh
From the floating length 2 it follows that along the yarn path every two cells can be merged together, and paper for type 2:2-1 can be used (Fig. 5.3). Seventeen yarns are required for the repeat.
7. Fill the cells along the path of each yarn.
In the current case, with a 2:2-1 structure (floating length 2, one yarn in a group) two cells have to be filled, then two kept empty, until the last ridge is reached. Turning back can be

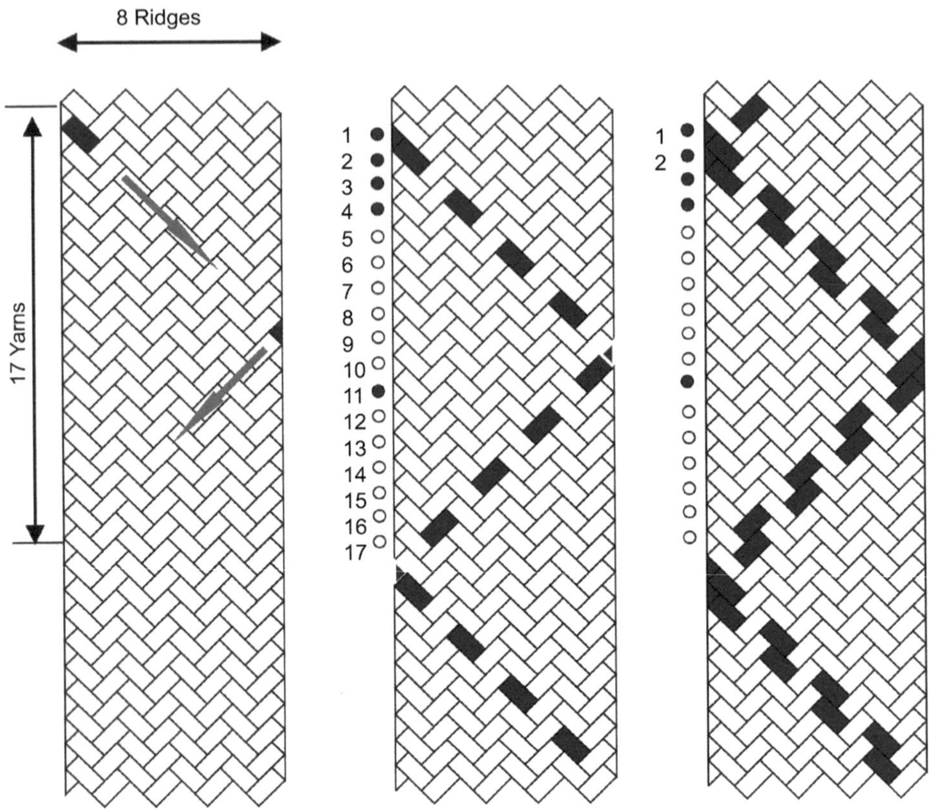

Figure 5.4 Selection of the mesh and marking visible cells for the first two yarns.

presented as a filled triangle and then again two cells filled and the two following skipped for the ridges in the opposite direction (Figs 5.4 and 5.5).

8. Extend the filling in both regions, both upper and lower, so that the complete interlacement is visible (Fig. 5.6).
   This is actually necessary for all carriers except the first, because they all have visible parts above their formal beginning on the drawing.

5.4 Popular patterns for flat structures with a floating length of 1

5.4.1 Track issues

A pattern with a floating length of 1, with one or two yarns in a group, usually requires a braiding machine with an even number of four-slot horn gears and two

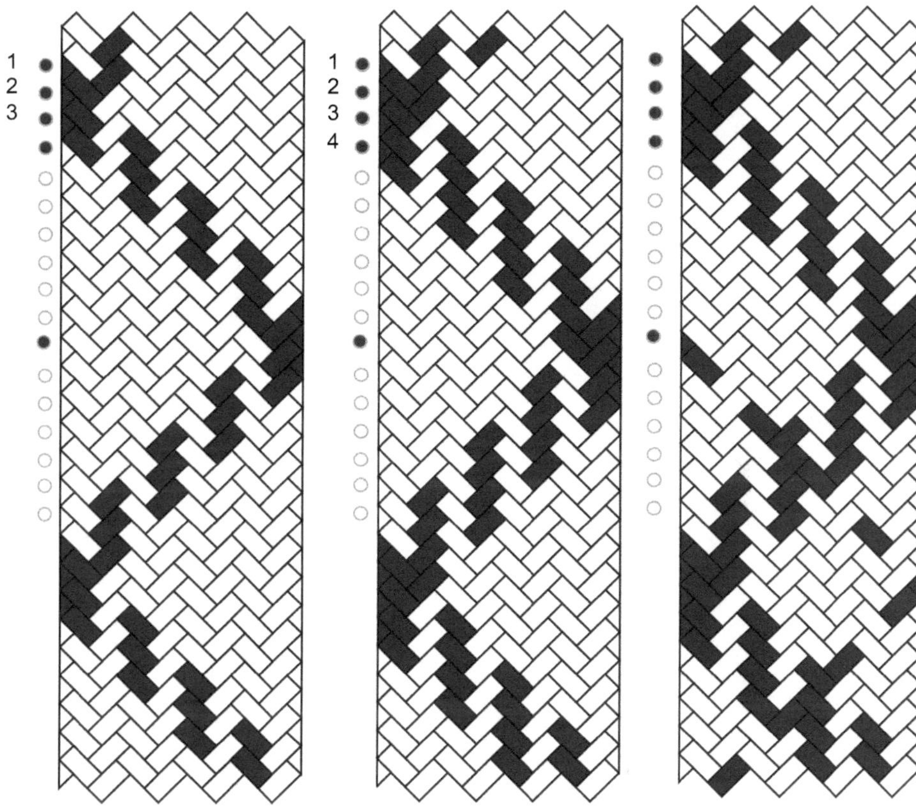

Figure 5.5 Filling the cells for the yarns nr. 3, 4 and 11. The other yarns between these are assumed here to be white. In the case of any other colour, their cells have also to be filled in the same way.

horn gears on the sides with six slots each (Fig. 5.7a). The full utilization of the machine is possible at a carrier arrangement "2 Full, 2 Empty." For this arrangement, after changing their direction of motion, each pair of carriers has to move into the empty positions between the other carriers. For this, the turning horn gears need to have two more sections than the normal horn gears. If any second carrier of the track marked on Fig. 5.7a as black is removed, then only half of the carriers will work, and a braid with nine yarns in a pattern 1:1-1 will be built on the same machine. This pattern can be identical to the 1:1-2 pattern with full occupation, but since each yarn is only presented once, the colour effect will be less effective than if two yarns in a group are used. The second disadvantage of the arrangement of Fig. 5.7b is that the product will have only half of the width, because of half the number of carriers and yarns. If such a machine (Fig. 5.7a and 5.7b) is an 18/1 machine, with 18 carriers at full occupation, and the last four-slot horn gear is demounted and the six-slot horn gear connected to the other horn gears (Fig. 5.7c),

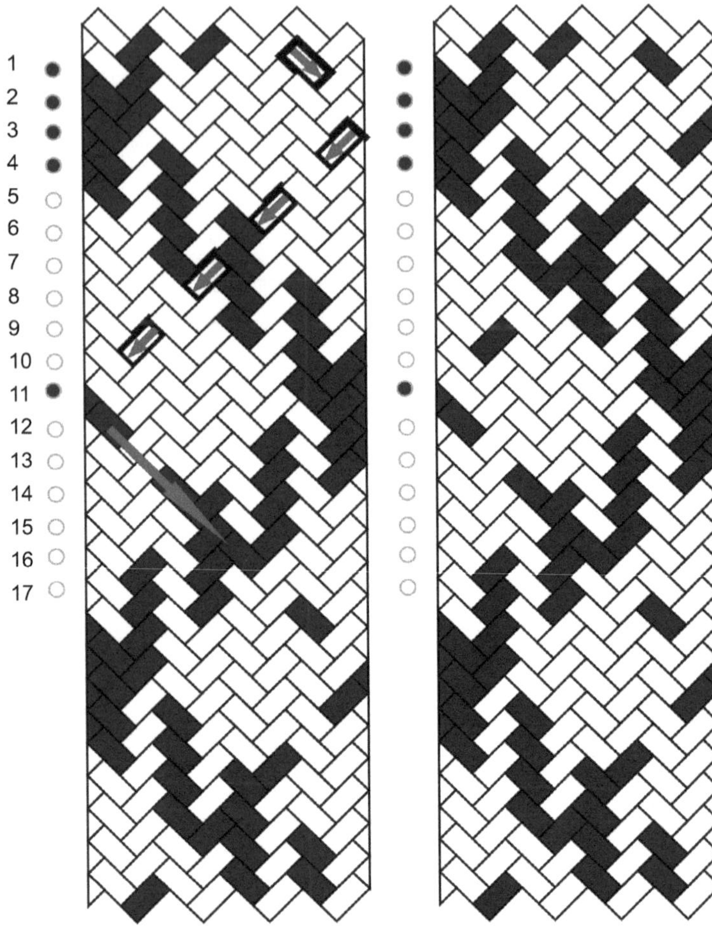

Figure 5.6 All visible cells of each yarn have to be filled. The most convinient starting point for the drawing of the yarn of carrier 11 is the place, where it changes its direction. This is in the middle of the repeat (a). All visible yarn pieces on the drawing have to be filled with the corresponding colour (b), both before and after the starting point.

then the machine will be turned into a 16/1 machine – with 16 carriers if fully occupied (Fig. 5.7d). Such machines, with an odd number of regular four-slot horn gears are seldom used, but they allow production of some different patterns to machines with an even number of regular four-slot horn gears. Examples are presented in a separate section as "Patterns on modified machines." Again, both 1:1-2 and 1:1-1 pattern types are possible, where the first one has full occupation with 16 carriers arranged "2 Full, 2 Empty," and a second pattern will be produced if every second carrier is removed, so that a "1 Full, 3" E arrangement is achieved (Fig. 5.7e).

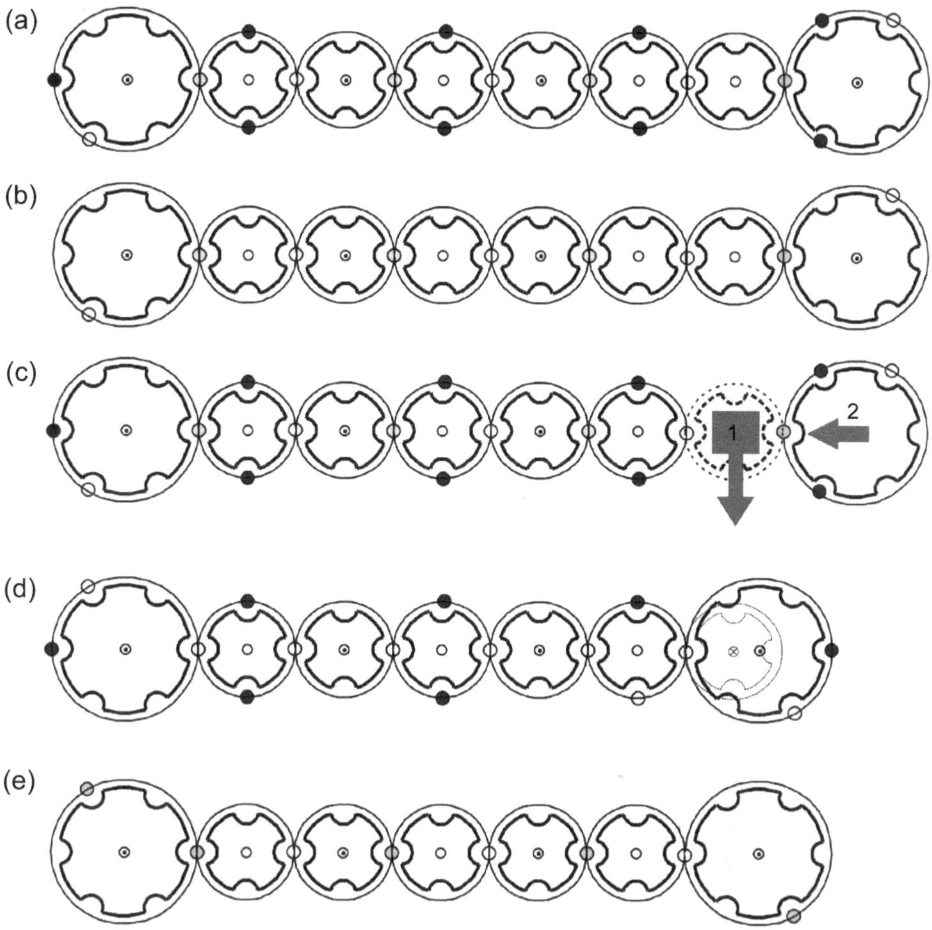

Figure 5.7 Horn gears and carrier arrangement for flat braiding machine with floating length of 1 and: (a) 18/1 machine for 1:1-2 pattern; (b) removing every second carrier the 1:1-1 pattern can be produced; (c) modification on the machine using reduction of one regular horn gear; (d) modified machine 16/1 for 1:1-2 pattern; (e) modified machine with partial occupation for 1:1-1 pattern.

5.4.2 Pattern with two yarns in a group (structure 1:1-2)

The following examples demonstrate popular patterns for 1:1-2 structures on a flat braiding machine. Figure 5.8 presents a pattern with angled colour stripes (Lepperhoff, 1953, p. 266; Engels, 1994, p. 184). The machine used for this pattern is configured as shown in Table 5.1.

Because the number of carriers (50) is not divisible by the colour repeat length of 4, at the end two carriers with the first colour have to be used. The repeat of 2 black and 2 white carriers is repeated 12 times, thus 12 angled stripes are visible. If the number of the repeats is reduced, and two black carriers are set between them, then

Table 5.1 Machine configuration for the pattern, presented in Fig. 5.8a

Machine number	50/1
Carriers	50
Horn gears (number × slots per horn gear)	1 × 6, 22 × 4, 1 × 6
Carrier arrangement	2 Full: 2 Empty
Colour arrangement	2 black: 2 white (repeat length 4)
Total bobbins	26 black, 24 white

Figure 5.8 Pattern with angled stripes (a) only stripes (b) interlaced in squared (Lepperhoff, 1953; Engels 1994).

the intersection points are changed, and squared or rectangular effects can be produced. The number of stripes in the square corresponds to the number of the repeated colour repeats. Here it has to be mentioned that at no configuration can a clear square effect be produced. In some cases, regions between the squares have a different filling. Figure 5.8b demonstrates the effect if the colour is repeated five times and there

Table 5.2 Machine configuration for the pattern, presented in Fig. 5.8b

Machine number	66/1 Sample suits as well for machines with 30, 42, 54, 66, 78, 90 carriers
Carriers	66
Horn gears (number × slots per horn gear)	1 × 6,40 × 4, 1 × 6
Carrier arrangement	2 Full: 2 Empty
Colour arrangement	(2 black: 2 white) × 5 times; 2 black colour repeat length 22
Total bobbins	36 lack, 30 white

Table 5.3 Machine configuration for the pattern, presented in Fig. 5.9a

Machine number	50/1
Carriers	50
Horn gears (number × slots per horn gear)	1 × 6, 22 × 4, 1 × 6
Carrier arrangement	2 Full: 2 Empty
Colour arrangement	1 black: 1 white Colour repeat length 2
Total bobbins	25 black, 25 white

are five stripes in any square. The same effect appears if the colour effect is repeated three times (not illustrated here). The arrangement for the machine in order to produce the sample in Fig. 5.8b is given in Table 5.2.

Another common possibility for colour patterning is the use of vertical stripes. These can be achieved if the two colours alternate, and because for the 1:1-2 pattern the group of carriers consists of two colours, the alternating colours are in the same group. Because of this, the stripes are as thin as the yarns. The pattern drawing is shown in Fig. 5.9a, and the configuration details are given in Table 5.3.

A more complex example, where different colours are used in the same group of yarns, is presented in Fig. 5.9b, and the corresponding machine and yarn configuration is given in Table 5.4. Slight changes in the middle area can lead to modifications of the pattern, and two more similar examples can be found in Engels (1994).

5.4.3 *Pattern with two yarns in a group (structure 1:1-2) on modified machines*

The most common flat braiding machines consist of an even number of horn gears. Adding or removing a four-slot horn gear, between the end six-slot horn gear and the

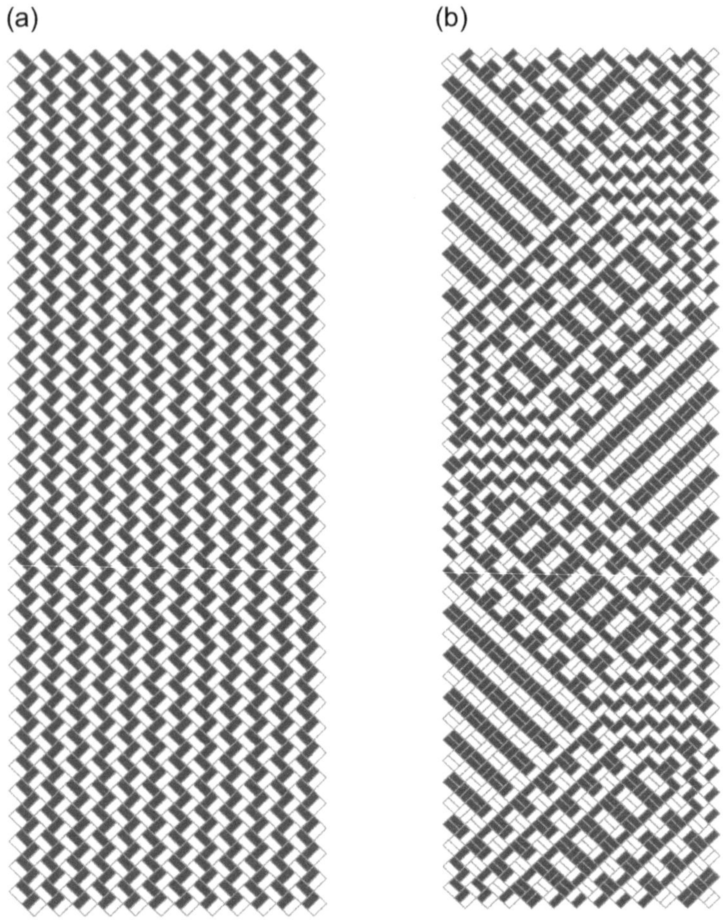

Figure 5.9 Using different colours in a group allows the production of vertical stripes (a) and some combinations (b).

Table 5.4 Machine configuration for the pattern, presented in Fig. 5.9b

Machine number	50/1
Carriers	50
Horn gears (number × slots per horn gear)	1 × 6, 22 × 4, 1 × 6
Carrier arrangement	2 Full: 2 Empty
Colour arrangement (here as colour 1 and colour 2 for black and white respectively)	(1 1 2 2) × 6 (1 2) × 6 1 × 1 (1 2) × 6 2 × 1
Total bobbins	25 black, 25 white

Table 5.5 Machine configuration for the pattern, presented in Fig. 5.10a and 5.10b (Lepperhoff, 1953; Engels, 1994)

Machine number	48/1, modified from 50/1
Carriers	48
Horn gears (number × slots per horn gear)	1 × 6, 21 × 4, 1 × 6
Carrier arrangement for Fig. 5.10a	2 Full: 2 Empty
Colour arrangement	(2 black: 2 white) × 12 times
Total bobbins 3 for Fig. 5.10b	24 lack, 24 white
Colour arrangement for Fig. 5.10b	(2 black: 4 white) × 8 times
Total bobbins 3	16 black, 32 white

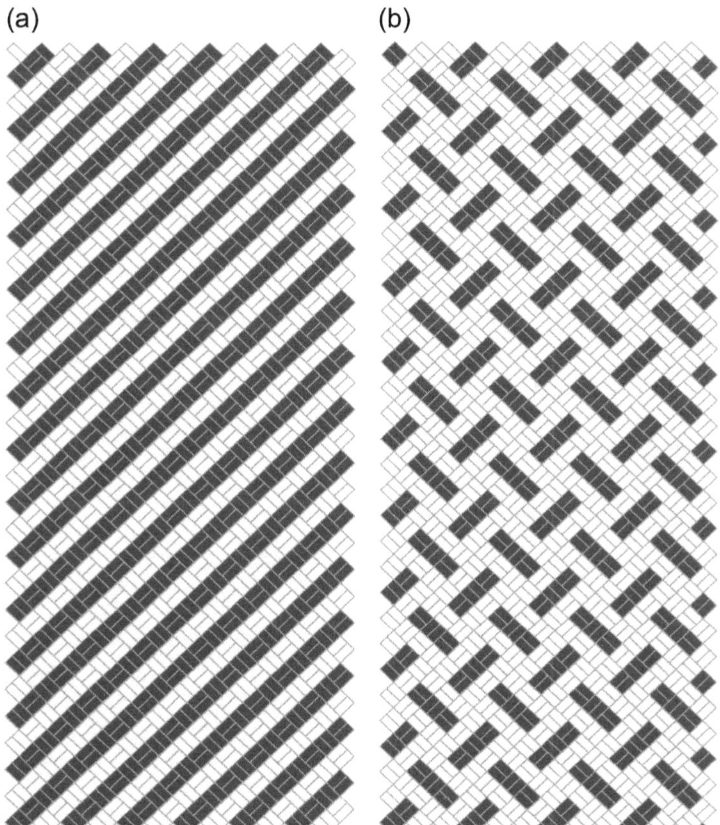

Figure 5.10 Pattern on modified machines (a) diagonal stripes (b) short lines.

next four-slot horn gear before it, changes the machine to one with an odd number of horn gears and an even number of carriers. This changed configuration allows the production of flat braids with similar effects to tubular braids – for example, diagonal stripes (Engels, 1994).

This machine modification has to be taken into account *before* the machine is produced. There are machines where the development engineers have considered this option and provided the required spare parts (horn gears, bearings, etc.), and there are machines where this was not planned for and is therefore not possible. Such possibilities extend the patterning capabilities on the machine, but can also make it significantly more expensive, because certain mechanical parts, such as the gears, have to be easily accessible, whereas in a standard machine they can be encapsulated in an oil bath.

Figure 5.10a shows an example of a pattern from such modified machine, where diagonal stripes are visible. The configuration for this sample is shown in Table 5.5.

5.4.4 Pattern with one yarn in a group (structure 1:1-1)

All colour effects produced from type 1:1-2 pattern, wherein the colour arrangement of the yarns in each group is the same (for instance 2 black, 2 white, several groups of 2 black and 2, 4 or 6 white, etc.), can also be produced from type 1:1-1 pattern, where only one yarn per group is taken. These samples require the same machines but at only half occupation (1 Full, 3 Empty), and because of this the braids will be of approximately half the width compared to the 1:1-2 patterns. Figures 5.11a and 5.11b show short diagonal lines and hound'stooth effects. The

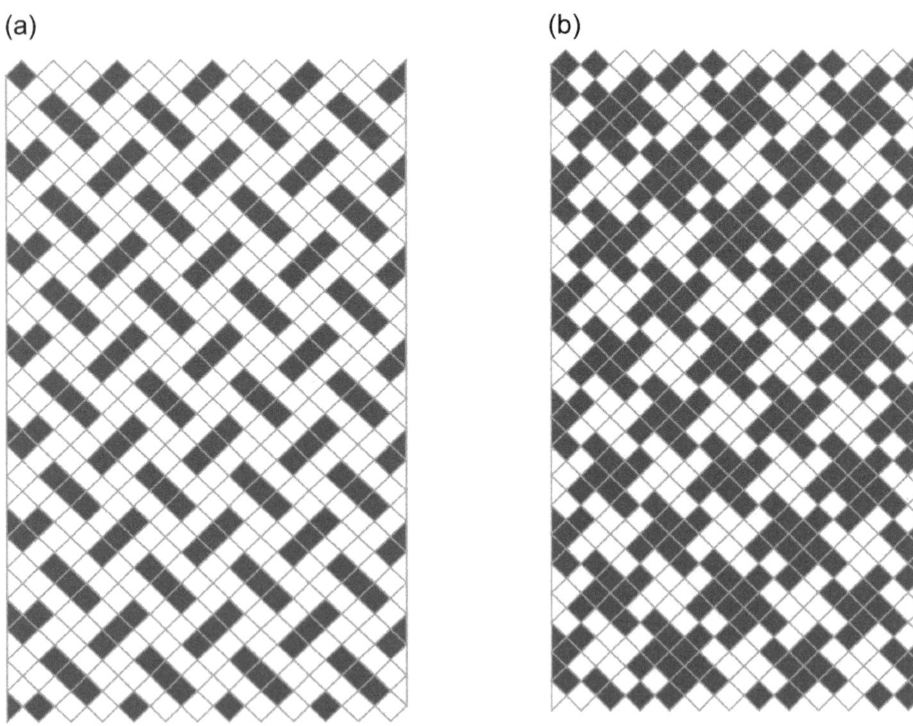

Figure 5.11 Short diagonal lines (a) and hound'stooth effect (b) on a machine with partial occupation (1 Full 3 Empty) with pattern 1:1-1.

Table 5.6 Machine configuration for the pattern, presented in Fig. 4b.11a

Machine number	48/1
Carriers	24
Horn gears (number × slots per horn gear)	1 × 6, 21 × 4, 1 × 6
Carrier arrangement	1 Full: 3 Empty
Colour arrangement	(1 lack: 2 white) × 8 times
Total bobbins 3	8 black, 16 white

Table 5.7 Machine configuration for the pattern, presented in Fig. 5.11b

Machine number	50/1
Carriers	25
Horn gears (number × slots per horn gear)	1 × 6, 22 × 4, 1 × 6
Carrier arrangement	1 Full: 3 Empty
Colour arrangement	(2 black: 2 white) × 6 times 1 black × 1 time
Total bobbins 3	6 black, 6 white

machine configurations are given in Tables 5.6 and 5.7. Long diagonal stripe effects are also possible, as shown in Fig. 5.12a and 5.12b, the configurations for which are presented in Tables 5.8 and 5.9.

5.5 Popular patterns for flat structures with a floating length of 2 (2:2-1)

5.5.1 Samples on machines with standard configuration

Flat braiding machines with a floating length of 2 are probably the most popular. Their horn gears are always in the configuration:

1 × 5 slots, N × 4 slots, 1 × 5 slots,

where N is an even number of horn gears. The arrangement is always "1 Full, 1 Empty", and there are always an odd number of carriers. Figure 5.13a and 5.13b show two simple colour effects, using two large groups of carriers of the same colour, here on a 33/1 machine with 16 horn gears, where a colour arrangement of 17 black and 16 grey yarn carriers is used. The interlacements of these large groups lead to the building of triangular zones of a single colour, with square zones between of the two colours, mixed. In the Fig. 5.13b in the middle of the two groups three carriers are used with white yarn.

Using shorter colour repeats, finer colour effects are possible. Figure 5.14a shows a variant of horizontal stripes on a 49/2 machine (horn gears 1 × 5, 22 × 4,1 × 5) and

Figure 5.12 Samples with diagonal stripes in a groups, realized by 1:1-1 pattern.

Table 5.8 Machine configuration for the pattern, presented in Fig. 5.12a

Machine number	50/1
Carriers	25
Horn gears (number × slots per horn gear)	1 × 6, 22 × 4, 1 × 6
Carrier arrangement	1 Full: 3 Empty
Colour arrangement	(1 black: 1 white) × 12 times 1 black × 1 time
Total bobbins 3	13 black, 12 white

Table 5.9 Machine configuration for the pattern, presented in Fig. 5.12b

Machine number	66/1
Carriers	33
Horn gears (number × slots per horn gear)	1 × 6, 30 × 4, 1 × 6
Carrier arrangement	1 Full: 3 Empty
Colour arrangement	((1 black: 1 white) × 5 times 1 black × 1 time) × 3 times
Total bobbins 3	18 black, 15 white

colour repeat (1 black, 1 white) × 24 and at the end 1 black. Figure 5.14b presents another pattern on the same machine, using the colour repeat (2 black, 2 white) × 12 and at the end 1 black, and Fig. 5.15a shows a hound'stooth effect on the same machine, with colour arrangement (4 black, 3 white) × 7. Fig 5.15b shows the same colour effect produced on a smaller machine – 21/2 with 21 carriers on ten horn gears. Another sample with a zigzag effect from a 49/2 machine, which can be produced on a machine with 25, 37,49, 61, 73, 85 and 97 carriers, is shown in Fig. 5.16a. The colour arrangement there is:

(1 black, 1 white) × 12
(2 black, 1 white, 1 black, 1 white) × 4
(1 black) × 1.

5.5.2 Samples on machines with modified configuration

If machines with a floating length of 2 are built with the possibility of changing some horn gears, then they can be modified for production of additional colour effects, as shown in Section 5.4.3 above. Figure 5.16b shows an example with a floating length of 2 and with a horizontal stripes effect, and Fig. 5.16c shows a hound'stooth effect. Both these examples can be produced on a 48/2 machine, modified from a 49/2, with horn gears (1 × 3 slots, 22 × 4 slots, 1 × 5 slots) and carrier configuration "1 Full, 1 Empty." For the horizontal stripes the carriers have the colour arrangement "1 black, 1 white," while for the hound'stooth effect "4 black, 4 white" is used.

Figure 5.13 Colour effects with two large groups of yarns with the same colour (a) and including third group in the middle of these groups (b).

5.6 Popular patterns for flat structures with a floating length of 3 (3:3-1)

For producing samples with a floating length of 3, horn gears with at least six slots are required (see Chapter 2, section 2.5.1). The configuration for flat braiding requires two horn gears at each end of the sequence of horn gears, which turn back

Figure 5.14 Horizontal stripes (a) and variant of diagonal stripes (b).

the direction of motion of the carriers and which have one (or three) slots more than normal horn gears (in this case with seven slots – Fig. 5.17). The most common horn gear configuration is 1 × 7 slots: N × 6 slots: 1 × 7 slots, where N is again an even number (Lepperhoff, 1953; Engels, 1994).

Figures 5.18a and 5.18b show example colour patterns for such samples, prepared for a 49/3 machine with horn gears (1 × 7, 14 × 6, 1 × 7), carrier arrangement 1 Full, 1 Empty, and 49 carriers. Of the samples: (a) has the colour arrangement (1 black, 1 white) × 24, and at the end 1 black; and (b) has (3 black, 3 white) × 4, (1 black, 1 white) × 12, 1 black at the end.

Figure 5.18c shows a sample from a modified 48/3 machine rearranged from a 49/3 machine to a horn gear configuration (1 × 5 slots, 14 × 6 slots, 1 × 7 slots). The

Figure 5.15 Hound'stooth effect patterns (a) colour arrangement for 49/2 carriers (4 black, 3 white) × 7 (b) on a smaller machine – 21/2..

carrier arrangement is again 1 Full, 1 Empty and all 48 carriers are equipped with alternating 1 black and 1 white yarn. Such machines are very seldom produced, but they allow clear, almost horizontally-orientated zigzag stripes.

5.7 Conclusions

As shown in this chapter, the colour patterns for flat braids are also limited, as for tubular braids. Mostly diagonal lines, vertical and horizontal stripes and hound'stooth

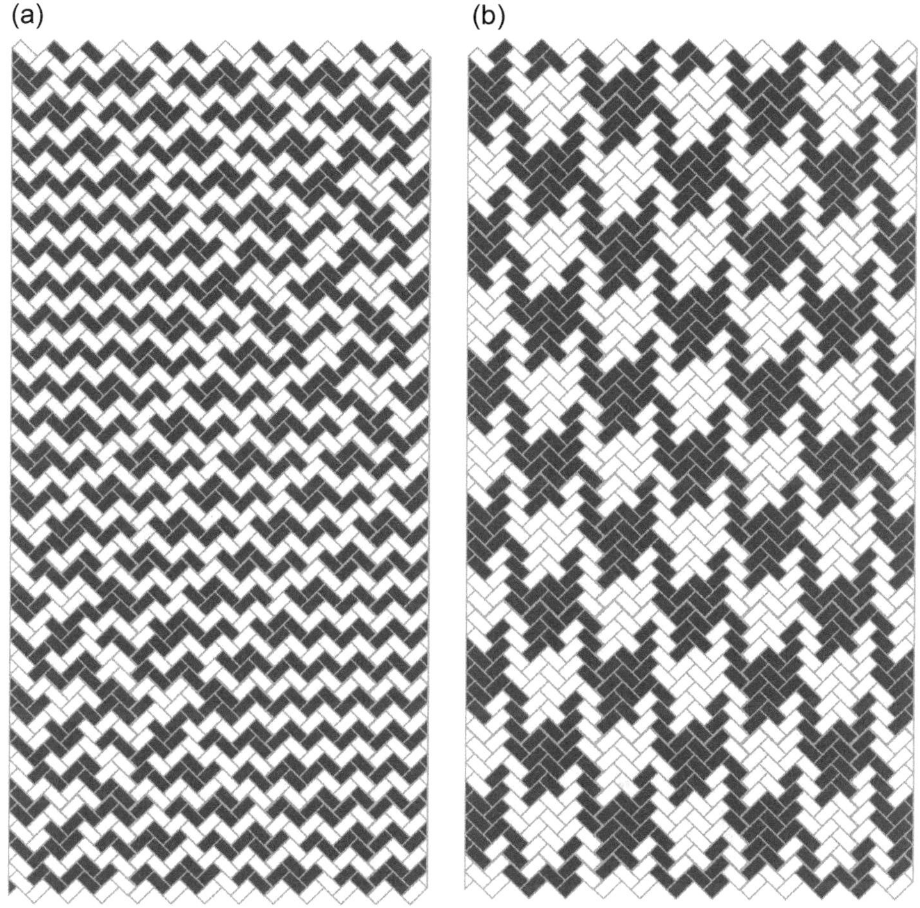

Figure 5.16 Samples with floating length of 2 on machines with modified configuration. (a) zigzag effect (b) horizontal stripes (c) the hound'stooth.

effects can be produced. For several of the patterns, special configurations of the machines are required, which is dependent upon either the possibility of changing the size of the horn gears and their gearing, or on building a completely new machine.

5.8 Sources of further information and advice

The colour patterning of flat structures in this chapter follows the explanation in Lepperhoff (1953) and Engels (1994), although some of the examples from these books are skipped, because they represent only slight modifications of other examples. According to Engels (1994), the patterns were given in the first edition of Lepperhoff in 1914, so they were published almost 100 years ago. Since this first

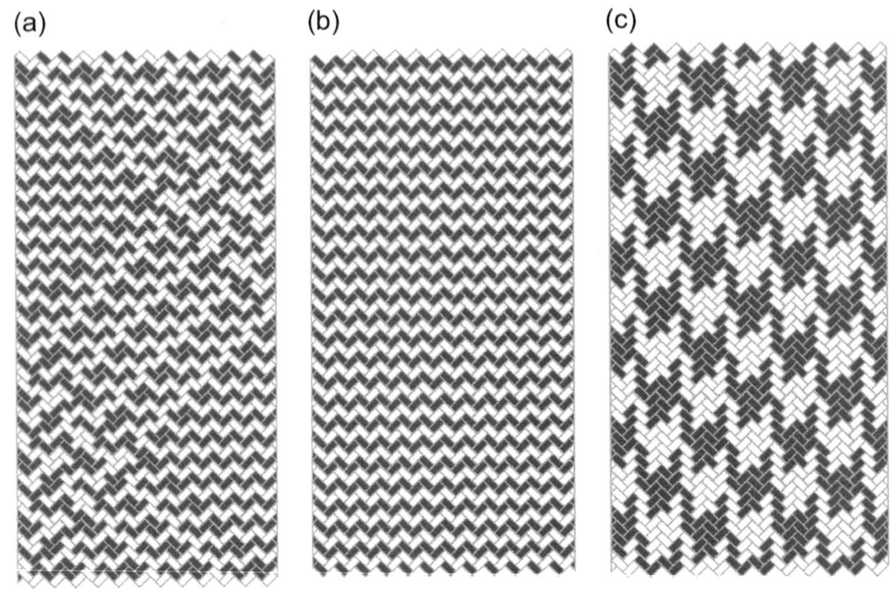

Figure 5.17 Example of horn gear configuration for producing of flat braids with floating length of 3.

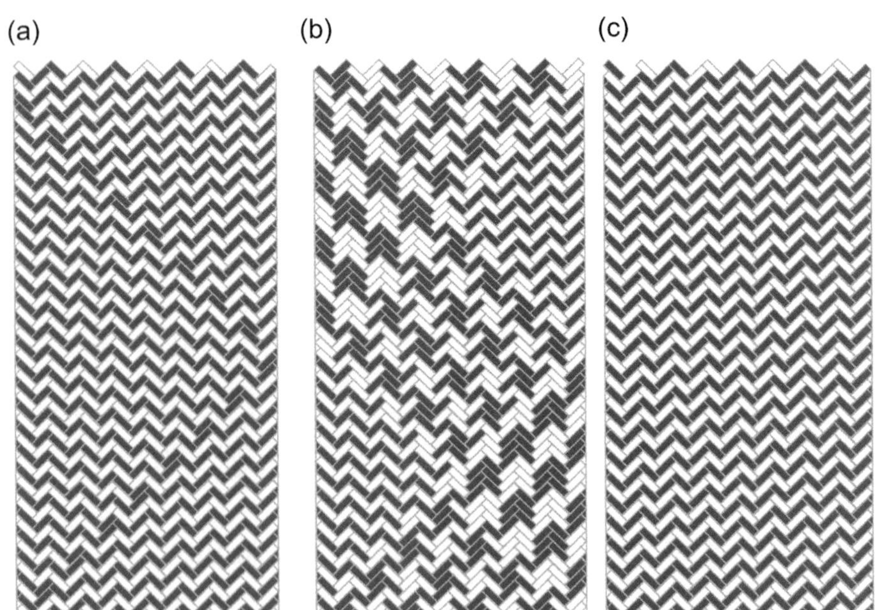

Figure 5.18 Examples of colour pattern of flat braids with floating length of 3 (a) colour arrangement (1 black, 1 white) × 24, and at the end 1 black; (b) has (3 black, 3 white) × 4, (1 black, 1 white) × 12, 1 black at the end. (c) modified 48/3 machine rearranged from a 49/3 machine with 1 Full 1 Empty carrier arrangement with and colour arrangement 1 black - 1 white.

edition is no longer available, both the other books are given as references. Several technical errors in these books have been corrected, since all patterning pictures in this chapter were constructed with the software TexMind Braider (Kyosev, 2012), described in Chapter 13, which allows fast picture generation at high resolution and fast checking of the colour sequences.

Except for the above-mentioned books, and the book by Goseberg *et al.* (1981), which again cites Lepperhoff's work, there is no other literature known to the author where the patterning of flat braids is described.

References

Lepperhoff, B. (1953): *Die Flechterei.* 3rd edn.: Eugen G. Leuze Verlag Saulgau/WTTBG.

Engels, H. (1994): Handbuch der Schmaltextilien. Die Flechttechnologie. Teil I: Maschinen und Verfahren zur Erzeugung konventioneller Geflechte, Mönchengladbach.

Goseberg, F., Brabender, K. and Moeller, P. (1981): *Textiltechnik Maschinengeflechte. Ausbildungsmittel Unterrichtshilfen.* Frankfurt am Mai: Arbeitgeberkreis Gesamttextil.

Kyosev, Y. K. (2012): TexMind Braider. Version 1.1. Mönchengladbach: TexMind UG. www.texmind.com.

Part Two

Maypole braiding machines and mechanics

Braiding machine components

6

6.1 Introduction

In this chapter, the basic modules and principles of maypole braiding machines are presented. The mechanical realization of the machine elements differs between producers, and some details of these differences are discussed in appropriate places, if they represent important developments during the history of braiding machines, or if they bring special advantages.

The yarn material is wound on bobbins, which are mounted on carriers (Fig. 6.1.). Every carrier usually holds one bobbin, and adjusts the yarn length (the yarn tension) during the braiding process. Every carrier follows a certain path, called a *track*. The track (for flat braiding machines) or a combination of tracks (for tubular and form braiding machines) determines the structure of the braid. The track determines the path for the motion of the carriers, but the carriers are moved forwards (or also backwards in 3D braiding) by the *horn gears*. Every horn gear is a disc with several cuts (slots) where the carrier foot fits. Each horn gear is mounted on a shaft, which has a corresponding gear. This transmits the motion from the motor to the mounted horn gear and the neighbouring gear. At the beginning of the braiding, all outer ends of the yarns go through the braiding point. Once the carriers start their motions, the yarns interlace and build the next pieces of braid at the braiding point. The produced braid is pulled out by a take-off system. The distance between each carrier and the braiding point does not remain constant during the braiding process and, as a result, the required length of yarn between the carrier and the braiding point changes. The differences in the required yarn length should normally *not* be compensated by yarn elongation (in most cases yarn elongation is not enough for such compensations), and a yarn length compensation mechanism has to be available on each carrier.

Because the carriers are the most important assembly on the braiding machines, the principles of how they work and the most common constructions are presented in Chapter 7, and will be not be discussed in this chapter. In this chapter, the carrier motion systems, the track plate, the take-off and some additional elements of the braiding machines will be discussed.

6.2 Carrier motion systems

The carriers move, following the track, which is a channel in the bottom plate of the machine. The bottom plate and the horn gears together make the *deck* of the machine. In classical braiding machines, two basic deck constructions exist: for sliding carriers

Braiding Technology for Textiles. http://dx.doi.org/10.1533/9780857099211.2.115

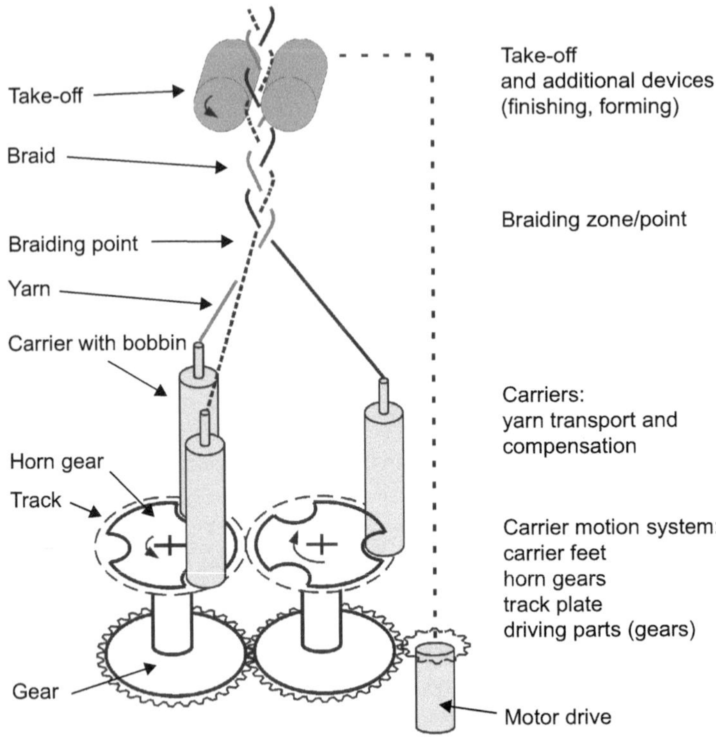

Figure 6.1 General principle of maypole braiding machine.

and for carried carriers. The difference between these two systems is in the way of fixing and moving the carriers along the plate.

6.2.1 Sliding carriers

The guiding foot, b, (Fig. 6.2) of *sliding carriers* (Douglass, 1964; Lepperhoff, 1953) slides along the track. Vertical motion of the foot of the carrier is limited by two small flat plates, a and a_1, (Fig. 6.2), which need to be well lubricated so that the friction between these plates and the main plate of the machine is reduced (Fig. 6.3). The horn gears move the carrier forwards by pushing the pin, b, (Fig. 6.2) but they do not carry the carriers. (According to Douglass, these machines are of English or American origin, but in several sources they are referred to as the 'Barmer System', which confirms their origin from the village of Barmen, now part of the city of Wuppertal in Germany (Douglass, 1964; Goseberg *et al.*, 1981)). Because the channel for the track is open, as the carrier moves along (Figs 6.3 and 6.4), fibres and dust from the (mainly natural) yarns fall down and mix with the oil lubricating the space between the plates. This is a serious problem, and the machines consequently require frequent cleaning. Another disadvantage of sliding carriers is the high friction

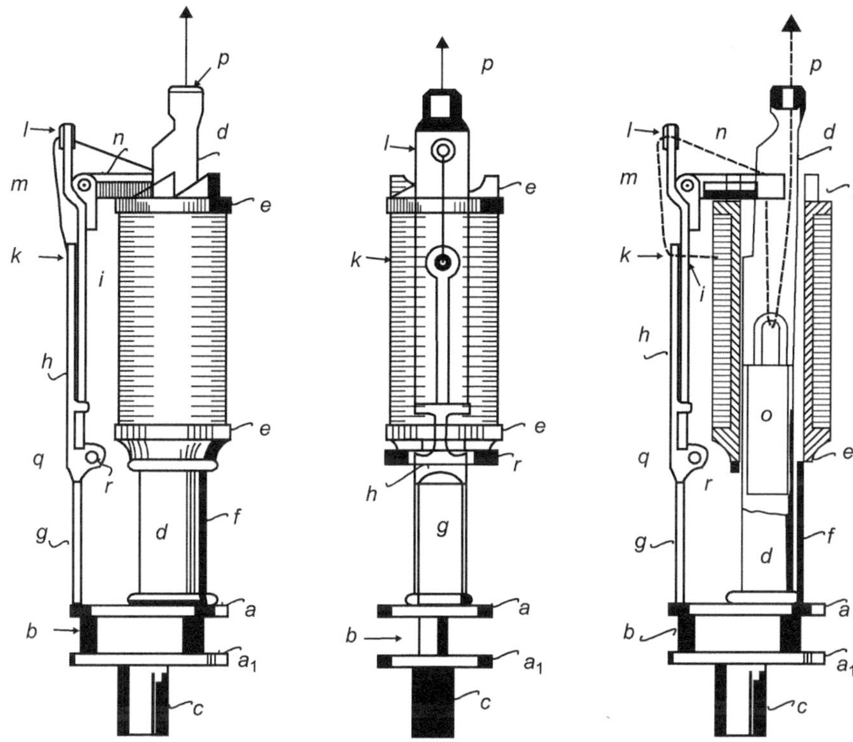

Figure 6.2 Sliding carrier 'Barmer System' (Lepperhoff, 1953, p.7). The foot c is moved from the horn gears and the guiding foot b is sliding through a channel (track).

Figure 6.3 Historical machine with sliding carriers. The feet of the carriers slide over the plate following the track. The oil and the dust can go inside of the track (author's own photo).

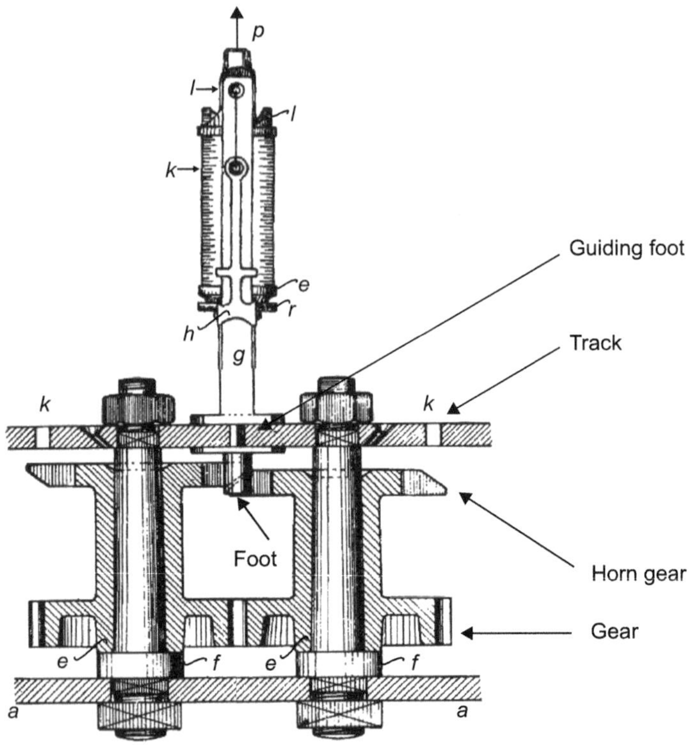

Figure 6.4 Motion system of gears, horn gears, track plate and carrier foot for sliding carriers.
Source: Picture created by combination of two figures from (Lepperhoff, 1953).

between the small plates, a and a_1, and the guiding foot, b, and the machine plate. Any vibration of the upper part of the carrier can cause a stick–slip effect, which can lead to irregularity in the carrier motion.

Sliding carriers are nowadays seldom used in practice, but they are mentioned at appropriate points to give the reader an idea of the historical development of the machines and also to understand the reasons for the popularity of carried carriers.

6.2.2 Carried carriers

In contrast to sliding carriers, *carried carriers* enclose the horn gear with small plates and are carried by the horn gears during motion (Fig. 6.5). The channel in the track plate determines only the direction of motion of the carrier (and does not support the carriers, as in the sliding carrier system). The vertical position of the carrier is not determined by the channel, and the carrier does not lie over the track plate but hangs on the horn gear, as can be seen in Fig. 6.6. Two guiding feet (Fig. 6.7) are mounted on the small bottom plate, and they determine the direction of motion of the carrier,

Figure 6.5 Elements of the motion systems of carried carriers. The carrier foot goes into the horn gear slot, and the small plates around the carrier foot are keeping the carrier on the horn gear, so that it does not stay on the track plate.

Figure 6.6 Elements of the carrier motion systems of carried carriers (author's own picture from machine of the company Spiraltex). This shows that the carrier hangs on the horn gear and does not lie on the track plate. In this example, the horn gears have three cuts per slot.

following the track. Since they are connected to the carrier only on the upper part, the track plate can be closed, in contrast to the track plate for sliding carriers. This allows the box for the gears to be completely closed, so that dust and fibre contamination is not possible, meaning that no cleaning of the gears is necessary (although the horn gears have to be cleaned!).The simple way to recognize the type of carrier system is to check where the sliding elements are – if there are two rotational guiding feet, or

Figure 6.7 Elements of the carrier motion system of carried carriers. The two guiding feet are under the small plate – they only determine the direction of the carrier motion.

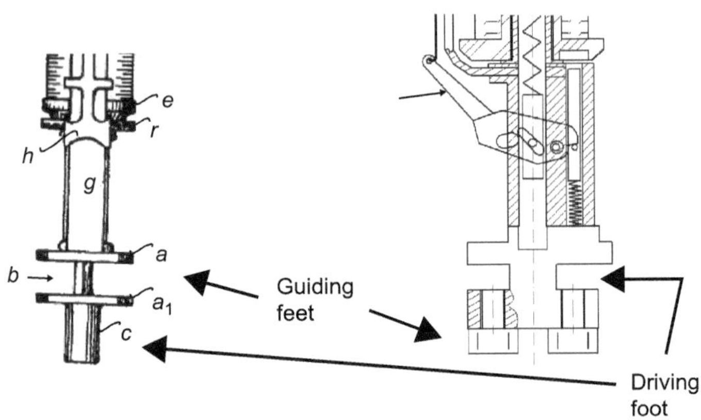

Figure 6.8 Comparison of the carriers for the two type of motion systems – sliding carrier on the left-hand side and carried carrier on the right-hand side.

only one fixed sliding element on the bottom part of the carrier. If the bottom part is in the form of a cylindrical shaft (the foot), it is a carried carrier, and if the guiding foot is above, then it is a sliding carrier (Fig. 6.8).

The company Körting Nachfolger Wilhelm Steeger of Wuppertal, Germany (Rehrmann and Ludwig, 2013), has its own solution forfixing and guiding the carriers – they use two plates for each horn gear and an additional plate on the carriers (Fig. 6.9, left-hand side). Because the gap between the plates and the horn gears is very small, the double fixation of the horn gears leads to more stable seating of the carriers and reduces vibration. This permits higher working speed at lower noise levels.

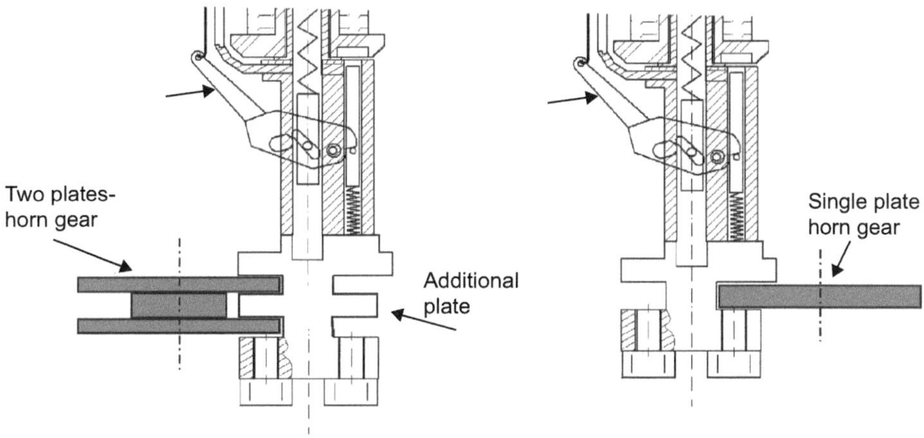

Figure 6.9 Two-plate horn gear system of the company Steeger (on the left-hand side) versus the classical system on the right-hand side.

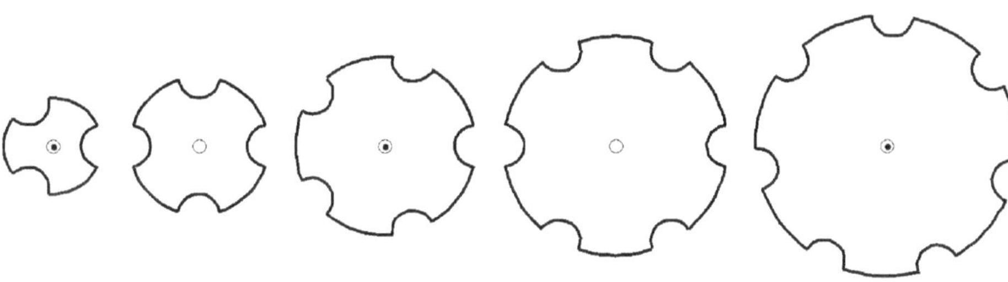

Figure 6.10 Schematic representation of horn gears with different numbers of slots.

6.2.3 Horn gear issues

Each slot of the horn gear has one cut (in some special constructions, three) where the carrier foot is accommodated during motion. The most commonly used braiding machines have horn gears with four slots. Flat braiding machines additionally have two horn gears with five slots at the turning points. Additionally, in several cases and applications, horn gears with three, six and more slots can be used (Fig. 6.10). The number of sections on the horn gear determines the type of pattern, as described in Chapter 2.

Theoretically, the horn gear radius is equal to the radius of the curvature of the arcs of the track around the horn gear (Fig. 6.11a). Actually, in several instances, the horn gears have modified geometry, with larger radii than the arcs (Fig. 6.11b). The deeper cut in the horn gear allows a larger contact surface to support the carrier and also guarantees that the applied driving force is in the same direction as the velocity of the carrier. Another reason for making the track radius smaller than the horn gear radius is that the two channels of the track between two horn gears meet at a greater angle.

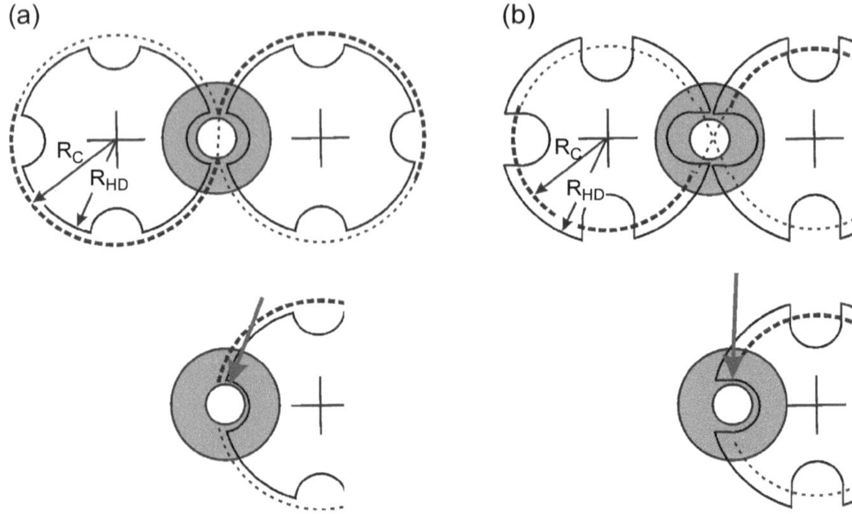

Figure 6.11 If the track radius is smaller than the radius of the horn gear, the leading of the carrier can be more stable (b) than in the opposite case (a).

An almost tangential contact point can be better for the dynamics of the process, but can also lead to problems with the motion of the guiding foot; the guiding foot can fall into the wrong part of the channel because both channels represent a continuation of its motion. For simplicity, all horn gears here are drawn as in case (a) in Fig. 6.11, because this gives a clearer representation.

At the transfer points, where the carriers change direction at the leading horn gear, the direction of the contact forces changes and the load on the contact area of the sections increases. The wear of these areas also increases with increased working velocity of the machine. To reduce repair costs and down-time following wear of these areas, some companies build their horn gears with exchangeable polymer sections (Fig. 6.12). The use of the polymer material at the contact area has a further advantage in reducing impact forces and the noise produced by the machine.

6.2.4 Guiding feet

Both sliding and carried carriers have parts – guiding feet (other words used for it are foot or shoe) – which follow the track and thus determine the motion of the carrier. Without the guiding feet, carried carriers cannot move from one horn gear to the next; the guiding foot enables this by following the track through contact with one or two walls of the channel. To do this successfully, the guiding feet need to have a particular form and dimensions. As shown in Fig. 6.13a, small elements of circular cross-section will not be able to 'select' the proper direction for the carrier themselves, because for these elements both directions are possible and allowed. The guiding feet have to be long enough so that, on exiting from one curve, they have the correct orientation to enter the next curve which is dedicated for the motion. In most

Figure 6.12 Exchangeable contact elements on the slots of the horn gears reduce the repair costs and time on the modern machines.
Source: Photo of machine, produced by the company August Herzog, Oldenburg, Germany.

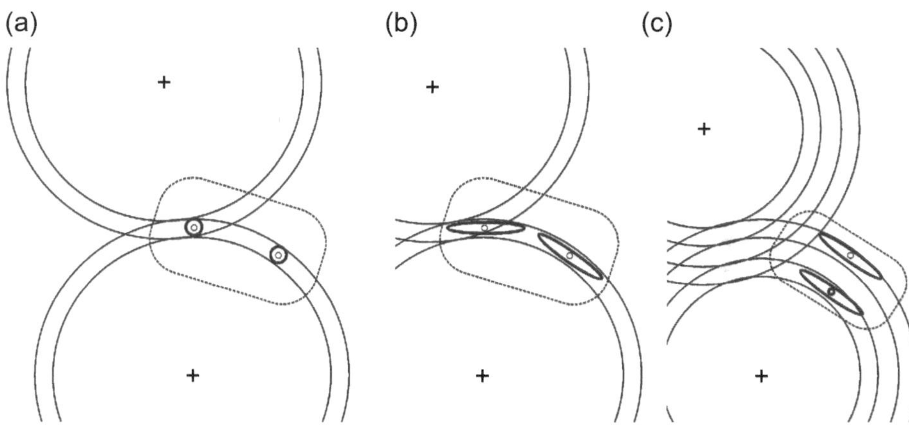

Figure 6.13 Some configuration of the guiding feet: (a) will not work – there are two possible ways for the foot and so the motion law is not more determined; (b) common case with two guiding feet (c) double tracks – for heavy metal wire braiding machines.

cases, two guiding feet per carrier are used (Fig. 6.1b) to achieve a smooth change of orientation of the carrier driving along the curve. Some heavier carriers use two track channels to similar effect (Fig. 6.13c). It must be mentioned that there are very good machines with only one guiding foot per carrier, which are not depicted in Fig. 6.13, because they can be seen as a special case of the variant in Fig. 6.13b.

The minimum length of the guiding foot depends on the radius of curvature of both the connected tracks. The guiding foot has to be longer than the length of the common part of the channel of both curves (Fig. 6.14a) in order to be sure to enter,

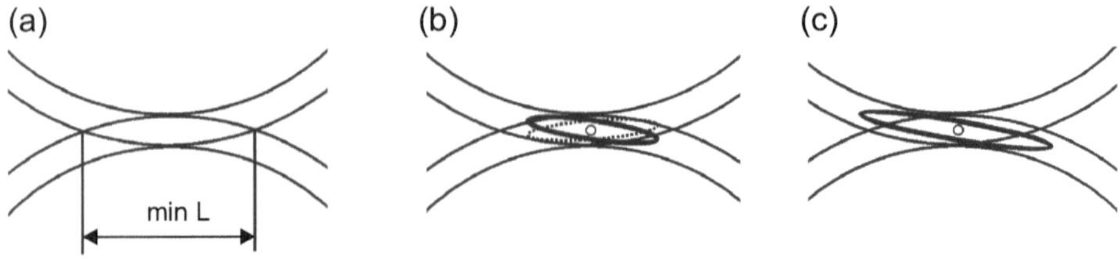

Figure 6.14 Minimal size of the guiding foot: (a) the foot has to be longer then min L (b) if it is shorter the motion is undetermined (c) correct length, the foot follows the proper curve and cannot switch to the false track.

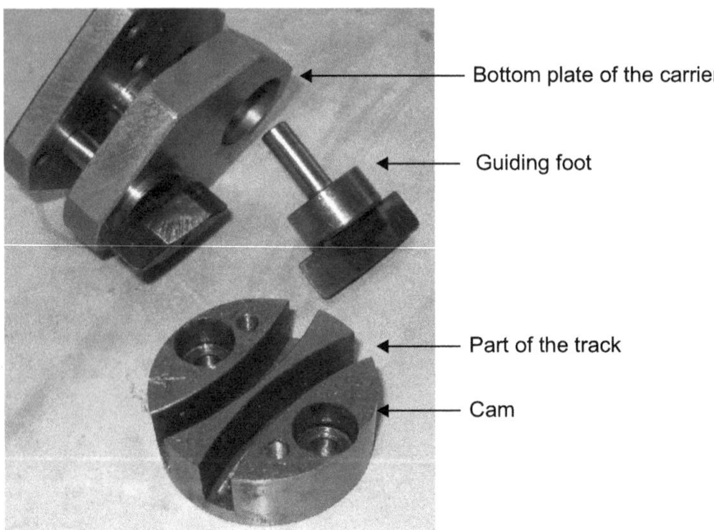

Figure 6.15 Exchangeable cam with a part of the tracks, and bottom part of a carrier with two guiding feet. The guiding foot can free rotate around its own axis during the motion.

without collision, the opposite curve of the track (Fig. 6.14c). If it is shorter, it can rotate, and its position is therefore geometrically undetermined (Fig. 6.14b). As well as the minimal radius of curvature on the entire track, the minimum length of the foot depends also on the width of the channel. It can be shown that for larger horn gears the feet have to be longer. One common way to reduce the length of the foot, and also the size of the carriers and of the machine, is to replace tangential arcs for connections between the tracks with linear ones, as explained above (Fig. 6.11) and also in Chapter 11 on Spiral braiding.

Figures 6.15 and 6.16 demonstrate one variant of a realization of guiding feet, together with a cam for part of the track. The guiding feet in this case are not fixed and can freely rotate inside the bottom plate of the carrier. Some companies use a longer fixed single foot per carrier and in this way reduce the number of carrier parts and thus the costs.

(a) (b)

Figure 6.16 Two guiding feet and cam as a part of the track: (a) separated for better visibility; (b) usual working configuration.

6.3 Track plate

6.3.1 *Classical braiding machines*

The gears of modern braiding machines with carried carriers are mounted between two plates – the base plate and the track plate. The common shafts for the tooth and horn gears go through the track plate, on which are formed the channels of the tracks. Flat braiding machines have one single track (Fig. 6.17), tubular braiding machines have two tracks (Fig. 6.18), and other machines (form braiding) have more tracks.

In addition to the main tracks, opening plates or service channels for each track are required, from where the carriers can be inserted or removed. These service channels are normally sealed by special elements during the normal working of the machine, and only in the case of rearrangement or exchange of the carriers are they opened. Each track needs such a channel because the carriers do not change their tracks, and can be added or removed to only one track (except for machines with switchable cams). Several producers do not form the track as a channel, but keep only the side of the channel where the guiding foot makes contact.

At the point where tracks pass between two horn gears the loading over the walls of the channel is usually higher because the centrifugal force of the carriers changes direction. In modern braiding machines these parts are built as separate, replaceable cams (Figs 6.15 and 6.16), so that only the cam and not the entire track plate needs to be replaced. The replacement of such elements takes only minutes, compared to several hours to replace the plate; the costs of the parts and the logistics are also not comparable.

6.3.2 *Modifications on machines with fixed track*

The track of a braiding machine determines the type and structure of the product. To extend the range of products for one machine, different options for rearrangement of the tracks of braiding machines are available – but only if the manufacturer has

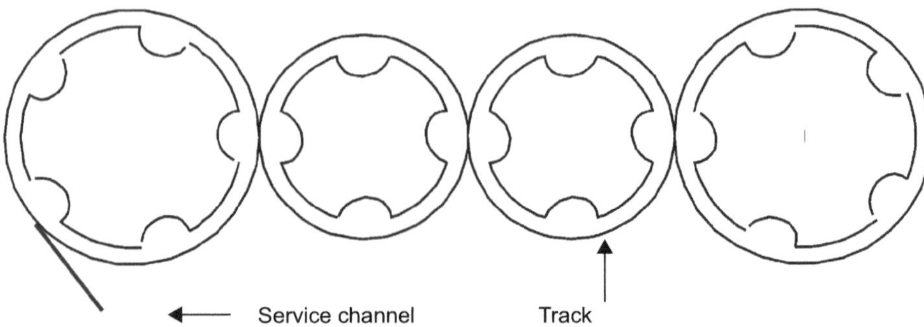

Figure 6.17 Track of flat braiding machine and the service channel.

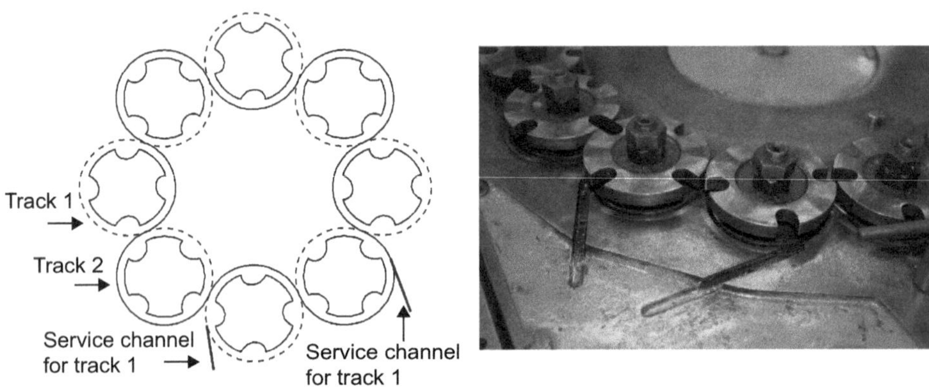

Figure 6.18 Tracks of tubular braiding machine with their service channels.

constructed the machine to allow these options. One of the most common changes is the modification of a flat braiding machine to tubular braiding, and vice versa (Fig. 6.19). A tubular braiding machine (Fig. 6.19a) can be modified to a flat braiding machine if the manufacturers have designed it for this change and provided the necessary parts for modification. This option will make the machine more expensive compared to a normal one, but in the case of machines with more then 24–32 carriers the added value for the machine is significantly higher than the price paid. For machines with only a small number of carriers, this modification is not cost efficient. (In Fig. 6.19 the machine illustrated has a small number of carriers for better visibility). The change from tubular to flat and vice versa is not only connected with the change of the track. Flat machines have horn gears at both ends which have one slot more (or less) than the normal horn gears, so that the carrier from one half of the track goes into an empty slot before starting on the opposite half of the track. When horn gears with different numbers of slots have to be added to an existing machine, their dimension and speed have to be synchronized with those of the existing horn gears – a horn gear with five sections for a machine (Fig. 6.19b) has to have a larger diameter (it has to carry more carriers, and the path of the carriers between the slots

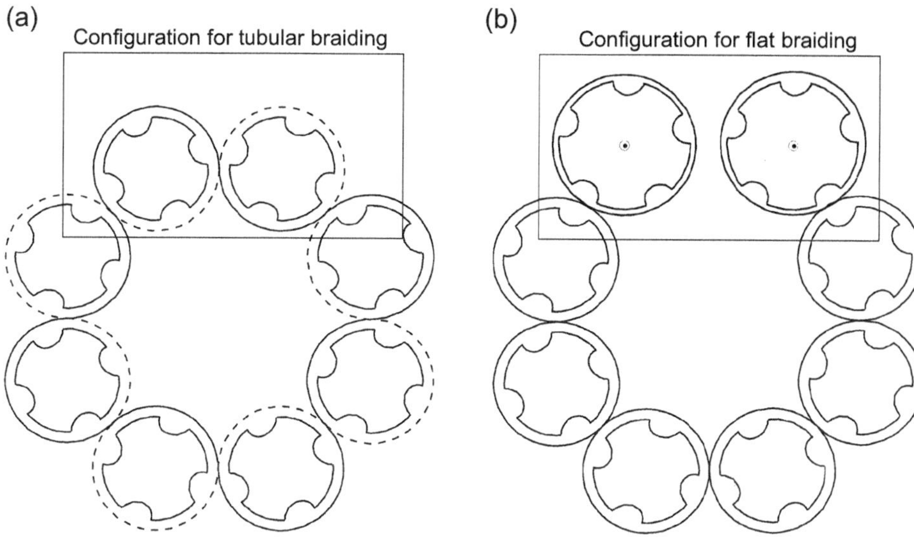

Figure 6.19 Modification of the track from tubular (a) to flat (b) braiding machine. For this modification the horn gears, the track plate segment, the gears and their bearing inside the marked rectangular area have to be replaced.

has to be equal). On the other hand, the velocity of the carriers has to remain constant (in order to have a stable braiding process without vibration or noise), so the larger horn gear has to rotate more slowly, and so has to be driven by a larger gear. The bearings of the larger horn gear and drive gear will thus have a different position to the smaller ones, and to allow for the difference a special eccentric bush is required. Because of these considerations, all the elements mentioned here – inside of the rectangular areas in Fig. 6.19 – have to be replaced to modify the machine from tubular to flat and vice versa. Often, instead of changing the complete track plate, only the cams at the crossing place and at the areas where horn gears are removed have to be replaced. The new five-slot gears for the flat configuration are known as Keleher gears (Douglass, 1964). The four-slot horn gears which are used in the configuration for tubular braiding have slots in the form of a trapeze – they are wider outside so that the carrier has enough time for the transition from the smaller horn gear. Depending on the manufacturer, the size of the machine (number of horn gears), the motion system (carried or sliding carrier) and the required working speed, these machine modifications can differ significantly. For this reason, only the principle is mentioned here, without construction details. Such details, for older machines, can be found in Goseberg *et al.* (1981) and Douglass (1964), but it has to be remembered that the working speeds at the time these books were written were significantly lower than they are today, and so the requirements for machine construction and dynamics today are greater.

In the same way, a set of two five-slot horn gears can be inserted in the middle of the track of one larger flat braiding machine to produce two smaller flat braids at the

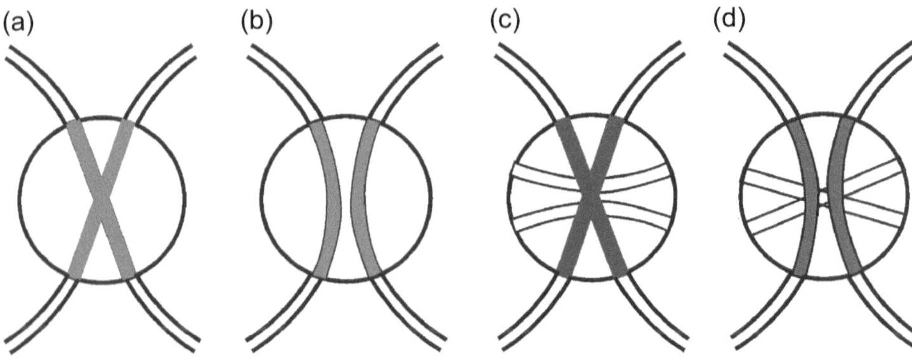

Figure 6.20 Cam between two horn gears (a) 'transmit' cam – all carriers will be transmitted to the next horn gear; (b) 'keep' cam – all carriers will remain on the same horn gear (c) combined cam, consisting of the both positions in a state 'transmit'; (d) combined cam in a state 'keep'.

same time, instead of one larger flat braid. Depending on the type of machine, the size of flat braiding machines can be reduced if more horn gears are removed and two end horn gears with five slots are mounted at the end of the track. In this way, one machine can be modified to produce smaller braids with fewer carriers.

One more simple way of modifying the track plate is demonstrated by the VarioBraider produced by the August Herzog company (August Herzog, 2013). In this machine, all horn gears have four slots and remain in their places; only the cams between the horn gears can be rotated between two positions – 'transmit' or 'keep' (Fig. 6.20). Such combined cams have two systems of channels, one for transmitting the carriers (Fig. 6.20a) and one for keeping the carriers (Fig. 6.20b), positioned at a certain angle, one to another, so that with a single rotation of the cam it can be switched between these two states (Fig. 6.20c and 6.20d). The angle in Fig. 6.20 is selected as 90°, but actually it can be lower to reduce the necessary rotation of the cam. The VarioBraider machine consists of a set of 4 × 4 horn gears, with a switchable cam mounted between each pair (Fig. 6.21). All the switches have to be set to one of the two positions before the machine is switched on.

6.3.3 Dynamically changeable tracks

Changing the track during the operation of a braiding machine allows the realization of products of different cross-section. Such machines, simple jacquard machines with control cards and two tongue-like switches, are used for producing of buttonhole braids. A buttonhole braid consists of two regions – one wide flat braid without a hole and one region with a hole – which is created as two smaller flat braids (Fig. 6.22).

The advantage of a buttonhole braid, compared to all other techniques for producing holes, is that there are no yarn cuts for producing the hole, which means that the hole is significantly more stable. For such products a modified flat braiding machine is

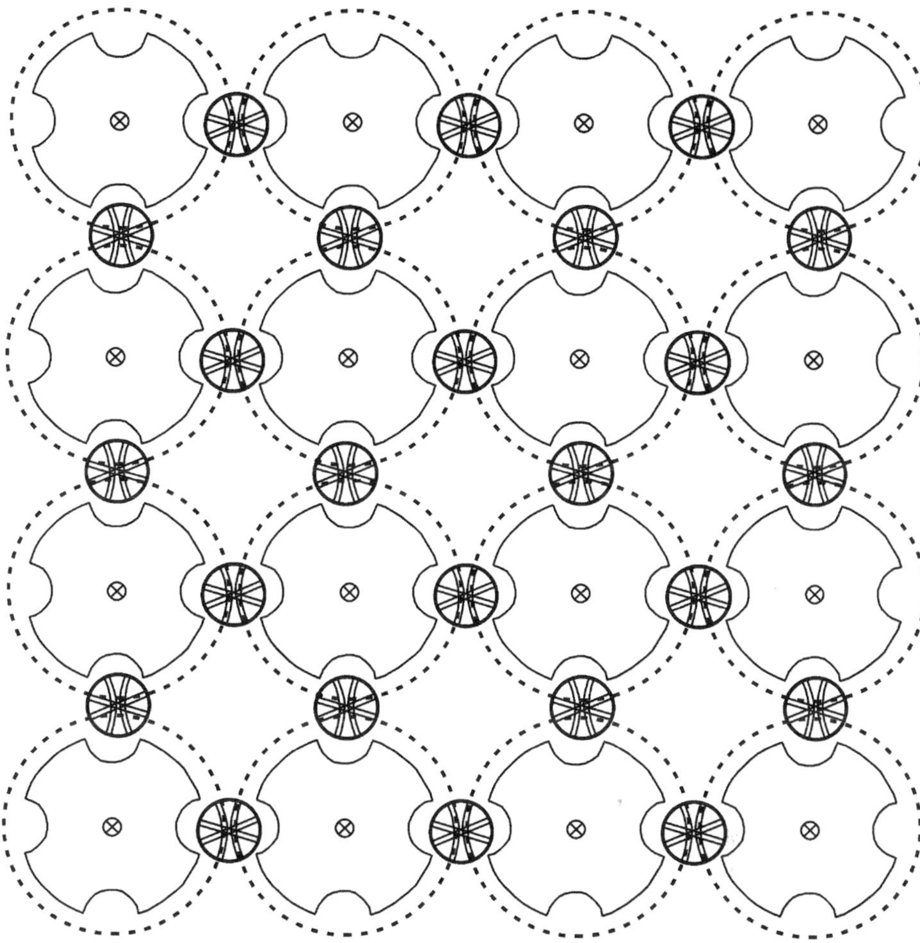

Figure 6.21 Configuration of 4 × 4 horn gears with cams between them for changing the track. The current set-up shows only the idea and do not show working configuration of the machine. Different such configurations are presented in more detail in Chapter 12 (Square and other types of form braiding).

used, where a central special arrangement of tongue-like switches is used (Goseberg *et al.*, 1981) (Fig. 6.23). These switches are thin at the top (and so 'tongue-like') and at one position (Fig. 6.24) they are part of the track, so that the carriers move through the entire machine and build one large flat braid. After the required length of the wide part is braided, the switches are rotated so that the normal channel is closed and an additional one opened for the carriers (Fig. 6.25). Following this track, the carriers go around two small horn gears and return to their normal track, but split into two groups, braiding two flat braids. After the hole length is achieved, the switches are turned back to their original position, and the process continues.

Figure 6.22 Button hole braiding sample.

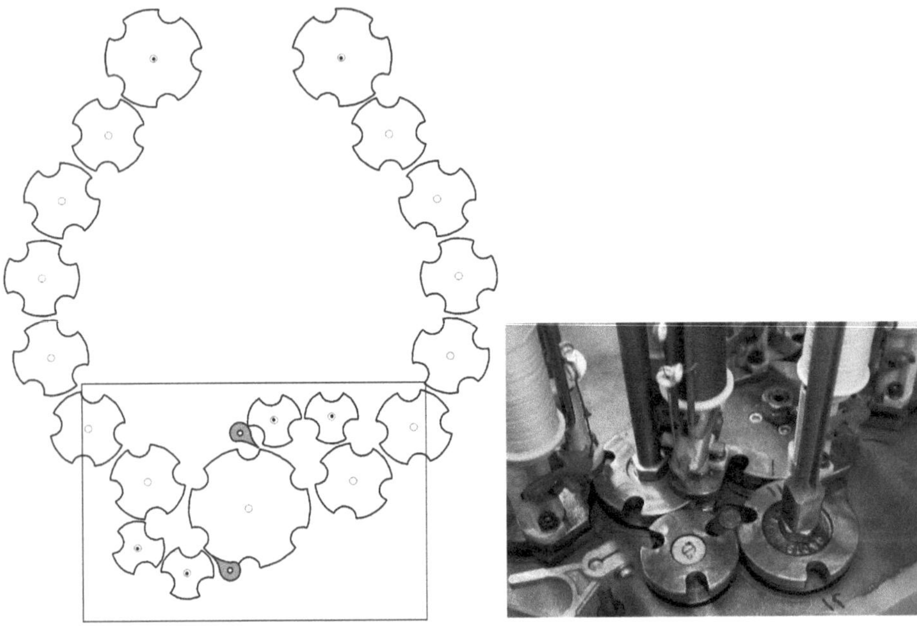

Figure 6.23 Example configuration for button hole braiding machine (following Goseberg *et al.*, 1981, photo the author). In the area, marked with the rectangle the machine is split into two flat braiding machines or switched again to one larger flat braiding machine during the work (see Figs 6.24 and 6.25).

With this construction the track changes dynamically while the machine is running. Another variant of dynamically changing tracks is the modified, newer generation of the VarioBraider VF, described in the previous section, where some of the switches are rotated while the machine is running, using step motors or pneumatic or hydraulic cylinders. This allows, for example, the production of ropes with up to three branches for medical braids (August Herzog, 2013). Detailed investigations into the uses of the machine for medical applications have been undertaken in recent years by the team of Dr Markus Milwich at ITV-Denkendorf, Germany (Müller, 2013).

The best known case of a machine with changeable track during braiding is the 3D braiding machine developed by the August Herzog company and the Institute for Textile Technique Aachen at the RWTH Aachen University (ITA) (Büsgen, 1993). This machine has the same configuration as the variable braiding machine shown

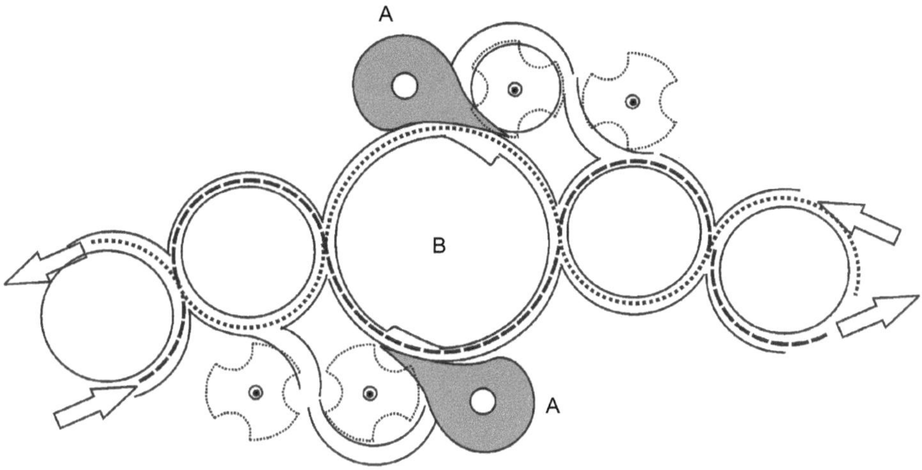

Figure 6.24 The tongue-like switches A are in a position such that the carriers are moving around the large horn gear B and the machine works as a one flat braiding machine. Note that the carriers have to move only in the demonstrated direction, and reversing of the motion direction can lead to machine damage because the guiding feet of the carriers can crash at the edges in the switching area (Goseberg *et al.*, 1981).

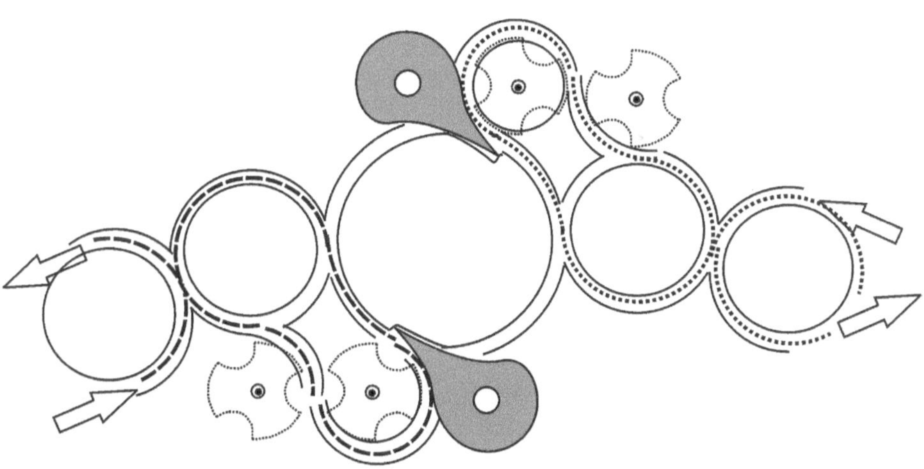

Figure 6.25 The tongue-like switches close the normal channel and forwards the carriers though two additional horn gears back. In this way the machine works as two flat braiding machines.

in Fig. 6.21, but all switches and all horn gears are individually controlled by computer (Fig. 6.26). During a two-step braiding process the track can change after any movement of the carriers and allow the production of parts with changing geometry and properties (Büsgen, 1993; Schneider, 2000; August Herzog, 2013). More details about the 3D braiding processes are given in Chapter 12.

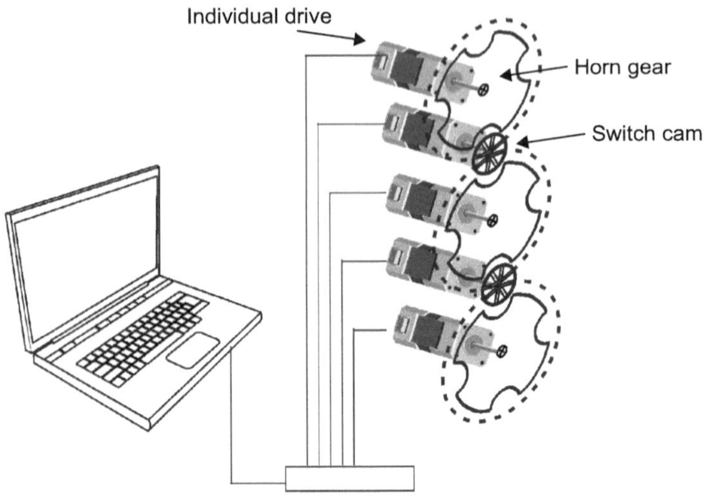

Figure 6.26 Principle of the 3D braiding machine with programmable track.

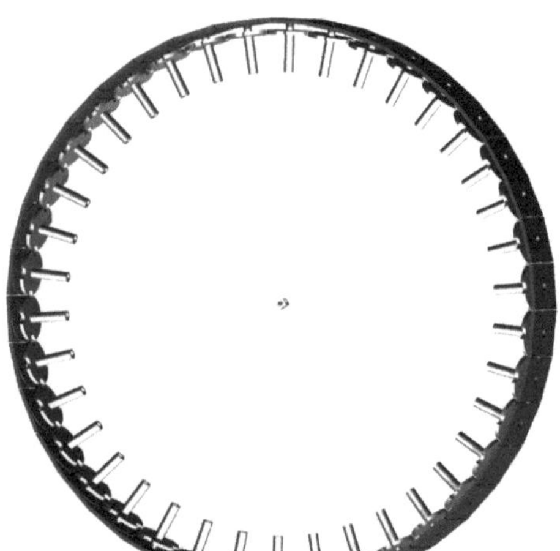

Figure 6.27 Principle configuration of the track plate and the horn gears of a radial braiding machine.

6.3.4 Track on radial braiding machines

Each of the braiding machines described above has a flat track plate. There is a modified type of tubular braiding machine, created mainly for overbraiding with high performance fibres, called a radial braiding machine, where the track plate is in the form of cylinder and the horn gears and the carriers are mounted on the inner part of this cylinder. Such a machine is shown in Fig. 6.27. For such machines, each carrier moves in one plane if carried from one horn gear, but when the carrier changes to the next horn gear, the plane changes as well. For machines with a small number of carriers the angle between the planes is higher, and the passing of the carrier from one plane to another can require additional changes to the construction of the horn gears and the carrier feet.

6.3.5 *Carrier motion systems in the future*

The track on braiding machines helps carrier guidance, but also limits the range of products from one machine. It is likely that in the near future, machines without tracks will be created where each carrier has its own independent drive and motion, programmed like a robot. Construction of such a prototype is possible even with simple robots today, but they have some significant limitations, which will have to be solved before such machines can be used for industrial production:

- production speed has to be high enough
- the machine has to be as small as possible
- yarn tension has to be kept constant, but high enough for the normal braiding process.

If the first point above is now ceasing to be a problem, the second and third points still remain. Each self-moving carrier (let us think of it as a 'car') needs energy and space for a motor and a battery, and this leads to larger geometries. Larger geometry leads to greater machine size and greater yarn length differences for compensation (see Chapter 7 Carriers for braiding machines).

Yarn tension is another factor – the yarns need sufficient tension in the braiding zone, which means that the carriers (or 'cars') will need to resist this tension. But how can a 'car' resist the force created by a rope connected to a 'helicopter'? Either with greater weight, which again means greater size and a greater energy requirement due to higher friction forces, or some other force, such as a magnetic force – which leads to the same problem.

The horn gears solve these problems very elegantly – they hold the carriers, but they also fix them and stabilize them in both directions on the vertical axis, so no yarn tension can take the carrier from the horn gear. These considerations show why machines with track and horn gears are still the easiest stable solution.

6.4 Braiding zone and the take-off

The take-off mechanisms of the braiding machine determine the braiding angle of the product and thus influence significantly the properties of the braids. The braiding angle is briefly introduced here, but more details about the relation between the braiding angle and the take-off speed are given in Chapter 8 'The mechanics of the braiding process'. The configuration of the braiding zone is very closely connected with the type of take-off device. For this reason the special elements around the braiding point or braiding zone and the take-off are presented together.

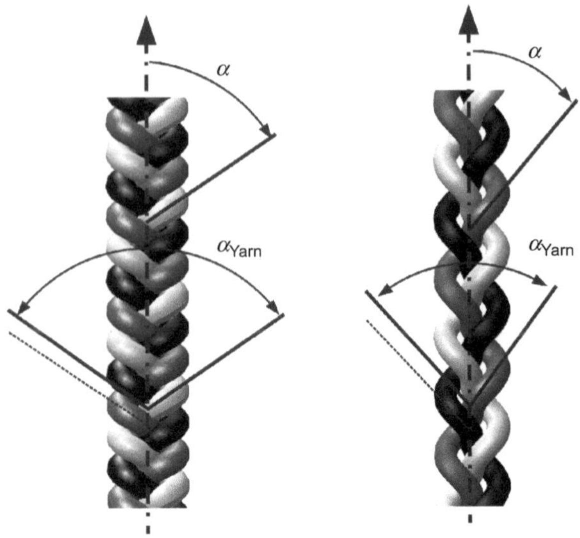

Figure 6.28 Definition of the braiding angle and angle between the yarns.

6.4.1 Braiding angle

The braiding angle is one of the most important parameters of braided structures and is defined as the angle, α, between the braid axis and the yarn axis (Fig. 6.28).

In the case of complex three-dimensional forms, the braiding angle is defined as the angle between the vector of the take-off velocity at the investigated braiding point and the yarn axis at that point. Normally, in symmetrical structures with homogenous yarns, the braiding angle, α, is half of the angle between the yarns, α_{Yarn}:

$$\alpha = \frac{\alpha_{\text{Yarn}}}{2} \quad [6.1]$$

In some cases the *helix* angle is used as the braiding angle (in the case of tubular braided structures)

$$\alpha_{helix} = 90^{\circ} - \alpha \quad [6.2]$$

The measurement of the helix angle is more complicated, because it depends on some 'virtual' horizontal axis, which is not directly recognizable in the braided products.

The braiding angle is a main geometric characteristic of the braided structure and it determines several of the mechanical and geometrical properties of the product, especially the elongation of ropes and the cover factor for overbraided structures.

The braiding angle in one structure can be changed or adjusted by changing the take-off velocity. The general rule is that the higher the take-off velocity, the lower the braiding angle – so the take-off velocity is the main machine parameter which can be used to change the braiding angle.

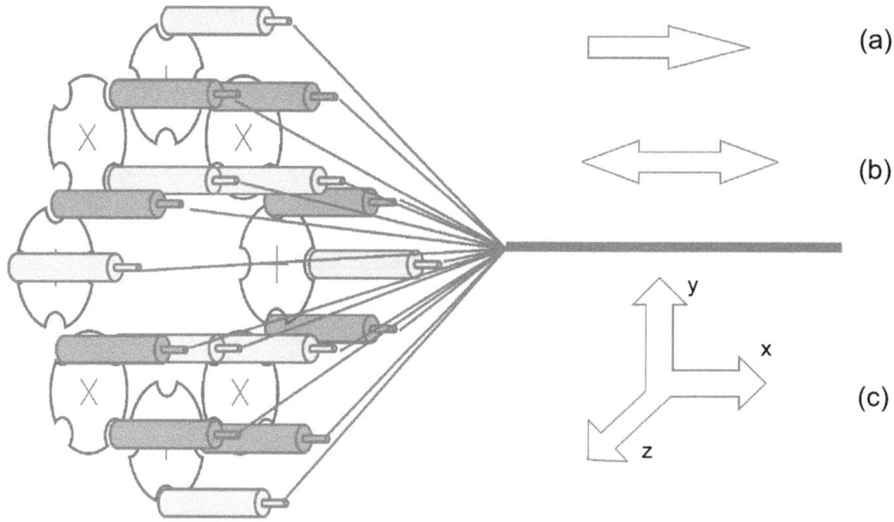

Figure 6.29 Commonly used variants for the take-off directions; (a) uniaxial–unidirectional; (b) uniaxial – bidirectional; (c) triaxial and with usually all six degrees of freedom–three translations and three rotations.

6.4.2 Take-off basics

In classical applications of braiding, the take-off mechanism takes the braided product from the braiding zone in only one direction and along one coordinate axis (Fig. 6.29, case (a)) with constant velocity. For the braiding of electric cables and overbraiding of some products, reversible take-off is necessary (Fig. 6.29 case (b)) in order for some regions of the product to be covered many times by the yarns. In this case changes of velocity can also be required, and the mechanisms are uniaxial, but bidirectional. The most complex case – overbraiding of curved mandrels with non-linear profiles – requires that the parts move during the braiding process in all three axes (Fig. 6.29c), including rotations. As the requirements for the velocity are different in these three cases, the mechanisms which do this are also different.

One additional criteria for constructing the take-off mechanism is the form of the final product. If the product is linear and not harmed by bending (textile ropes, shoe laces, etc.), simple systems with belt friction around cylinders can be used. Continuous products with constant cross-section can be taken off using caterpillar devices without bending or damage to the parts. In the case of overbraiding, curved profiles have to be moved through the braiding zone by special robot, since each part is unique.

6.4.3 Uniaxial, one-directional take-off mechanisms

In order for the take-off velocity to be well controlled, there has to be very good contact between the driven elements of the take-off mechanisms and the braid. For thin ropes and laces a system of several cylinders is used, combing pressure between

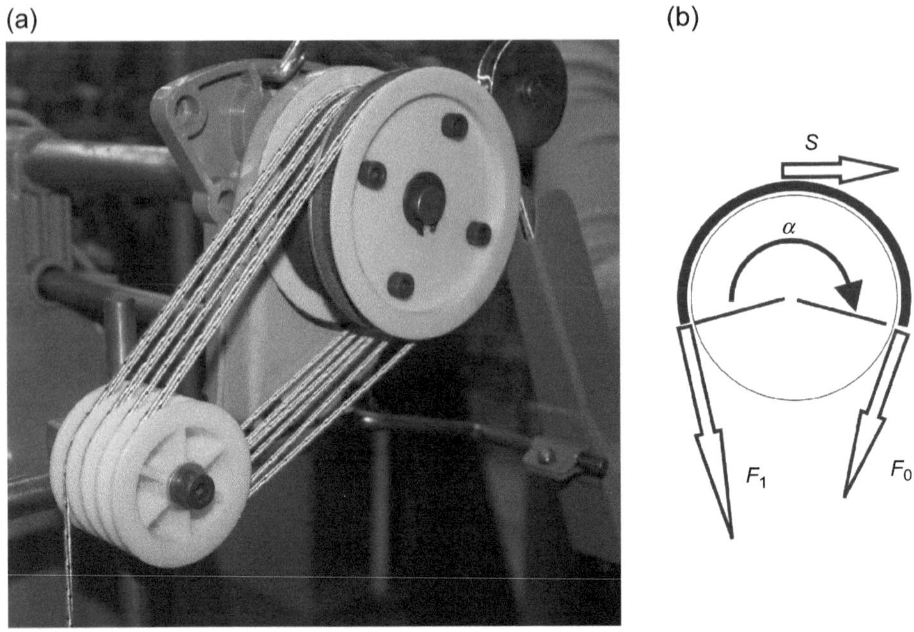

Figure 6.30 Take-off mechanism with one take-off cylinder and one guiding wheel (a); (b) forces in the rope around the wheel.

some of them and friction around the cylinders to avoid slippage between them and the braided product. Figure 6.30 demonstrates one of the common systems for thin ropes, where the braided product goes around the larger wheel and the smaller wheel five or six times. For the forces (Fig. 6.30) on one single partial winding around any cylinder, the well-known equation of belt friction, also known as the capstan equation or the Euler–Eytelwein equation, is valid:

$$F_1 = F_0 \cdot e^{\mu\alpha} \qquad [6.3]$$

where F_0 and F_1 are forces after and before the wheel (looking from the braiding zone), is the friction coefficient between the rope and the wheel, and μ the warping angle. The friction force, which tries to stick the rope to the wheel, is:

$$S = F_0 \cdot e^{\mu\alpha} - F_0 = F_0 \cdot (e^{\mu\alpha} - 1) \qquad [6.4]$$

It can be shown that, for a friction coefficient $\mu = 0.3$, after three windings around the wheel force F_0 has to be less than 10% of F_1. In the normal case, the force F_0 is the self-weight of the rope, and four to five windings have to be made in order to avoid slippage from vibrations. Another issue is that if the rope is wound around one take-off cylinder, the single windings can intersect and disturb the normal take-off process. To prevent this, a second wheel, usually with channels, is used as a rope guide. This second wheel reduces the total contact force of the take-off device, because it

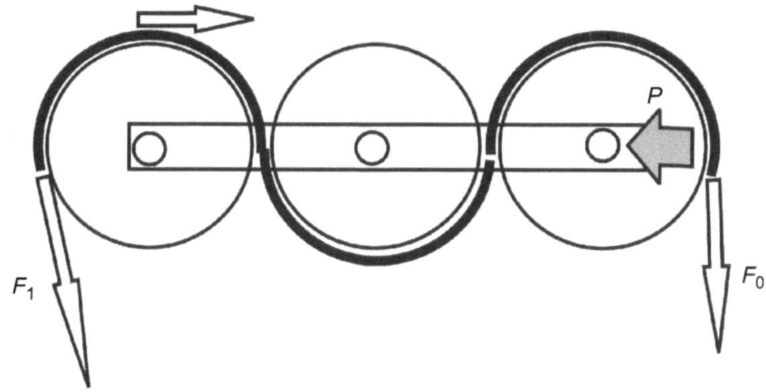

Figure 6.31 Take-off mechanism with three rollers using belt friction and contact friction from the normal force P.

often has a lower friction coefficient and is normally not driven, but is technologically necessary for a stable take-off process. To increase the contact force between the take-off wheel and the rope, the take-off wheel is usually covered with a special rubber material with a high friction coefficient, and at the end some additional pressure roll is used.

On braiding machines with more carriers the required take-off force is higher, and the grip between the rope and the take-off mechanism has to be better. In such cases, in addition to the belt friction around the cylinders, pressure between them creates an additional friction force proportional to that pressure, P (Fig. 6.31).

The take-off device pulls the ropes out of the braiding zone, which also has to be suitably designed for the dimensions and the types of rope. Since for ropes and laces the braiding zone is very small, it is commonly referred to as a braiding point. Here we will consider the braiding point to be generally the place where yarn interlacing is happening (Fig. 6.32). Usually in the area of the braiding point, a braiding die is mounted to control the yarns or to change the geometry of the interlacing zone. In some machines the braiding die has a special form and is called the braid former. The selection of the proper form, size and position of the braid former is very important for the normal braiding process. The position of the former can usually be adjusted using a system of three interconnecting beams, which allow change of the position in the three coordinate directions (Fig. 6.32).

The size and the form of the braider have to be selected depending on the product. The former opening has to have a sufficiently small opening so that it keeps the braid fixed and prevents movement in the horizontal direction (in the plane normal to the braiding axis). But the opening also has to be large enough so that the friction between the former and the braid does not affect the braiding process or the operation of the motor. For tubular braids the opening has a circular cross-section. In this case, several openings can be drilled on one plate, so that a change requires only simple rotation of this element.

In the case of braiding machines for thicker ropes, a spitted, two-element former is provided, as, for example, on the machines of the August Herzog company (Scheider,

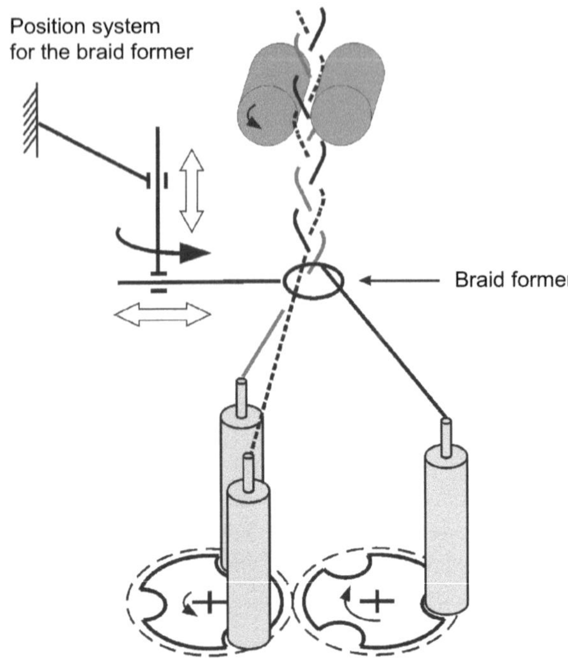

Figure 6.32 Position of the braid former.

1991). This allows easy change of the former element during the adjusting process at the set-up of a new product, without cutting of the expensive rope material, which could have a thickness in some cases of more than 10 cm.

Often the formers are also combined with equipment which controls the yarn tension. If a yarn is broken, the machine stops automatically. The working principle of such equipment is described in more detail in Chapter 8.

For flat braiding machines, the braid former is different depending on the width and form of the braided structure. Usually, such form braiders are of open form, and the opening has to be on the side of the horn gears, where the carriers change direction. Because there are more yarns in the other direction, the resulting force keeps the braid flat to the surface of the forming element. The open profile also allows more convenient operation with the yarns during breaks and in the loading process. Figure 6.33 shows some types of formers used for different flat braided products.

6.4.4 Uniaxial, bidirectional take-off mechanisms

If the braided products are sensible to lateral pressure or bending, caterpillar take-off devices are used (Fig. 6.34). These distribute the pressure on the product over a large contact surface, which allow take-off at a stable speed without slippage, without increasing the lateral load on specific areas and without bending. The distance between both tracks of the caterpillar is adjustable and has to be set anew for each rope. Because of the larger contact area, the caterpillar devices can work in both

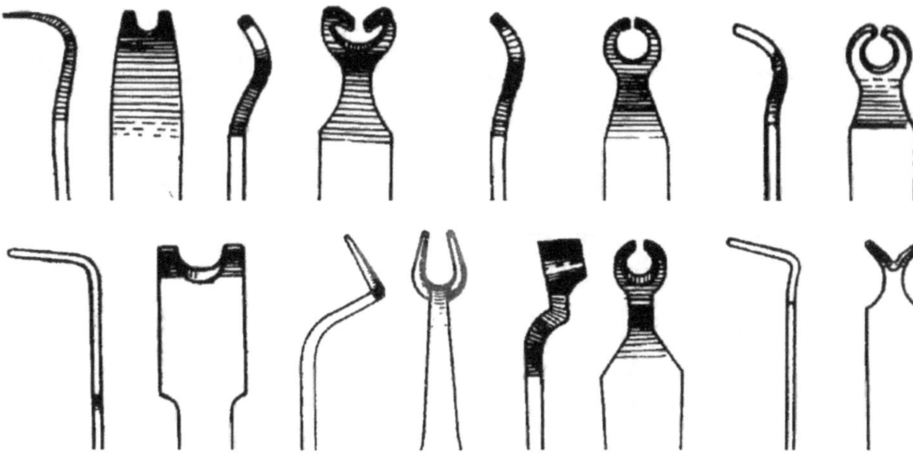

Figure 6.33 Formers with different forms and sizes, used producing flat braids (Lepperhoff, 1953).

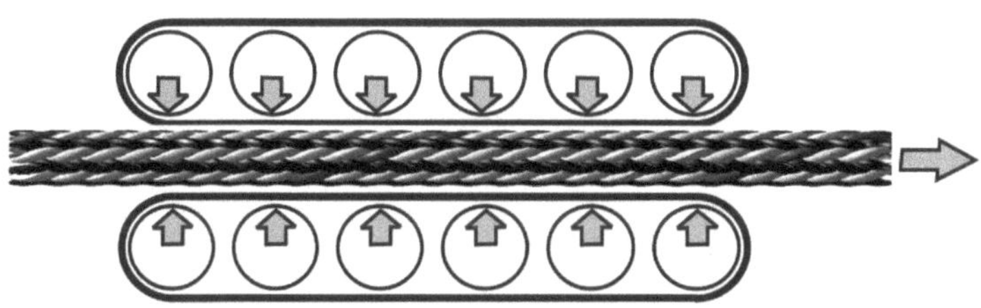

Figure 6.34 Caterpillar devices can apply higher forces with distributed, adjusted for the product pressure and without its bending.

directions – forwards and backwards, if this is programmed during their development – and they can transfer larger forces. The caterpillars are usually (but not only) used in a horizontal position, as shown in Fig. 6.35.

On machines where reversible motion is possible, the geometry of the braiding zone (here it is no longer a 'braiding point', because usually thicker and more complex parts are overbraided) is usually determined by two braiding rings. The size of the rings depends on the diameter of the mandrel (Fig. 6.36)).

6.4.5 Take-off devices with position control

The overbraiding of profiles of complex form requires precise positioning of the mandrels during the overbraiding process. Depending on the profile of the mandrel, it may have to be moved forwards and backwards, it may in some places have to stay fixed for a certain amount of time, or it may have to rotate around some axis; for this a robot arm is used (Figs 6.37 and 6.38). The process is usually discontinuous, because after each part the braiding process has to be stopped in order for yarns to be separated from the part, and the robot to load the next mandrel for overbraiding.

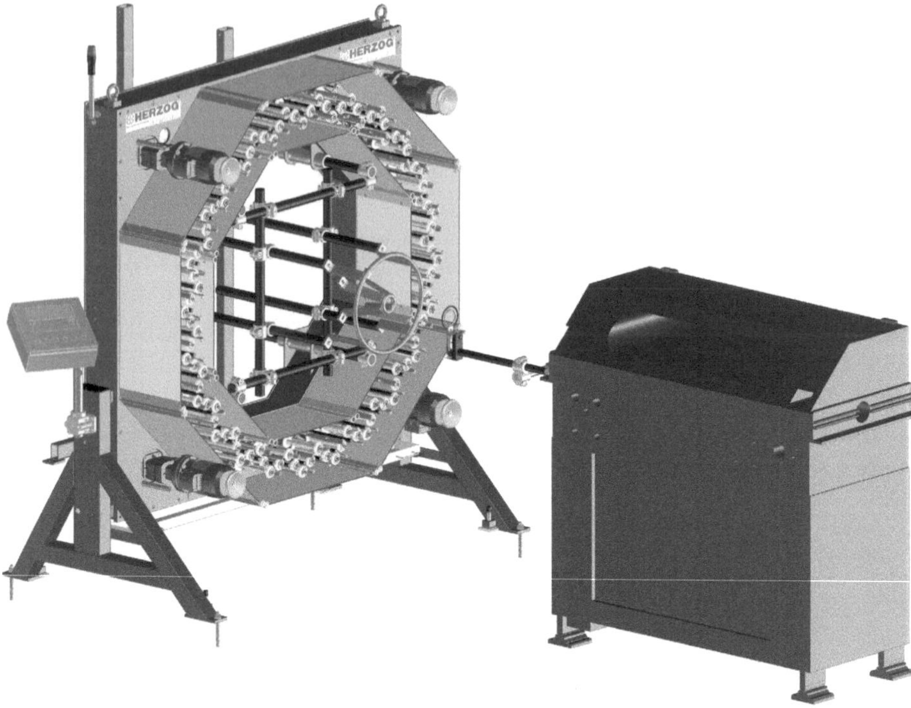

Figure 6.35 Machine with horizontal take-off. The braiding plate is placed vertically so that the braided product can be taken off horizontally. The capsuled take-off machine is on the right-hand side on the picture.
Source: Courtesy of the Company August Herzog Textilmaschinenfabric GmbH & Ko KG, Oldenburg (August Herzog, 2013).

6.4.6 Manual take-off

Manual movement of the cables during the overbraiding process is used during the overbraiding of electric cables for airplanes, cars and so on. In this case the operator can use the take-off device for overbraiding the regular regions, but after that he/she has to stop the machine and move some of the cable parts between the yarns to cover the area where the parts of the cable are split. Figure 6.39 shows the main points of such work. There is a take-off device available, where the worker can fix the cable during the processing of longer parts of the cables where there are no branches. When a branch has to be processed, the machine has to be stopped to move some part of the branches in or out of the braiding zone. During the braiding of the areas with branches, the cable is moved by hand. Because both hands are necessary for this work, stopping and starting the machine is usually performed with a foot pedal.

The braiding zone on machines with manual take-off has a different configuration from normal braiding machines. The braiding ring is significantly larger and is usually mounted on a special cylinder in the central part of the machine. In this

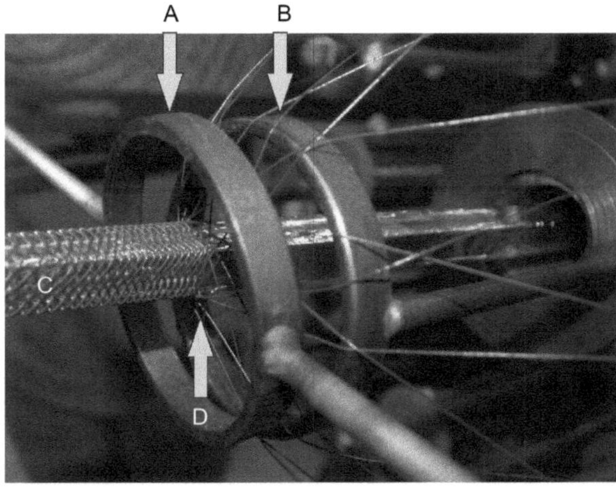

Figure 6.36 Two braiding rings for support the overbraiding process in two directions. The ring A is necessary if the take-off speed is high and the braiding zone D is moved to the left hand side. Ring B is in contact with the yarns for lower take-off speeds, where the braiding zone tries to move near to the carriers and for motion of the mandrel C in the opposite direction (to the right-hand side on the picture).

way, contact and entanglement between the cable which has to be overbraided and the moving carriers is prevented (Fig. 6.40). The ring also allows overbraiding in the reverse direction.

6.5 Additional elements in the braiding machine

Several braided products require special configuration of a braiding machine to achieve extended features or properties. Here we discuss some of the most commonly used.

6.5.1 Core guiding

A core is used in the middle of tubular (and some square or similar) structures. As already mentioned in the previous section, it may be an electric cable, which has to be covered by cut or heat resistant material, or a rope which has to be stabilized for working under high pressure, or just cotton threads, used for giving volume to shoelaces. Flexible cores are guided through the machine plate via a specially mounted pipe. To ensure the stability of the braiding process, the soft core materials have to be kept under constant tension. Different kinds of brakes are used to control the tension, depending on the type, geometry and properties of the material. For classical textile materials, normal disc brakes with springs or weights can be used. An alternative solution is to mount some braking system such as, for example, a belt friction device, around a bobbin flange (Fig. 6.41).

Figure 6.37 Configuration of radial braiding machine with robot arm as take-off device. *Source*: Courtesy of the Company August Herzog Textilmaschinenfabric GmbH & Ko KG, Oldenburg (August Herzog, 2013).

'Cores' which are not of linear form are called mandrels. These forms are used in composite production to give the basic form of the new product. They are produced from light material which can be removed (or not) after the overbraiding, and which has to have enough strength in order not to deform under the yarn tension during braiding. Since they are moved by a robot arm, there is no other special equipment on the machine necessary for their motion.

6.5.2 Piles for middle ends (inlays)

Middle ends (in braiding for composites they are known as 'inlay yarns') are another kind of yarns, similar to the cores, but they are fixed between the braiding yarns (Fig. 6.41) (Engels, 1994). Because of this they can be used on any of the braiding machines – flat, tubular and form braids. Middle ends are yarns which are guided through the entire machine table including the tooth gear, shaft and the horn gear. Additionally, a special pile has to be mounted on the horn gear in order for the middle end to be guided over the level of the carriers so as not to disturb the normal braiding process. The tension of the middle ends has to be controlled and kept constant during the braiding process, and normally devices identical to those employed for the core

Figure 6.38 Overbraiding of complex profile. The part is moved with the robot arm. The braiding rings from the both side allows control of the braiding geometry in both directions. *Source*: Courtesy of the Company August Herzog Textilmaschinenfabric GmbH & Ko KG, Oldenburg (August Herzog, 2013).

tension are used. The tension of the middle ends normally has to be higher than the braiding yarns; otherwise they will interlace with them rather than remain in the middle between the braiding yarns.

Both core yarns and middle ends can be used to produce hyper elastic materials (gummy tapes for underwear) or, alternatively, to limit elongation. In composites, braids which have middle ends are usually called 'triaxial' braids, because they have yarns in all three directions: 45°, −45° and 0°.

6.5.3 Braiding needles

The braiding needles are special thin, long, metal parts which are mounted on the top of the piles for the middle ends. Their task is to regulate the moment of change of the yarn tension direction, usually for flat braided structures and other braided structures where the carriers are reversing direction (Fig. 6.42). The needles are very important for making flat braided products of good quality, especially materials with a low friction coefficient, where the slippage between the yarns means that the interlacing points can get shifted in one direction or the other.

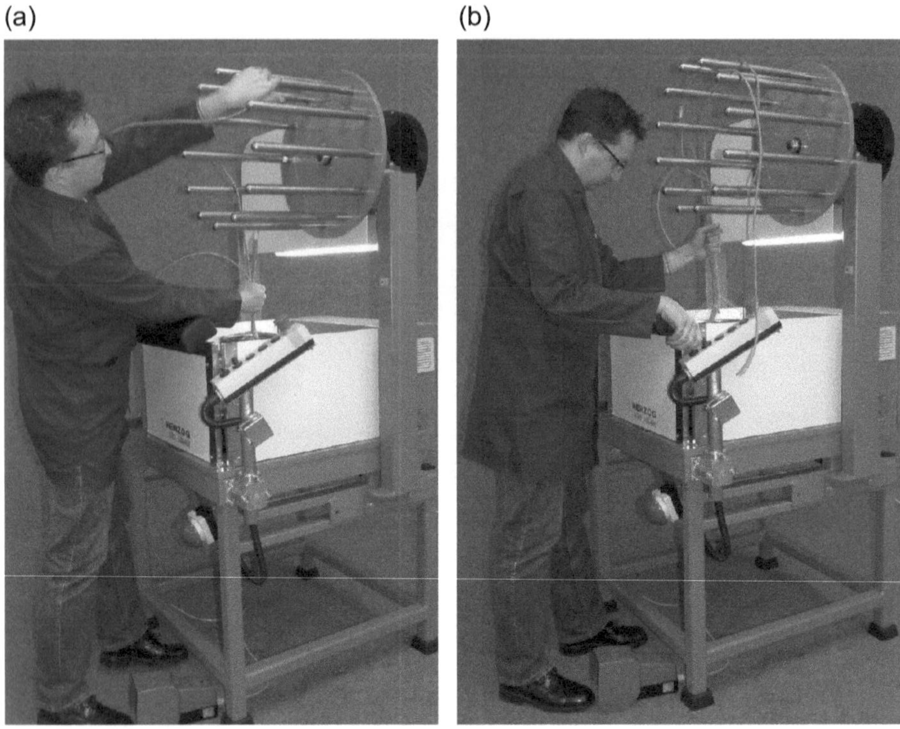

Figure 6.39 Overbraiding of cables. (a) preparation step (b) overbraiding.
Source: Courtesy of the Company August Herzog Textilmaschinenfabric GmbH & Ko KG, Oldenburg (August Herzog, 2013).

6.5.4 Depot for particles for the core

It is common to use a core or mandrel presented as discontinuous material, such as metal spheres or other particles. These are inserted into the braid at certain, defined distances, and are used to keep the rope stable, for instance, for jalousie ropes or in some fishing meshes. In such a case the take-off of the machine is changed, and the braided product goes down through the middle of the machine. The particles are supplied from a pipe, B, usually using vibration transport or similar. A mechanism, A, (Fig. 6.43) allows only one piece at a time to go into the braiding area, where it is overbraided from the yarns, and the ready-made product is taken off under the machine through the pile, D.

6.5.5 Special parts for fancy braids

Several machines can have additional, usually program-controlled needles, which fix and hold only some yarns to build loops from one or side or the other. These are com-

Figure 6.40 Braiding zone during the overbraiding of cables on the machine KBB 1–24–100. The large braiding ring and the cylinder below it are protecting the cable three from contact with the carriers and braiding yarns outsize the braiding area.
Source: Courtesy of the Company August Herzog Textilmaschinenfabric GmbH & Ko KG, Oldenburg (August Herzog, 2013).

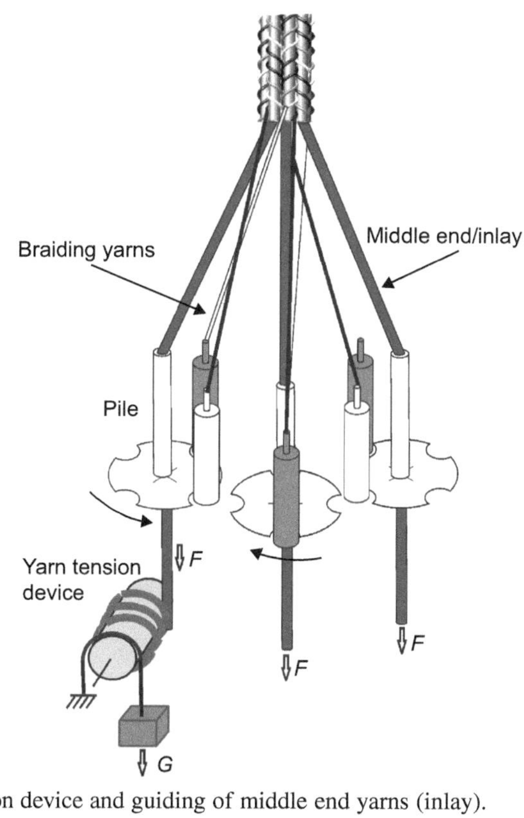

Figure 6.41 Tension device and guiding of middle end yarns (inlay).

Figure 6.42 Braiding needles on flat braiding machine. On the right-hand side are visible three yarns, whose upper part remains in the same orientation, as during the moment of the building interlacement with some other yarn even the bottom part, is already moved into the opposite direction because the carrier changes already the direction of its motion.

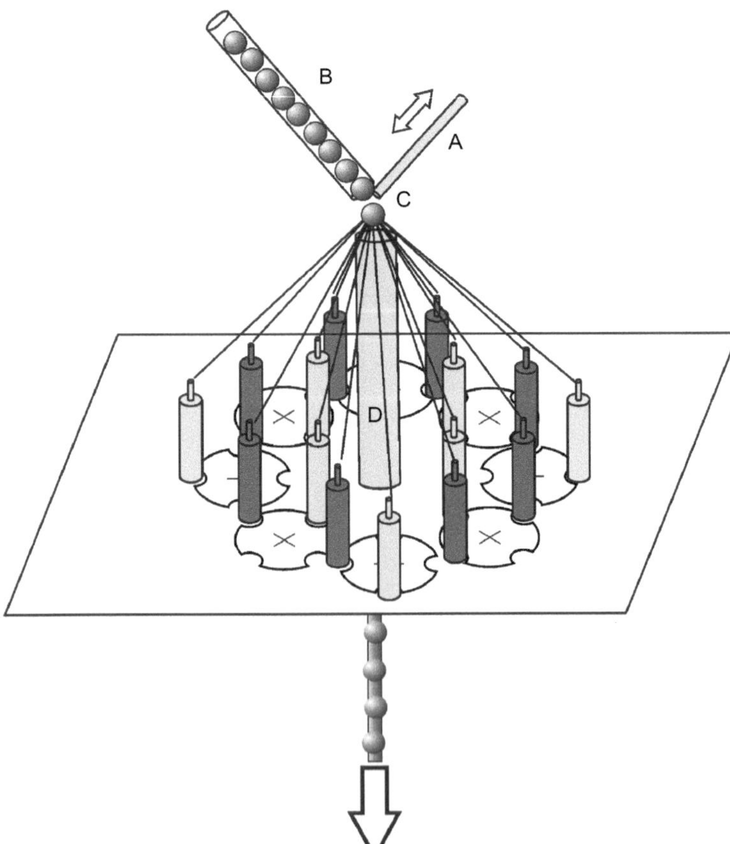

Figure 6.43 Inserting particles for overbraiding from a depot. The take-off in such machines is reversed, so that the product, guided from a rope (like a guide for cores) goes through the entire machine table and is taken off to the bottom side of the machine.

mon techniques for fancy braids for clothing, where the principle is that only at certain times is one braiding needle moved into the braiding zone, and then taken out again.

6.6 Some special configurations

6.6.1 Production of endless ropes

In some types of horizontal take-off machines, the complete plate of the braiding machine is constructed in two parts; one part can be rotated around the connection point (Fig. 6.44), and on the other side a small opening appears. Such machines are used for the braiding of multi-layered endless ropes. The rope is braided, and after take-off is guided by a system of rollers so that it comes again into the braiding zone for overbraiding with the next layer. After the final diameter and number layers are achieved, production stops and the track plate has to be opened (Figs 6.45 and 6.46), so that the endless rope can be taken out of the machine without cutting. A state-of-the-art machine of such a configuration is shown in Fig. 6.47.

6.6.2 Form braiding

In most standard cases the take-off mechanism is mounted on top of the braiding machine. In some cases, however, it is more practical for the product to be taken off from the bottom and to be immersed immediately in a coating bath (as in the production of gaskets, seals and other form-braided products), formed and then taken off and wound (Fig. 6.48).

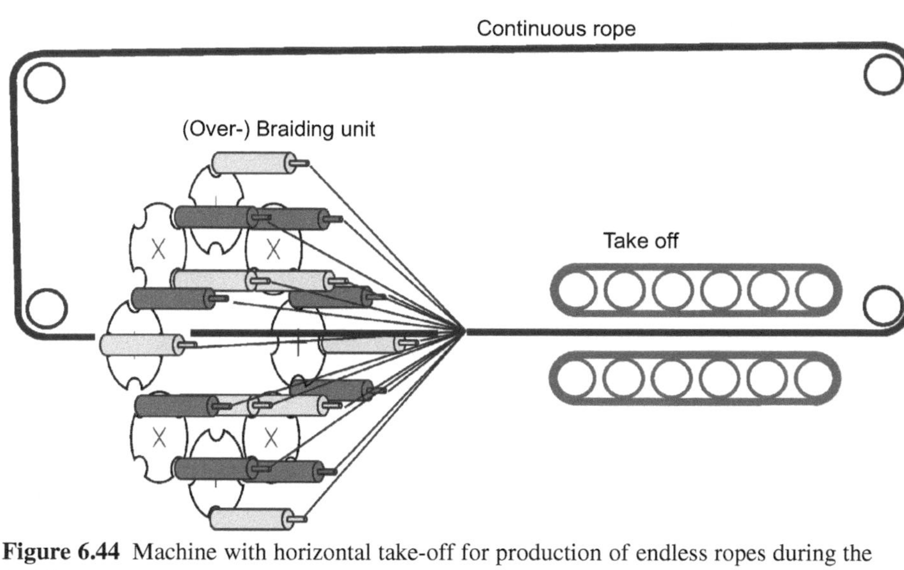

Figure 6.44 Machine with horizontal take-off for production of endless ropes during the braiding process.

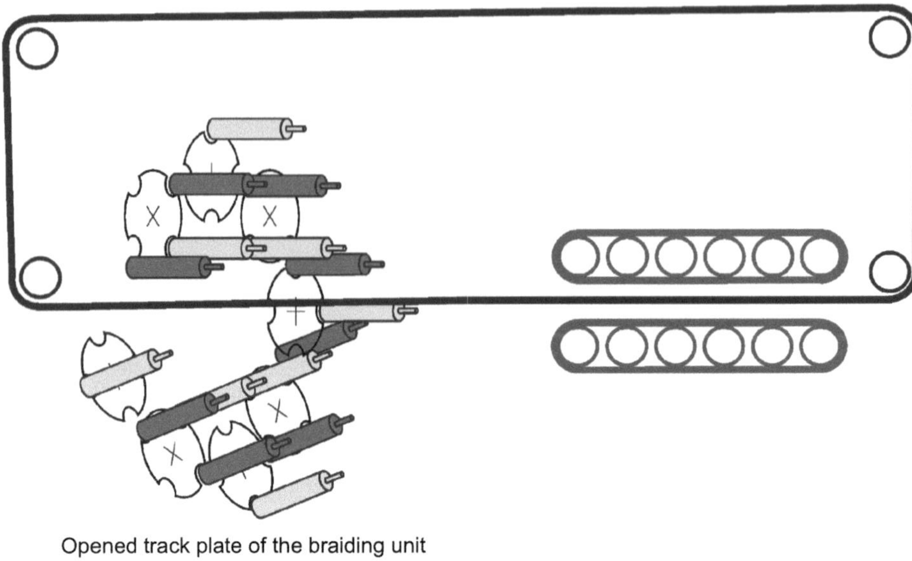

Figure 6.45 Machine with horizontal take-off for production of endless ropes; during the removing of the rope from the machine the braiding plate is opened into two parts.

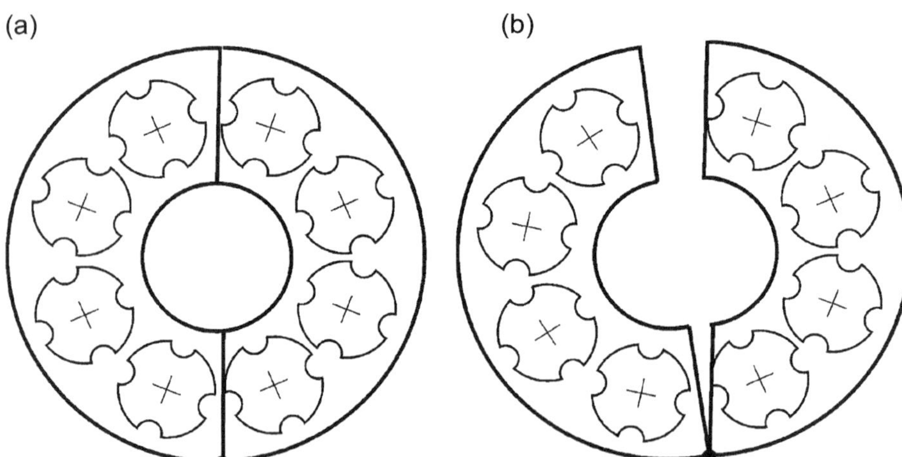

Figure 6.46 Machine with horizontal take-off. The plate can be opened in the upper side remaining on the bottom part fixed with a joint (a) normal state (b) open configuration.

(a) (b)

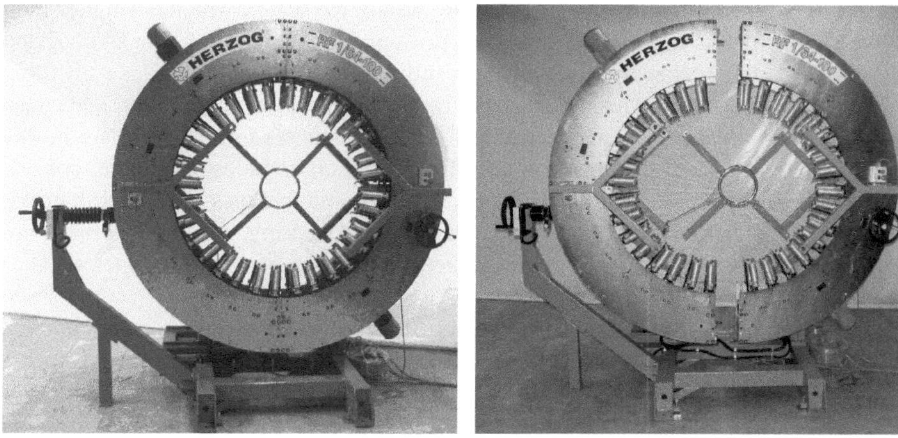

Figure 6.47 Radial braiding machine, Radial Braider RF 1/288–100, with set-up for 64 braiding bobbins and 32 0°-threads, with capacity for production of endless ropes (a) in normal state (b) open configuration.
Source: Courtesy of the Company August Herzog Textilmaschinenfabric GmbH & Ko KG, Oldenburg (August Herzog, 2013; Grave *et al.*, 2009).

Figure 6.48 Packing braiding machine with coating bath and take-off mechanism on the bottom side.
Source: Courtesy of the Company August Herzog Textilmaschinenfabric GmbH & Ko KG, Oldenburg (August Herzog, 2013).

After the braiding process and coating, the new product is usually pulled though an opening with a certain form (for example, square for seals), which is smaller than the braid. During this process the product is pressed and compacted to the given shape by the former. The higher friction and the resulting significantly higher take-off force has to be considered (or at least mentioned to the machinery manufacturer!), because a stronger take-off mechanism is required. The complete take-off mechanism in this case has to be designed to transfer higher forces, which includes, of course, consideration of the proper motor drive, shafts, tooth profile and thickness of the tooth gears, etc. More details about the configuration and types of packing braiding machines are given in Chapter 12. The same is also valid for the braiding of climbing ropes or other ropes where higher forces are expected.

The advantage of the take-off on the bottom side lies in reduced space requirements, storage and costs, because three processes – braiding, coating and forming – and eventually winding – are combined in one machine.

6.7 Sources of further information and advice

The best place to look for information about innovations in the area of braiding machine constructions are fairs such as the International Textile Machinery Association (ITMA) Fair (www.itma.com), or events and meetings with the machinery producers. Also patents for new devices and machines. There are other sources available, but they are mostly research investigations or older books, which do not encompass the state of the art in the area of braiding machine construction.

References

August Herzog (2013), ‘Product Catalogue’, available at: www.herzog-online.com (accessed 01.2013).

Büsgen, A. (1993), *Neue Verfahren zur Herstellung von dreidimensionalen Textilien für den Einsatz in Faserverbundwerkstoffen: Dissertation.*

Douglass, W.A. (1964), *Braiding and Braiding Machinery*, N.V. Uitgevesmaatschappij Centrex, Eindhoven.

Engels, H. (1994), *Handbuch der Schmaltextilien. Die Flechttechnologie. Teil I: Maschinen und Verfahren zur Erzeugung konventioneller Geflechte*, Mönchengladbach.

Goseberg, F., Brabender, K. and Moeller, P. (1981), *Textiltechnik Maschinengeflechte: Ausbildungsmittel Unterrichtshilfen*, Arbeitgeberkreis Gesamttextil, Frankfurt am Mai.

Grave, G., Zimmermann, K., Drechsler, K., Rathjens, A. and Kyosev, Y. (2009), ‘Simulation of 3D overbraiding – solutions and challenges’, paper *Presented at Second World Conference on 3d Fabrics and their Applications*, 6–7 April, Greenville, South Carolina, USA.

Lepperhoff, B. (1953), *Die Flechterei,* 3rd edn., Eugen G. Leuze Verlag Saulgau/WTTBG.

Müller, L.E. (2013), ‘Biomimetisch optimierte verzweigte Faserverbundstrukturen mit hoher Tragfähigkeit’, *Melliand* (in printing).

Rehrmann, R. and Ludwig, A. (2013), 'Product documentation', available at: www.steeger-online.com.
Scheider, H. (1991), *Flechtscholle* EP 0498 944 B1.
Schneider, M. (2000), *Konstruktion von dreidimensional geflochtenen Verstärkungstextilien für Faserverbundwerkstoffe, Berichte aus der Werkstofftechnik*, Shaker Verlag, Aachen.

Carriers for braiding machines 7

7.1 Introduction

Carriers are the most important parts of the braiding machine. They have the largest number of functions and have to be expertly designed and produced. They are mounted over carrier feet, and these are part of the carrier motion system, described separately in Chapter 6. Carriers are moved along the track and they transport the bobbins while keeping the yarn under as constant a tension as possible (Fig. 7.1). After a yarn is interlaced with other yarns in the braided structure, a new yarn length from the bobbin has to be supplied for the interlacing process to continue. If there is a yarn break at the end of the yarn, the carrier must identify it and stop the machine. Different elements of the carriers are responsible for these functions. These are discussed in more detail in this chapter.

7.2 Carrier features

There are some basic features which each carrier has to be able to satisfy and without which the braiding process will not work. These features are not so important for the braiding process itself, but are important for the users of the machine and can also improve production levels by reducing service breaks. Optional features can make the life of machinery builders easier, allowing the use of the same carriers in different applications.

7.2.1 *Basic features*

The basic features of carriers are all related to the yarn (or wire) and the braiding process. Independent of their construction, carriers have to be able to perform a number of basic operations simultaneously, which is demonstrated in Fig. 7.2.

First, the carrier has to hold the bobbin with the yarn (or wire). It is due to this that in several patents it is misleadingly named the "bobbin holder." Holding the bobbin, the carrier moves along the track and has to perform several additional tasks, so it is not simply a bobbin holder. The distance L1 (Fig. 7.2) between the carrier and the braiding point is not the same as the distance L2, when the same carrier is under the braiding point. This distance changes periodically during the braiding process and depends on the machinery stitch, configuration of the track plate, and the position of the braiding point. As a result of these variations, different yarn lengths are required between the braiding point and the carrier. To guarantee a stable braiding process with constant conditions, the yarn tension has to remain as constant as possible, and

Braiding Technology for Textiles. http://dx.doi.org/10.1533/9780857099211.2.153

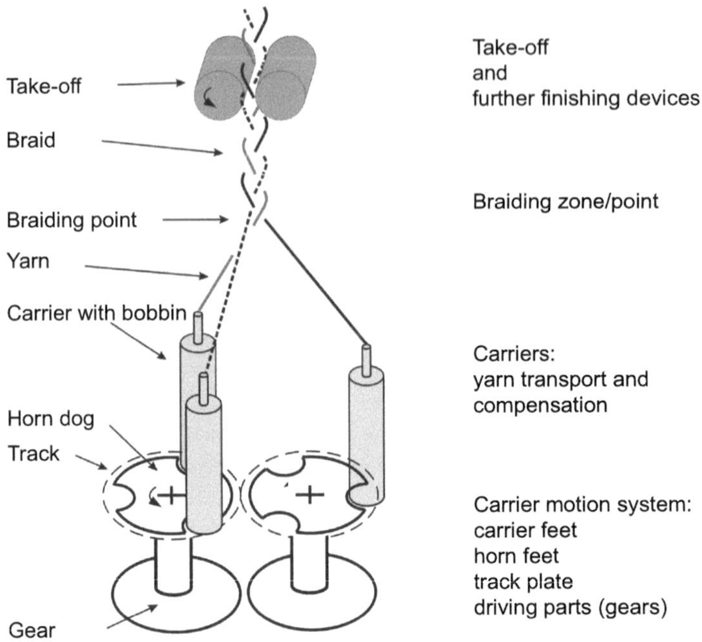

Figure 7.1 Main parts of the maypole braiding machine.

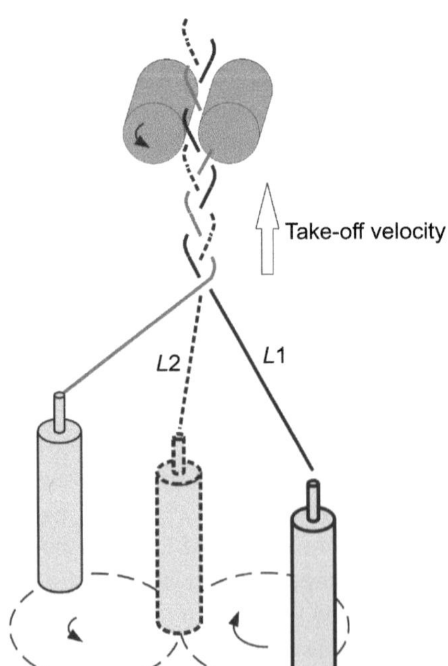

Figure 7.2 Basic features of the carriers. The distance between the carriers and the braiding point changes during the braiding process.

thus the carrier has to be able to compensate for differences in yarn length. Each carrier has to release more yarn if required, but has to retract it if the distance between the carrier and the braiding point becomes shorter. Furthermore, because the braided product is taken off, a new piece of yarn has to be released for the next interlacing. Finally, if a break occurs, the carrier has to produce a signal which stops the machine. The carrier is the most appropriate device on the braiding machine to the control yarn breaks, since it controls the tension as well. The tension increases rapidly before the yarn breaks and decreases to zero after the break, so if these changes can be detected, the machine can be stopped in time.

In summary, the basic features or operations of the carrier can be listed as follows (Douglass 1964):

Feature 1. Hold a bobbin with yarn material (as much material as possible).
Feature 2. Compensate for differences in length during the braiding and maintain yarn tension.
Feature 3. Pay out the correct quantity of yarn as required.
Feature 4. In the event of the yarn running out or becoming broken, the carrier should produce a signal to ensure that the machine stops immediately.

7.2.2 Self-evident features

Since the amount of yarn material on the bobbin is limited, it has to be replaced after a certain period. To reduce delays in setting up the machine, the construction of the carrier needs to allow for:

Feature 5. Easy replacement of the bobbin.
Feature 6. Simple, light construction with a minimum number of working parts
Feature 7. Rigidity, so that the carrier does not deform due to yarn tension

7.2.3 Optional features

In certain machines, such as packing braiders and radial braiding machines, the carrier has to:

Feature 8. Keep its functionality independent of its orientation in the space.

From an economical point of view, the machine has to run as fast as possible. However, running fast is usually inversely proportional to the quality of production. To formulate this as a feature, the carrier has to:

Feature 9. Keep its functionality independent of its velocity.

To keep the machine size in a suitable case the carrier has to:

Feature 10. Be as small (compact) as possible.

7.2.4 Ideal carrier

The ideal carrier is a goal engineers can work towards, improving existing carriers or creating completely new constructions. It has to:

1. carry a bobbin *with the maximum possible amount* of yarn.
2. maintain *constant* tension on the yarn during the braiding process.
3. pay out the correct quantity of yarn as required *without causing peaks in tension*, and taking into account the optional features.

Meeting these requirements usually involves compromise. More yarn means bigger, heavier bobbins and a bigger carrier. This then is not suitable for faster machines, because of inertia forces, and also bigger carriers means larger braiding machines.

7.3 Yarn length compensation devices

There are different ways to classify carriers. In the praxis the most popular classification is based on the type of compensation device – "slider" or "lever" carriers. "Slider" and "lever" are the two main types of carrier used which transmit the force between the yarn and some force element such as a weight or spring.

Actually, there are carriers where the differences in yarn length in the braiding process are compensated by winding and unwinding the yarn directly from the bobbin, using neither slider nor lever. The bobbin can be fixed and moved by a spring, or it can be moved by a motor mounted on the carrier. These two cases are still the subject of research rather than large, scale production, but they will be of interest in the near future.

Most carriers currently work with bobbins which do not participate in the compensation process. They are referred to as carriers with "bobbin independent" compensation, because an additional length compensation device is required (Fig. 7.3).

This additional device applies tensile force through a lever or slider on the yarn so that the excess yarn length can be retracted into the carrier. The force itself can be created by a weight or spring.

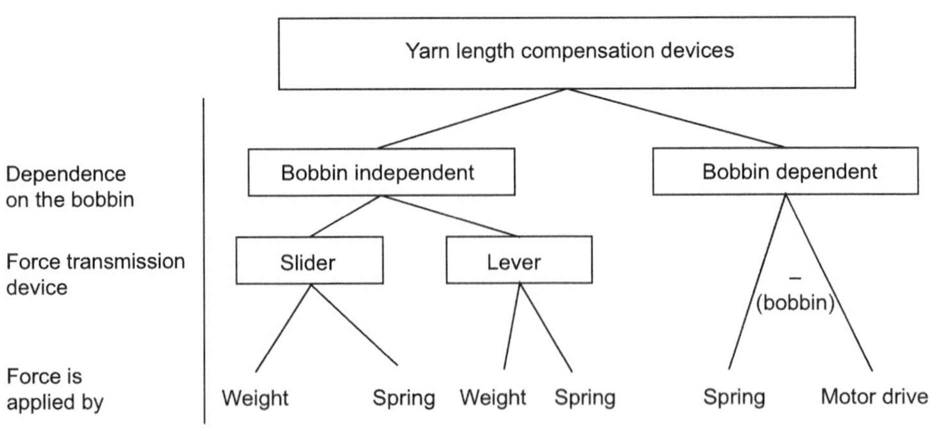

Figure 7.3 Classification of the carriers, depending on the working principle for the yarn length compensation device.

The most popular solutions are:

weight, connected to slider (carriers with weight compensation);
spring, connected to slider (carriers with slider compensation);
spring, connected to lever (carriers with lever compensation)

The names in brackets are the popular labels and describe completely the type of compensation device of the carrier.

7.3.1 Carriers with weight compensation

The first machine carriers used weight(s) for yarn length compensation.

As mentioned at the beginning of this chapter (Fig. 7.3), the required yarn length between the carrier and the braiding point varies during the braiding process. The variations in length have to be compensated to some degree inside the carriers. The weight-balanced carriers use one or two weights, which create the yarn tension and keep it constant. Under the weight forces, the weight moves down if surplus yarn length outside the carrier is available and retracts it. The bobbin is mounted on the carrier using a special bobbin adapter, so that the teeth on the upper part of the bobbin are in contact with the bobbin fixation lever (Fig. 7.4). This lever is pressed down onto the teeth by the yarn tension force or by a special spring inside the carrier, and it does not allow the bobbin to rotate during the braiding process. Having a fixed bobbin, one end of the yarn is fixed as well. The yarn goes from the bobbin through the middle eye (Fig. 7.5) so that the outer weight is hanging on it. If the distance between the carrier and the braiding point increases, more yarn length is needed. In this case the yarn pulls the outer weight up and the required length is taken from the

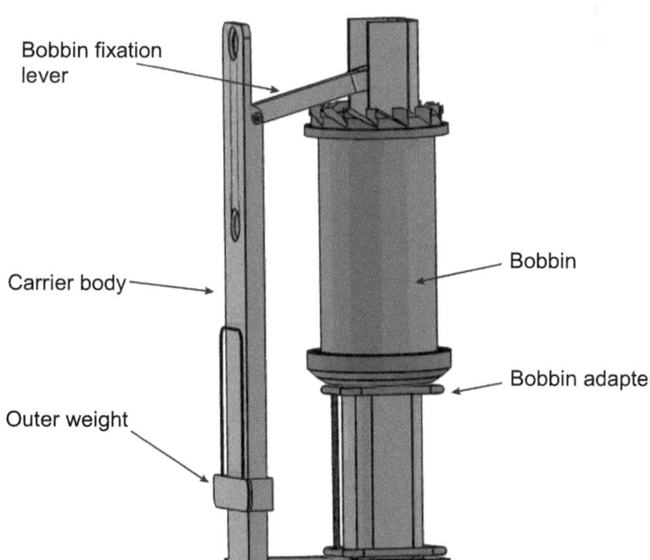

Figure 7.4 Weight-balanced carrier.

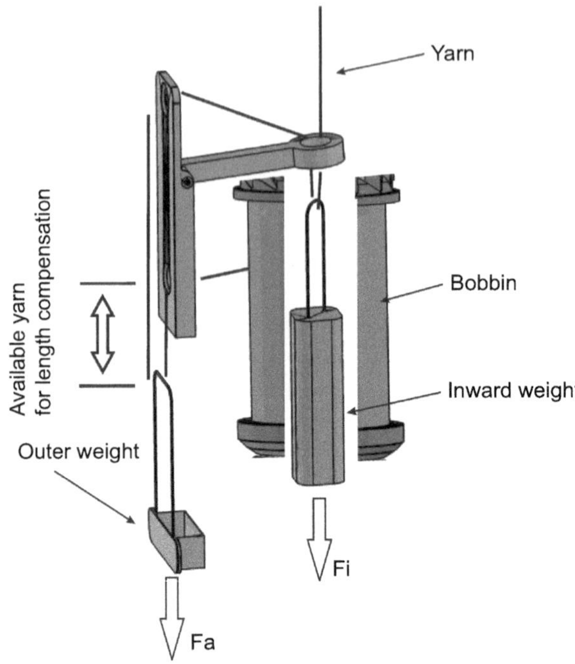

Figure 7.5 Yarn compensation elements in a weight carrier.

available yarn length in the compensation area. After the entire yarn length from the outer compensation area is used in the braid, the inner weight starts to compensate for the difference in length as well, moving up or down (Fig. 7.6). After the entire yarn reserve is used, the yarn tension increases, and the inner weight moves up and shifts the bobbin fixation lever up. This releases the bobbin, so that it can rotate and release some more yarn, until the inner weight takes the lever down and fixes the bobbin again. The outer weight is normally less than the inner one and so compensates for the differences first.

Since the weights are constant and their gravitational force is not dependent on their position, the yarn tension does not change during the braiding process. The weight can be easily adjusted by adding or removing mass (or a set of different weights) and so it is possible to guarantee constant and very precisely adjusted yarn tension. This advantage can be used only in slow-working machines with vertical carriers. This weight balancing principle cannot be used for horizontally orientated carriers or hanging carriers, which change their orientation (radial braiding). Additionally, weight-balanced carriers cannot be used in machines which are running faster than a certain speed (see Chapter 8, 'The mechanics of the braiding process').

There are also weight-balanced carriers in existence, where the inner weight is replaced by a spring with a very low spring constant – such as one of the classical carriers from the advent of braiding machinery, the so-called "Barmer Büchsen Klöppel," shown in Fig. 7.7.

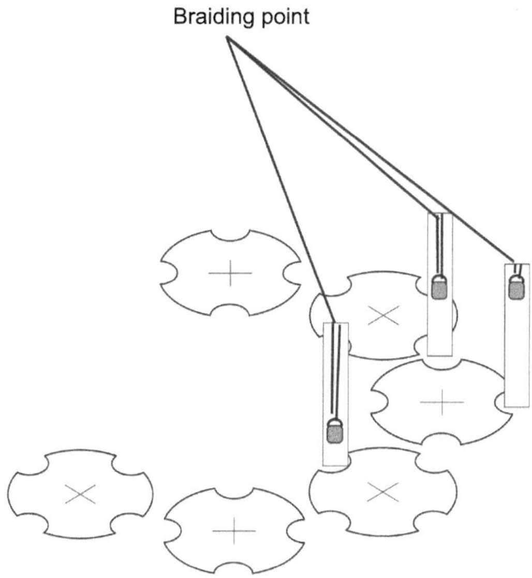

Figure 7.6 Position of the weight depending on the carrier position with respect to the current yarn length between the carrier and braiding point.

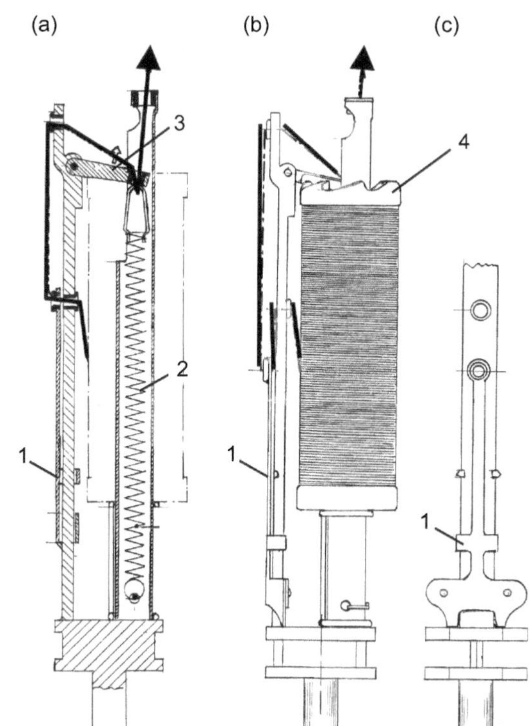

Figure 7.7 Weight-balanced carrier where the bobbin control is arranged with a spring, "Barmer Büchsen Klöppel" (Goseberg *et al.*, 1981). (a) section view (b) side view (c) view of the sliding weight 1.

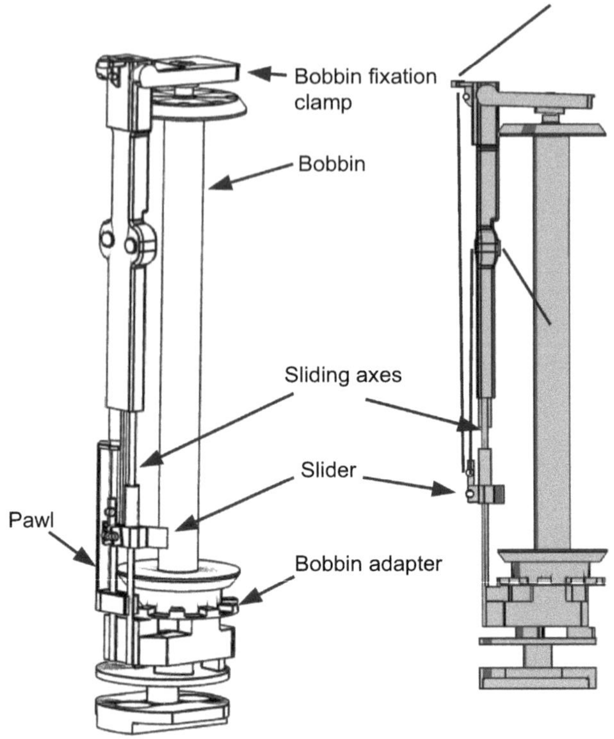

Figure 7.8 Slider-balanced carrier.

7.3.2 Carriers with slider compensation

The slider-balanced carriers generally work on the same principles as weight-balanced carriers, the one difference being that the sliding element is guided precisely, not just hanging in space as was the case for the previous weight-balanced carriers. Figure 7.8 shows a slider-balanced carrier, where the working force is created by a spring.

The spring presses the slider, directly, transmitting its force to the yarn. This saves space and allows the use of larger bobbins, making the braiding process more effective. The slider is supported in such cases at two axes, ensuring its stable motion. As the distance between the carrier and the braiding point decreases, the yarn tension also decreases and the spring moves the slider down until the tension forces and the spring forces reach a new equilibrium. When no more yarn is available in the compensation area, the slider reaches its upper position, contacts the plough and raises it (Fig. 7.9).

The tooth in the upper side of the plough releases the bobbin adapter, and then the bobbin can rotate freely until enough yarn length is unwound from the bobbin. After some new yarn length is available, that slider moves down under the action of the spring force, the plough moves down as well under the action of its spring force, and its tooth enters the next free position in the bobbin adapter and fixes it (Fig. 7.10).

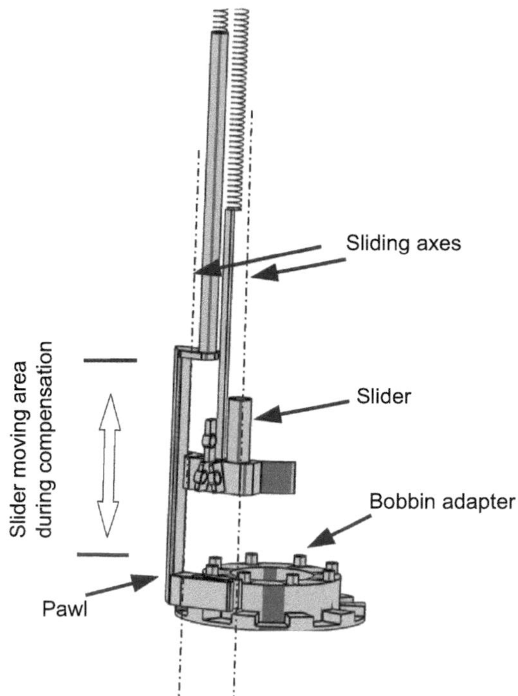

Figure 7.9 Elements of slider-balanced carrier – slider, plough and bobbin adapter.

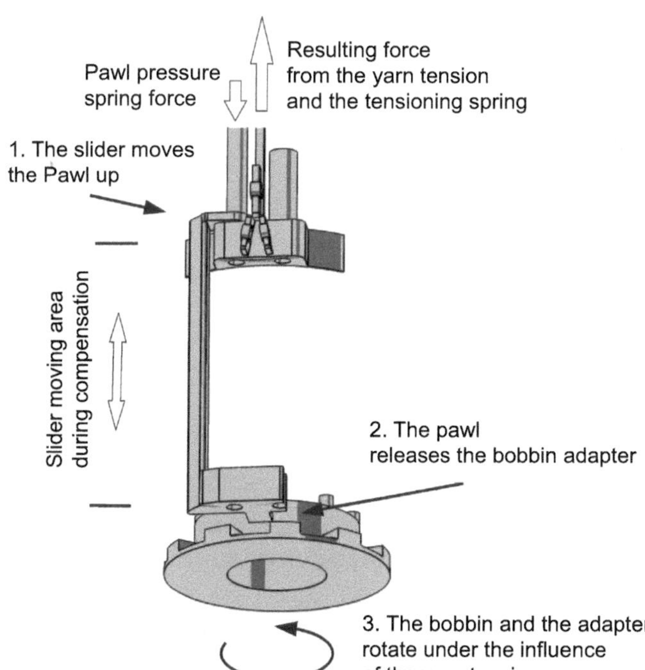

Figure 7.10 Sequence of releasing the brake.

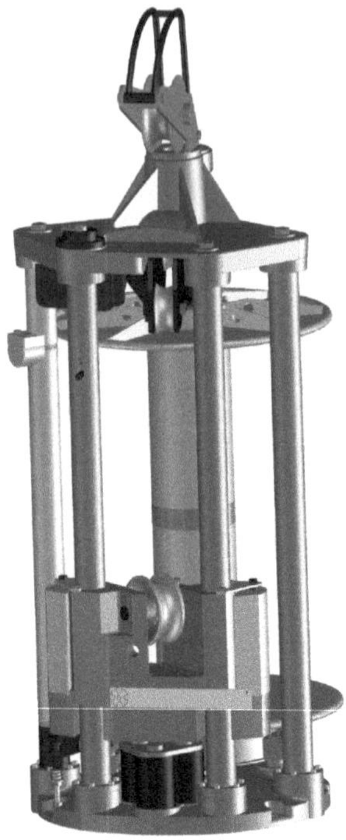

Figure 7.11 Slider-balanced carrier, which uses weight for creation of the yarn force.
Source: Courtesy of the company August Herzog GmbH & Ko KG, Oldenburg, Germany.

The advantages of slider-balanced carriers are:

- the slider needs less space, in comparison to a lever construction (described in the next section)
- the load of the connecting parts is lower (the lever transmits higher forces in the joints);
- there is more room for larger bobbins.

As a result, slider-balanced carriers are generally used for medium-sized to very large braiding machines. Figure 7.11 shows the construction of a slider-balanced carrier with sliding weight compensation for very thick ropes where the yarn tension (the "yarns" here are actually ropes, with diameters of several centimetres) can reach several kN.

Sliding has its disadvantages – for very fine yarns and delicate materials, the friction forces during the sliding process are higher, compared to lever-balanced carriers. These forces can reach the yarn breaking force, and this can cause plastic deformation in the yarns. The yarn tension regulation is also not sufficiently precise for such materials. In these cases, lever-compensated carriers are used.

7.3.3 Lever-balanced carriers

Lever-balanced carriers use a lever to transmit the spring force to the yarn. The spring force does not depend on the orientation of the carrier, which solves the problems of

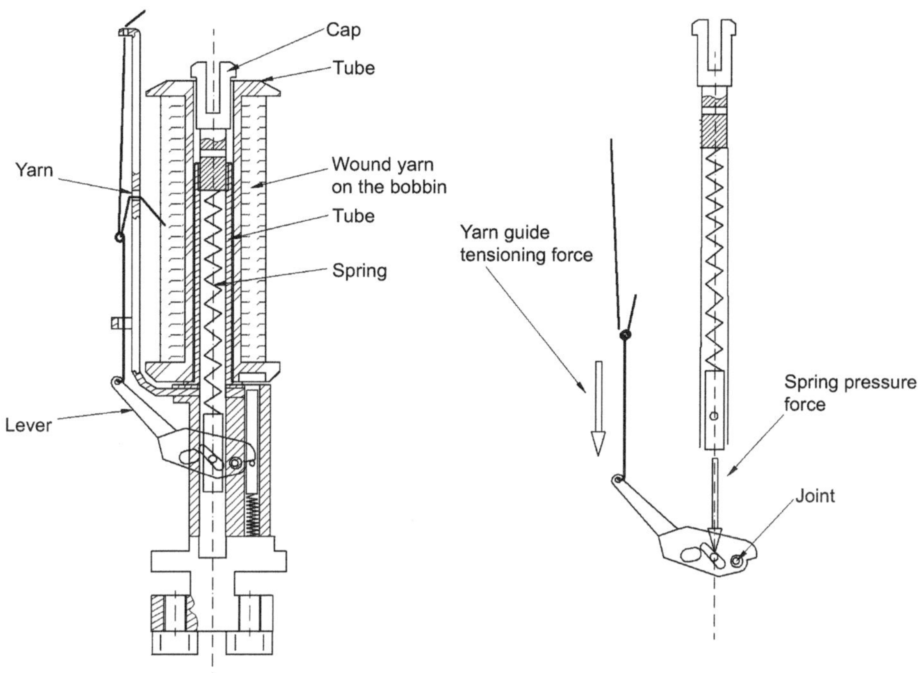

Figure 7.12 Lever-balanced carrier.

weight being used in weight-balanced carriers. Figure 7.12 presents the construction of a typical lever-balanced carrier.

The yarn tensioning spring is mounted inside the carrier body and presses the lever with a metal pressure pin mounted on a special cylindrical body. The lever rotates down under the action of this force and thereby also tenses the yarn, connected to its end by a wire. The lever actually reduces the force for the yarn, but it allows the application of the yarn force on the proper place.

If the yarn reserve is already used, the yarn tension increases, and the lever end will be taken to its upper position (Fig. 7.12). Its other end presses down the pawl so that the pawl end is hidden within the carrier body and releases the bobbin (Fig. 7.13). At that moment the bobbin can rotate free and the next length of new yarn is introduced into the compensation system. Once enough yarn is available, the lever rotates back and the pawl stops the bobbin (Fig. 7.14).

Lever-balanced carriers are the most popular carriers for textile braiding, where the yarn tension forces are not so high. They have a relatively simple construction and can work in all orientations. If a change is required in the working yarn tension, the tensioning spring has to be changed. Companies are usually able to provide a set of five to ten types of springs to cover the complete working range of yarn tensions for a given carrier. The springs are marked with a colour at the end (i.e. white, green, blue, black, silver, yellow, etc.) enabling users to recognize the different types easily. Table 7.1 presents an example of such sets, for the machines of the company August Herzog (Engels, 1996).

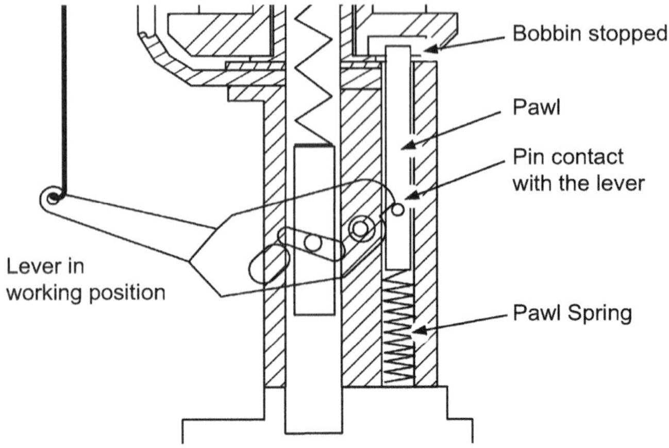

Figure 7.13 Lever carrier – the pawl released the bobbin, it can rotate and the next piece yarn can be unwounded.

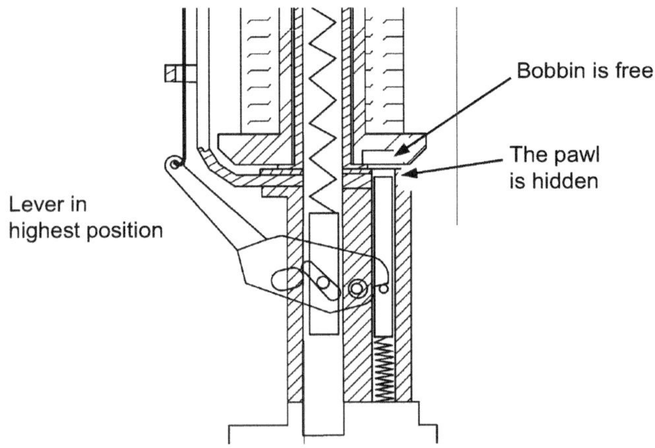

Figure 7.14 Lever carrier – the bobbin is stopped by the pawl.

If the yarn/wire tension has to be higher, then lever-balanced carriers are not suitable. The tensioning spring has to apply a force about three to ten times higher on the lever than the yarn tension because of the different arm lengths of these forces. In this case, slide-balanced carriers provide a better solution.

7.3.4 Single, double and multiple compensation

The available yarn length ΔL, which compensates for the differences in the distance between the carrier position and the braiding point, is equal to the possible path of the slider or the lever eyelet multiplied by two, because two pieces of yarn are available (Fig. 7.15a)

Table 7.1 Yarn tension in cN for single compensation for different springs (Engels, 1996)

Colour	Nr	Type A	Type B	Type C	Type D
Yellow-orange	17		900		
Light grey	16		160*		195
Rose	14		220		
Light brown	11		140	145	115
Dark red	10		115	135	105
Dark green	9		90	93	75
Silver	8	90	70	89	52
Light green	7	75	65	63	50
Dark blue	6	55	40	50*	40
Light blue	5	50	30	57	30
Black	4	40		40	
Light red	3	35		28	
Yellow	2	25		12	
White	1	15		3	

* These values are given in the literature (Engels), but because they do not follow the arrangement (from higher value to lower), and were not available for testing, they are marked as suspect values.

$$\Delta L = 2 \cdot \Delta S$$

This is the case of single or normal compensation. This available yarn reserve is not enough for some machine configurations with large carrier paths, where the distance between the carrier and braiding point are greater. If double eyes on the carrier body and the moving part (slider or lever) are mounted, it is possible to extend the length reserve, making double compensation (Fig. 7.15b)

$$\Delta L_{\text{Double}} = 4.\Delta S$$

Independent of the type of carrier (lever or slider, spring or weight) if carriers with eyelet for double compensation are available, these can be used for single compensation as well. The opposite is not possible. By the same principle, especially for very fine and delicate materials, more than double compensation can be used (e.g. triple or higher). Such cases are actually only applied to carriers with rollers as yarn guides, because otherwise the friction between the yarn and the yarn guides would increase significantly.

It is important to note that double compensation *doubles* the difference in yarn length, which can be compensated, but also *doubly reduces* the yarn tension (Fig. 7.16) since the spring force is distributed over double the number of yarn pieces. The same rule is valid in the case of multiple compensation – the higher the compensation, the lower the yarn tension.

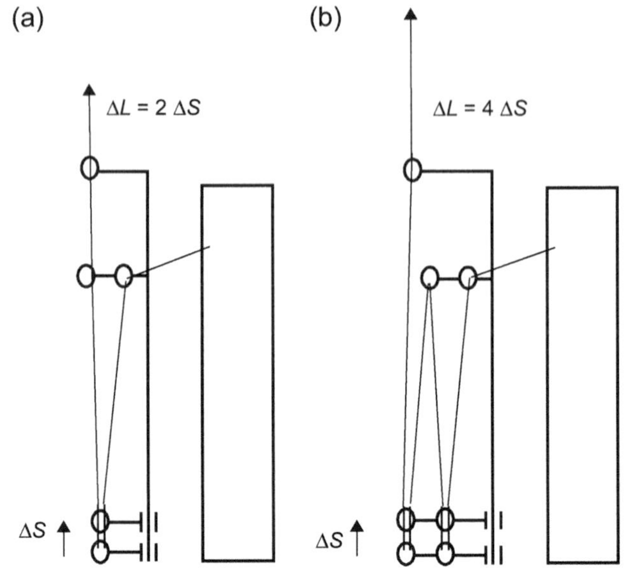

Figure 7.15 Single (a) and double (b) compensation.

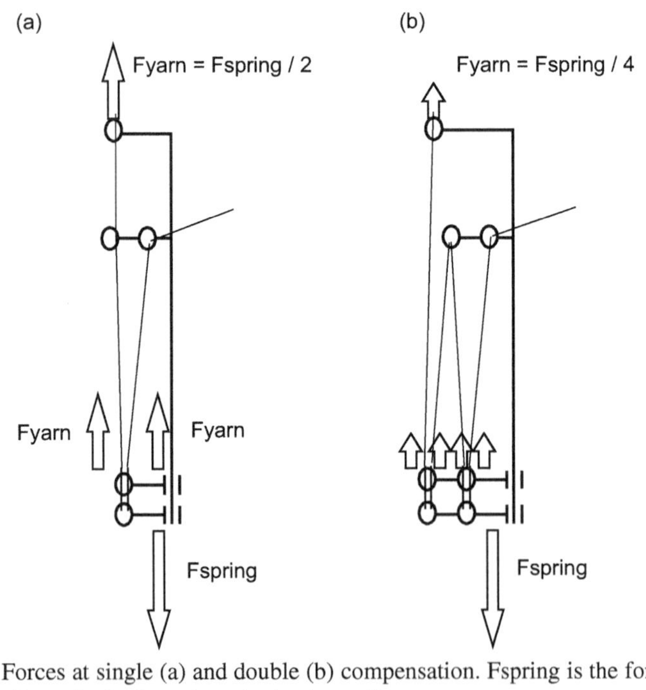

Figure 7.16 Forces at single (a) and double (b) compensation. Fspring is the force, created by the spring; Fyarn is the force in a single piece of yarn.

7.4 Bobbin brakes and bobbin form

7.4.1 Brakes

The bobbin brakes prevent the bobbin motion during the braiding process and in this way keep one of the yarn ends fixed. If the yarn in the compensation area is already used, then the brake has to release the bobbin to free some yarn length. The most commonly used type of bobbin brakes have been mentioned above in carrier construction.

Figures 7.13 and 7.15 present the common solution with normal pawl from the bottom side. Figure 7.10 demonstrates the working of a similar brake in slider-balanced carriers.

In larger carriers, where the bobbins weigh more than 100 kg, such a plough is not suitable, because the bobbin has a large moment of inertia and the rotation will give rise to significant force where the pawl tries to stop it. In such cases, friction brakes, similar to those in cars, are used – Fig. 7.17. The friction brakes convert the kinetic energy of the bobbin into thermal energy while allowing smooth release and stopping of large masses.

There are existing constructions for yarn compensation with friction brakes, based on belt friction on a disc connected with the bobbin and special metal wire (Fig. 7.18). In this case, the friction wire holds the bobbin and only if it is released, because of increased yarn tension, is the bobbin able to rotate. This construction is used mainly in carriers with horizontal mounted bobbins for wire or other such materials.

Figure 7.17 Friction brake for larger carriers.
Source: Courtesy of Company August Herzog GmbH & Ko KG, Oldenburg, Germany (Herzog, 2012).

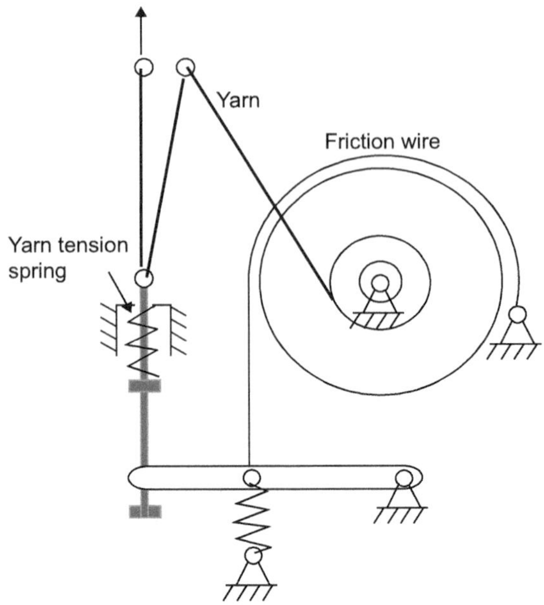

Figure 7.18 Brake based on the belt friction.

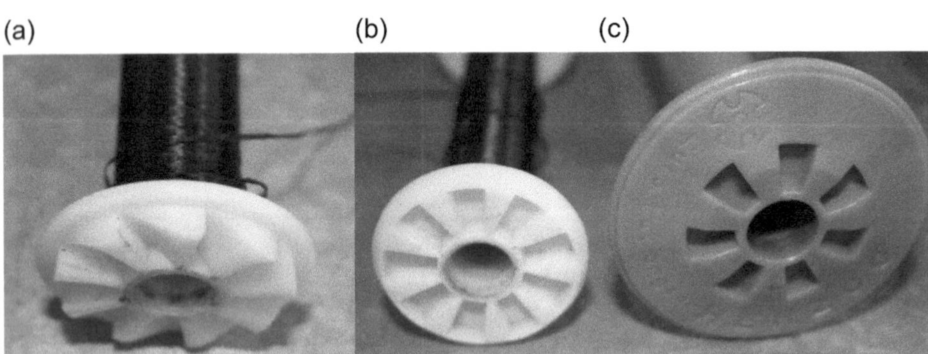

Figure 7.19 Tubes with different forms for direct contact with the brake. (a) with triangle teeth, usually used on one side of the tube only (b) and (c) with symmetric openings, usually applied on both the tube sides and working in both directions.

7.4.2 Bobbin tubes

The type of tubes used for yarn winding depends on the type of bobbin brakes. The carriers which still use the first invented braking system, as presented in Fig. 7.7, have on the upper side triangular-shaped dents (teeth) (Fig. 7.19a). The newer system, where the plough is on the bottom side (Figs 7.12–7.14), requires tubes with special openings as shown in Fig. 7.19b and 7.19c. All these tubes have direct contact with the plough during the stopping process.

Figure 7.20 Bottom sides of the tubes, which connect with the braking system using an adapter.

In some cases, companies use an adapter between the brake elements and the bobbins. The advantage of using an adapter is that the tubes do not come into direct contact with the elements and any deformation of the tubes cannot influence the braking process. Additionally, adapters can be designed and mounted more precisely, whereas with direct interaction with the tube, larger tolerances have to be considered. The disadvantage of adapters is their more complicated construction.

Figure 7.20 shows the bottom sides of the tubes which connect with the braking system using an adapter.

7.5 Yarn guides (eyelets, rollers, other elements)

Yarn eyelets have the function of leading the yarn if it has to change direction. The usual carrier requires at least three eyelets – one for unwinding, one on the lever/slider and one at the top of the carrier. The position and construction of the last eyelet has to ensure that during the braiding process the yarn does not come into contact with the other carriers. For this reason, this last eyelet is mounted at the highest position of the carrier (for vertical carriers) and is nearest to the braiding point. For double compensation, there are at least two additional eyelets. As the covering angle is between 90° and 180° at each eyelet, the friction is significant and special materials are used to reduce the friction coefficient during sliding. For this purpose, ceramic eyelets (Fig. 7.21) or specially coated metal eyelets (Fig. 7.22) are used. The producers of braiding machines are able to provide different sets of eyelets for better work and lower abrasion with any specific materials, but there is very little public information in this area.

Carriers which have to work with wire or thicker materials (ropes) use rollers instead of eyelets (Fig. 7.23). In this way, the sliding friction is replaced by rolling

Figure 7.21 Ceramics eyelet.

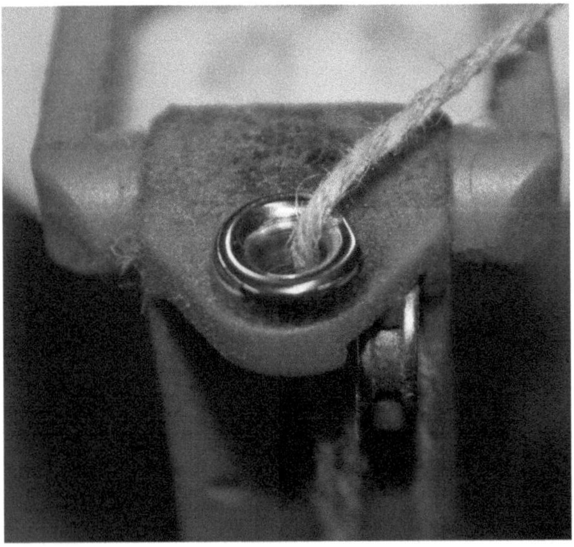

Figure 7.22 Eyelets from coated metal. Dust, built up during the processing of a flax yarn, is also visible on the picture.

friction. which is significantly lower and can be further reduced using bearings. Another effect of using rollers is the decreased curvature at the bending points, which is also very important for wire and other non-elastic materials. To reduce the number of changes of direction for wire carriers, their bobbins are often mounted horizontally, as described in the next section.

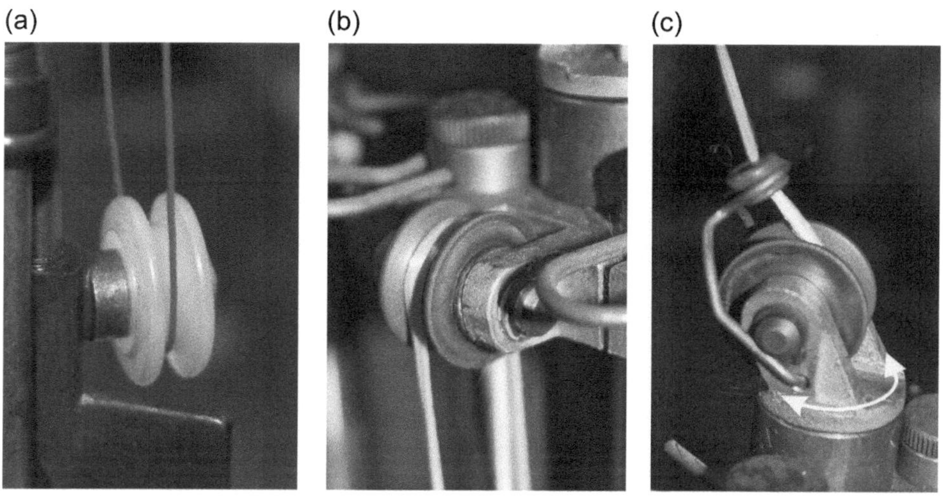

Figure 7.23 Roller eyelets: (a) from plastic material; (b) from metal; (c) upper eyelet from metal. Note that if the upper eyelet is produced as a roller, it (together with its bearing) has to be able to rotate additionally around the carrier axis.

7.6 Carrier modifications for special applications or materials

7.6.1 Carbon and glass fibres

Special care has to be taken when using carbon and glass fibres. These fibres break when bending around a profile of small radius of curvature, such as in normal eyelets. When processing such fibres, special carriers have to be used, which are designed so that the yarn is guided smoothly and with as little curvature as possible. More information about processing can be received from the producers, August Herzog GmbH & Ko KG, Oldenburg (Herzog, 2012) and Körting Nachfolger Wilhelm Steeger GmbH & Co. KG, Wuppertal (Steeger, 2012).

The proper sizing of the multifilament glass or carbon yarns can reduce several problems caused by single filament breaks, due to yarn bending at the eyelets. Reducing most problems does not mean that the filaments do not break if the yarn is properly sized; it simply means that the ends of the broken filaments do not go on the yarn surface and do not disturb the unwinding and braiding process, as can be seen in Fig. 7.24. However, bending the sized yarn in such cases can still cause filament breaks, which can be demonstrated by the reduced strength of the material after the braiding process.

7.6.2 Carriers with horizontal bobbins, for example for wire

A metal wire is used for overbraiding cables and high pressure tubes. The metal wire is a material with higher bending stiffness, contrary to the almost negligible stiffness

Figure 7.24 Broken glass filaments during the unwinding of unsized yarn.

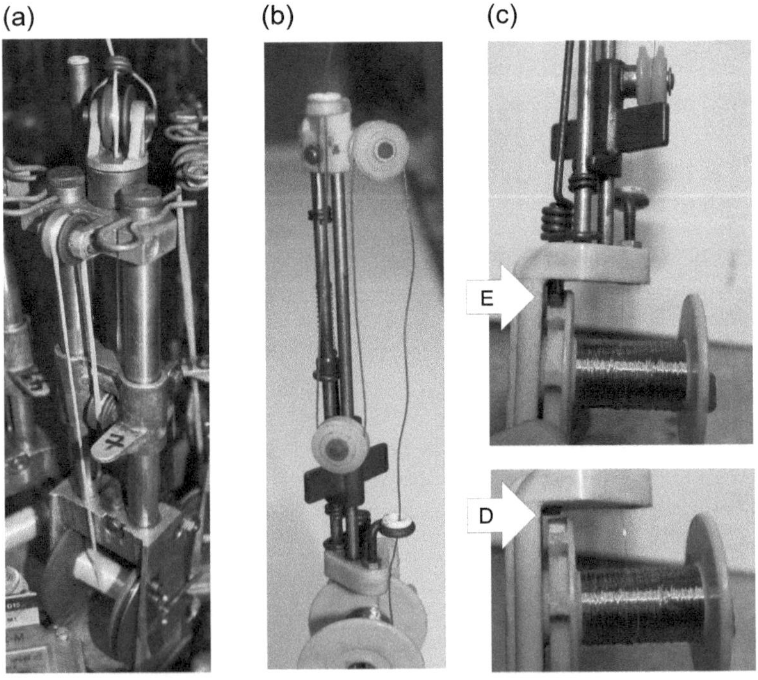

Figure 7.25 Carriers with horizontally placed bobbins for braiding of wire: (a) Carrier for heavy material; (b) carrier for fine metal wires; (c) break system for the carrier where at position E plough fixes the bobbin during the braiding process, and at position D the plough releases the bobbin and a length of wire can be unwound.

of textile yarns. Because of the stiffness of the material, it cannot be easily pulled through the eyelets and, as mentioned in Section 5.5, special rollers are used instead of eyelets to guide the material. Additionally, specially developed wire carriers are generally used where the wire's path is optimized to have as few directional changes as possible. The most common system uses horizontally placed bobbins (Fig. 7.25a and 7.25b).

The carrier presented in Fig. 7.25a uses brakes with belt friction (see Section 7.4.1), while the carrier in Fig. 7.25b uses a system with a slider and plough. The plough fixes the bobbin during the braiding process (Fig. 7.25c, position E) and if the material in the compensation area is not sufficient, the plough releases the bobbin (Fig. 7.25c, position D) and a length of wire can be unwound.

The disadvantage of the horizontally mounted bobbins is in their capacity – if larger bobbins have to be used the dimensions of the carrier will increase as well.

7.6.2.1 HP Carrier

The "HP" carrier is a special construction, which was used in the last 20–30 years in faster braiding machines. The name comes from "Herzog Patent." "HP" was associated with one special construction of carriers (Fig. 7.26). These carriers were built for a couple of years on machines which were able to run significantly faster than other braiding machines. Their construction is mainly welded iron sheet parts, which are light and stable even at higher working speeds. In order to save space, the compensation system is positioned on top of the carrier (Fig. 7.26), and these carriers can take very large bobbins, reducing the time needed to change the bobbins.

The yarn compensation system also uses a steel sheet spring (Fig. 7.27a) connected to the lever (Fig. 7.27b). The lever can be moved up as well by centrifugal forces and not only on the yarn tension, which allows braiding with very fine yarns. The tube with the yarn connects to the bobbin brake system using an adapter (Fig. 7.28). The brake system is also modified and designed to be able to stop larger bobbins, even with the tension force of fine yarns. HP carriers are not frequently used in braiding machines. Some of their disadvantages are the higher noise level and more limited possibilities for adjusting the yarn tension. An additional reason is that the handling during the bobbin change is not as comfortable as for classical carriers, and workers do not like working with them. Last, but not least, the modern braiding machines using lever- or slider-balanced carriers are already working at similar levels of productivity. Despite not currently being very popular, some renaissance for this construction can be expected because it is nearer to being an ideal carrier than the classical systems and would be able to work at even higher speeds if redesigned and improved at the mentioned points.

7.6.3 Carriers for ultra-fine braids

For ultra-fine braids lever-balanced carriers are commonly used. In some special medical applications where yarn tension has to be kept constant, weight-balanced

Figure 7.26 HP carriers for two different sizes of bobbins.

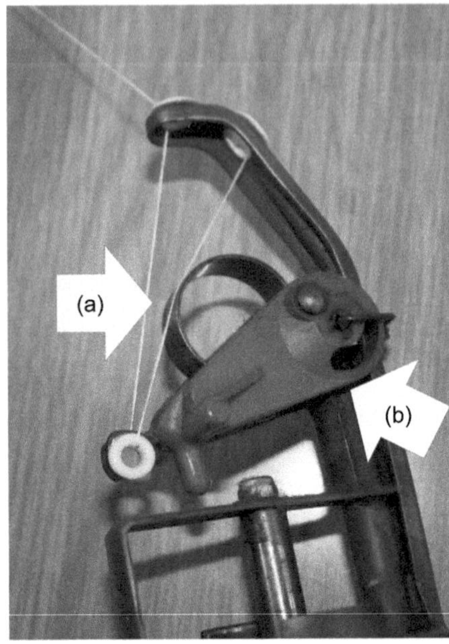

Figure 7.27 Compensation system of HP carriers (a) steel sheet spring (b) lever.

Figure 7.28 Tube form and adapter with the brake of HP carriers.

carriers are also used. In 2011, the company August Herzog introduced a new product line for ultra-fine braids in the medical industry – the braiding machine KB 1/12–56, with 56 mm pitch size with two new types of carriers FDr 56-T and FDr 56-D. The FDr 56-T is designed for wires such as stainless steel, NiTi, titanium and copper. They allow ultra-fine wire diameters of 0.025 mm/0.001 inch to be processed. With a horizontal bobbin and only one eyelet, the wire is guided in an optimum way, which also ensures easy handling. For monofile and multifile yarns the carrier FDr 56-D was developed, which is able to process materials with linear density of 4 denier. Braiding machines in the KB 45 series are available with 4–288 carriers in vertical or horizontal execution. All machines come with a PLC unit for programming, and can run in a clean room environment without any lubrication of the track plate (Herzog, 2012).

7.7 Sources of further information and advice

There are very few books and sources available about carrier construction. Those books where carrier construction is good and explained in detail date from 1953 to 1964 (Lepperhoff, 1953; Douglass, 1964). The latest book, which consists of some more detailed description of carriers, is written by the author's predecessor, Professor Harald Engels (1994). However, this book is actually written in German, and is not available in bookshops, so it cannot be used by any non-German speaking readers.

Information about innovative designs of braiding carriers can be found in some PhD theses and reports of research projects. Some journal papers can also be found, but usually carrier construction is handled very superficially, so these are not mentioned here. The producers of braiding machines usually give a part of the technical drawings to their customers, but only parts of the catalogues of the company Cobra LTD (Cobra, 2012) are available to the public, to analyse the entire construction. One of the best sources for understanding and following the development of carriers is found in the patents for carriers. However, due to their limited volume, developments and trends in patents are not discussed in this chapter.

References

Cobra Braiding Machines (2012) Part Catalogs, www.cobrabraids.com.

Douglass, W.A. (1964) *Braiding and Braiding Machinery*. Eindhoven: N.V. Uitgevesmaatschappij Centrex.

Engells, H. (1994) Handbuch der Schmaltextilien, Die Flechttechnologie, Teil 1, Mönchengladbach, Fachhochschule Niederrhein, FB Textil undBekleidungstechni.

Goseberg, F., Brabender, K. and Moeller, P. (1981) Textiltechnik Maschinengeflechte. *Ausbildungsmittel Unterrichtshilfen*. Frankfurt am Mai: Arbeitgeberkreis Gesamttextil (in German Language).

Herzog (2012) Prospects, Pictures and other information of company August Herzog GmbH & Ko. KG, Oldenburg (Herzog, 2012), www.herzog-online.com.

Lepperhoff, B. (1953) *Die Flechterei*, Eugen G. Leuze Verlag, Saugau/Wttbg.

Steeger (2012) Prospects, Pictures and other information of company Körting Nachfolger Wilhelm Steeger GmbH & Co. KG, Wuppertal (Steeger, 2012), www.steeger-online.com.

The mechanics of the braiding process

8

8.1 Introduction

This chapter presents the braiding process from the perspective of the laws of mechanics. Without these laws, events causing effects or defects, and some limitations, cannot be understood correctly. Readers should not be afraid of the equations. Understanding the process is the most important aspect, and the equations are only for those readers who wish to perform calculations. The theory is kept as simple as possible, so that in most cases a basic introductory course in mechanics should be sufficient to follow its content.

As the title of the chapter refers to mechanics, it incorporates various aspects of the braiding process. There are some static equilibrium relations about yarn tension (statics), and some kinematics (velocities).

8.2 Braiding point parameters

What occurs around the braiding point is the most important question in the braiding process. At that site the new braid is built, and any anomaly here will result in defects in the final product. There are several instances where minor errors at the braiding point can result in large losses, even financial losses. For instance, the braids used in the replacement of human organs are very fine, they have to be geometrically precise and, despite being small, are very expensive. One centimetre of such braid can cost more than 100 USD, and one defect during the braiding process can turn several centimetres into waste. Another instance where the losses can be ten or more times more expensive per centimetre of braid is the thick ropes for oil platforms, ships, etc. made from Dyneema®. In such cases, all parameters of the braiding process, such as tension, geometry, winding, position of the braiding point, etc., have to be fixed exactly, and normally the winding length of each bobbin is also precisely determined, to minimize any material losses. To calculate these parameters, an understanding of the relationship between take-off velocity, position of the braiding point, yarn tension, etc. is very important.

Braiding Technology for Textiles. http://dx.doi.org/10.1533/9780857099211.2.177

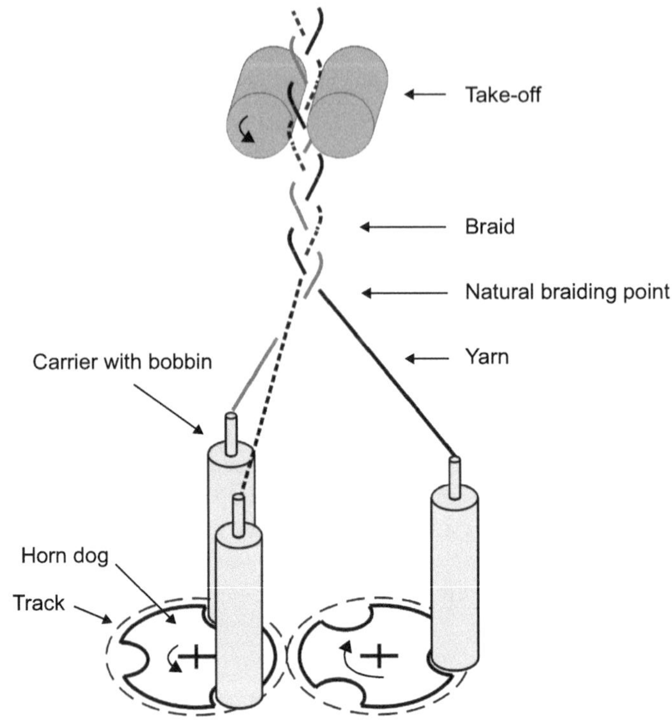

Figure 8.1 Schematic representation of the elements of the braiding process around the braiding point.

8.2.1 Natural braiding point

In classical braiding processes, for instance during the production of ropes or shoe laces, the braiding velocity and take-off velocity are considered to be one and the same, and only the take-off velocity is referred to. Actually, this is true only if the position of the natural braiding point remains unchanged during braiding (Fig. 8.1).

The *natural braiding point* is the place where the yarns interlace into a braid. The position under the natural braiding point is usually a vertical coordinate, perpendicular to the braiding plate, and therefore is referred to as the *height of the braiding point*. The workers in braiding companies know that the braiding point "moves up" if the take-off speed is increased, and vice versa. The explanation for this is shown in Fig. 8.2, with a single yarn participating in the tubular braiding process.

Let us simplify the problem and assume that the carrier moves in a circle around the machine axis, and some tube has to be over-braided (Fig. 8.2a, the tube is not shown). If the yarn between the carrier and the braiding point (AB), after contacting the tube at point B, remains in that place without any slippage or movement, then the braiding angle α will be equal to the angle β. The geometry of this case is identical

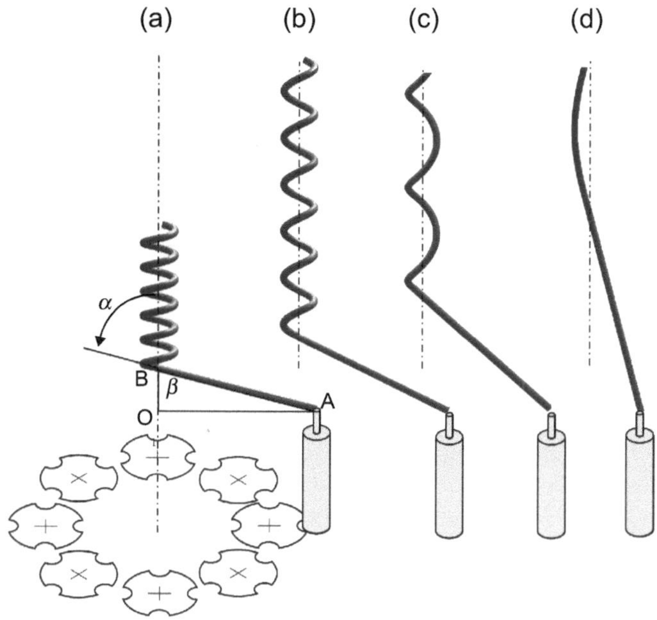

Figure 8.2 The position of the natural braiding point depends on the take-off speed and hence of the braiding angle.

to the geometry during the process of winding yarn around a cylindrical body, but in braiding the cylindrical body moves up (from the take-off velocity) not the yarn guide. If the cylindrical body (or braid) is taken off faster, (Fig. 8.2b, 8.2c and 8.2d) then the angle β becomes smaller, as will the braiding angle α, and the natural braiding point B will move up.

Figure 8.3 demonstrates the effect of using yarns of different thickness while applying the same geometry. If the yarn diameter is smaller or equal to the step between two windings, the angles α and β can remain equal; if not, slippage between the yarn and the winding body occurs. If the yarn is thicker than the theoretical space between the two yarn axes, the braiding angle α and the angle β will differ (Fig. 8.3c). In this case, the dimensions of the newly built body increase faster than the take-off speed removes the new part from the braiding point, or for the braiding – the take-off velocity will not be equal to the braiding velocity (Fig. 8.4).

Note that Fig. 8.3 does not have to be used for the definition of limiting conditions during braiding, because the second yarn system, usual for braiding, is not presented there. In the case of braiding, additional yarn from the opposite carrier system will be available between each two windings on Fig. 8.3. Figure 8.3 is intended only to give a simplified visualization of the fact that applying the same geometrical (actually kinematical, because velocities are also included) theories to the machine using different yarn thicknesses can significantly change the position of the natural braiding point and the process geometry.

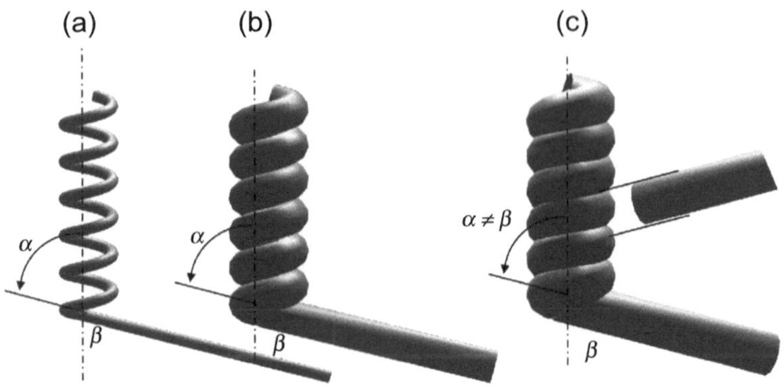

Figure 8.3 Take-off velocity is not always equal to the braiding velocity.

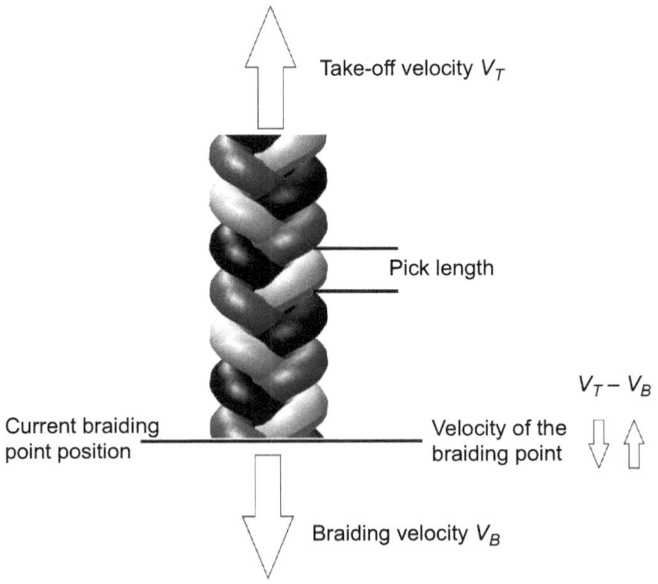

Figure 8.4 Braiding velocity and take-off velocity.

8.2.2 Braiding velocity and take-off velocity

On the braiding machine, there are two velocities which can be adjusted, but are normally fixed during the braiding process – the take-off velocity, and the angular velocity of the horn gears. The take-off velocity V_T (Fig. 8.4) is the velocity with which the already produced braided structure leaves the braiding point. It is a result of the activity of the take-off mechanism. The yarns interlace due to the motion of the carriers and build new braid elements at the braiding point. The velocity caused by creating a new length of braid is referred to here as the braiding velocity V_B. This velocity describes the rate of increase in the length of the braided structure. (Special case: during the over-braiding of non-symmetrical parts, the braiding velocity can be different at different places of the mandrel. In such a case, it will be separated at every vector point on the braiding zone around the mandrel.)

The take-off velocity and the braiding velocity normally have opposite directions; depending on the relationship between their magnitudes, different scenarios are possible, and these lead to structures with different parameters.

Case A: Braiding velocity equal to take-off velocity $V_T = V_B$

- This is the common scenario for the braiding of ropes and other conventional braids.
- The length of the newly built braid is equal to the length of the taken-off braid over the same time period. The natural braiding point remains constant, and the process achieves a steady state. Only in this way can the braiding former and the yarn-controlling devices around it work properly. For all calculations relating to the process, the take-off velocity is used and the braiding velocity is not referred to, because the two are the same.

Case B: Braiding velocity lower than take-off velocity $V_T > V_B$

- In this case the braid is taken off from the braiding zone before a dense structure is built. Strictly speaking, the braiding velocity in this case is still equal to the take-off velocity, but the yarns do not achieve their final position at the braid former, and they are not well fixed between the other yarns. Such braids are more flexible and, in the case of over-braiding, the structure is open, with spaces between the yarns. In the case of rope braiding (without core), the yarns can be moved easily.
- The natural braiding point in this case is somewhere after the take-off mechanism (Fig. 8.5b) and the former supports the braiding process, collecting the yarns at an ergonomically better place (Fig. 8.6). Such braiding is used if ropes with very low elongation are required. Due to the very high take-off speed, the yarns are almost parallel and the braiding angle is close to zero (Geo. Gleistein & Sohn GmbH).

Case C: Braiding velocity higher than take-off velocity $V_T < V_B$

- This case is unusual for rope braiding, because the braiding point moves "down" in the direction of the carriers with the velocity $V_B - V_T$. After a few moments, the braiding point will be under the level of the carriers' yarn guides, and the yarns will collide or wind around them and the process cannot work any more (Fig. 8.5a).
- In the case of over-braiding (Fig. 8.7) there are situations where the take-off velocity has to be equal to zero, or lower than the braiding velocity. In these cases, a special braiding ring is required to prevent collision between the yarns and the carriers. A radial configuration of the carriers on the radial braiding machine partially solves this problem, but for better yarn control and, especially, to allow for a change in direction of take-off, braiding rings are mounted from both sides of the braiding area.
- By over-braiding part of the core, the "braiding point" is extended to a braiding curve – it is a curve around the mandrel, where at a certain moment in time the yarns interlace to build new braids. The braiding velocity has a different direction at any point of that curve (Fig. 8.7).

8.2.3 Open braids and closed braids

When the take-off velocity is lower than the braiding velocity ($V_T < V_B$), stationary braiding is no longer possible. The minimum take-off velocity at which the braiding

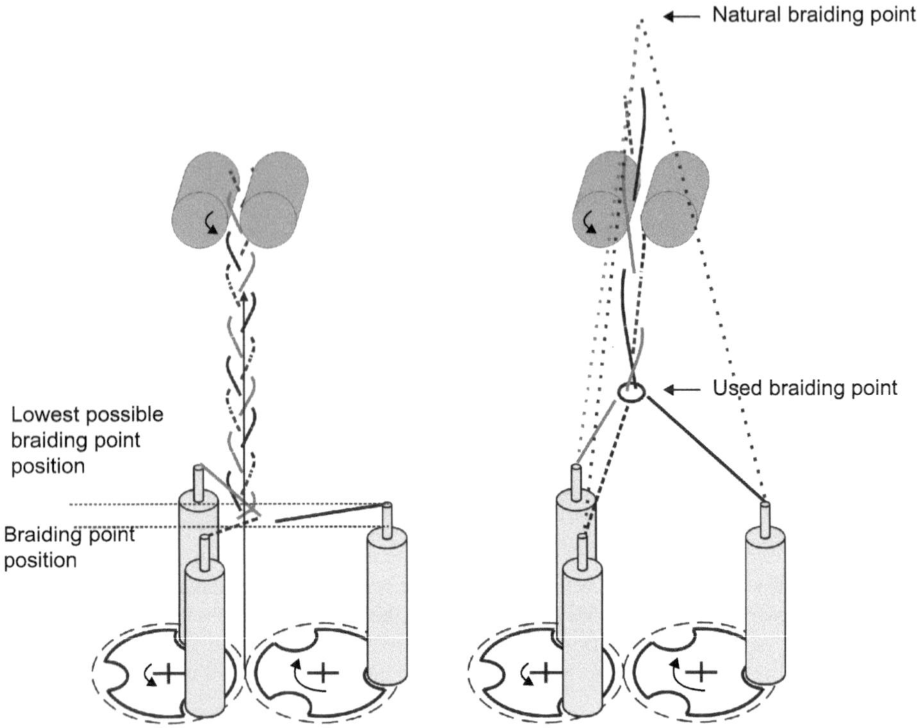

Figure 8.5 Natural braiding point to low and to high.

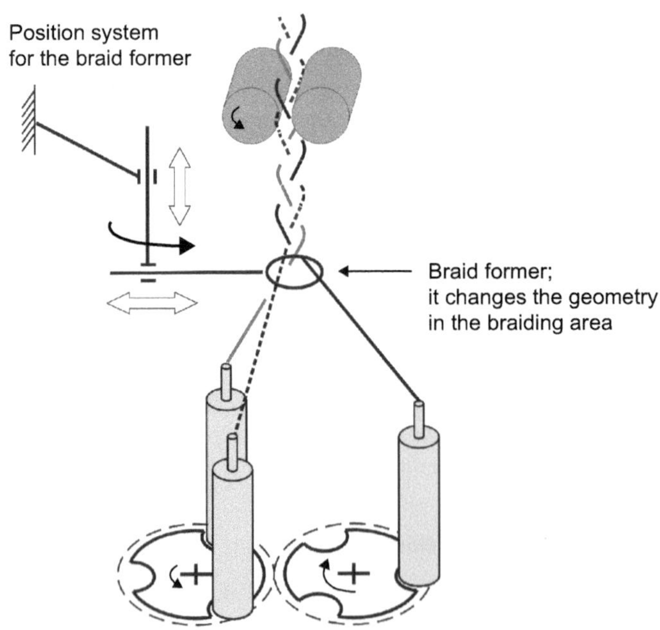

Figure 8.6 The braiding former determines the place of the real braiding point.

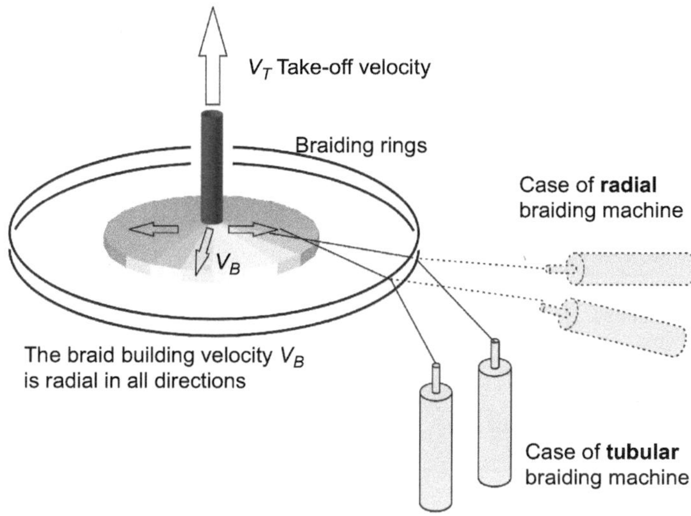

Figure 8.7 Braiding velocity in the case of over-braiding of core with changes in their diameter.

works without moving the braiding point is $V_{T,\min}$ and with this take-off velocity the densest structure which can be obtained with the given material and machine configuration (Fig. 8.8) is produced. If the take-off velocity is higher than this minimum value, but still low enough so that the yarns interlace close to the adjacent yarns, closed braids are produced:

$$V_{T,\min} \le V_T \le V_{T,\text{closed},\max} \qquad [8.1]$$

The exact values for both $V_{T,\min}$ and $V_{T,\text{closed},\max}$, depend on the geometry of the braiding zone (diameters, number of yarns) but also on the type of yarns used. For solid materials, such as wire or monofilament yarns, whose cross-section barely changes during braiding, the range, described in Equation [8.1], is very narrow. For multifilament yarns or textured yarns, this range is larger, because the yarns change their cross-sections to more or less flat without any empty spaces, so the structure is still "closed" (Fig. 8.10a). If the take-off velocity is larger than the $V_{T,\text{closed},\max}$

$$V_T \ge V_{T,\text{closed},\max} \qquad [8.2]$$

open braids are created, which means some pores remain between the yarns and the neighbouring yarns do not touch each other. Depending on the type of braiding and the type of yarn material, these open braids can differ slightly. If braids without core and middle ends (inlays) are produced, then the structure looks like a closed structure during the production process (Fig. 8.8).

After that, under minimal axial pressure the braid becomes shorter, thicker and looser – the yarns can change their position significantly without affecting their

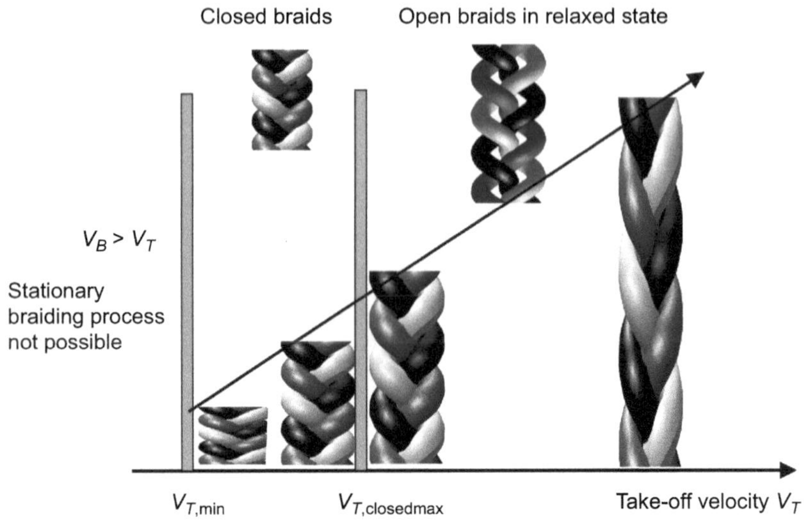

Figure 8.8 Open and closed braids for braiding without core (here flat braid).

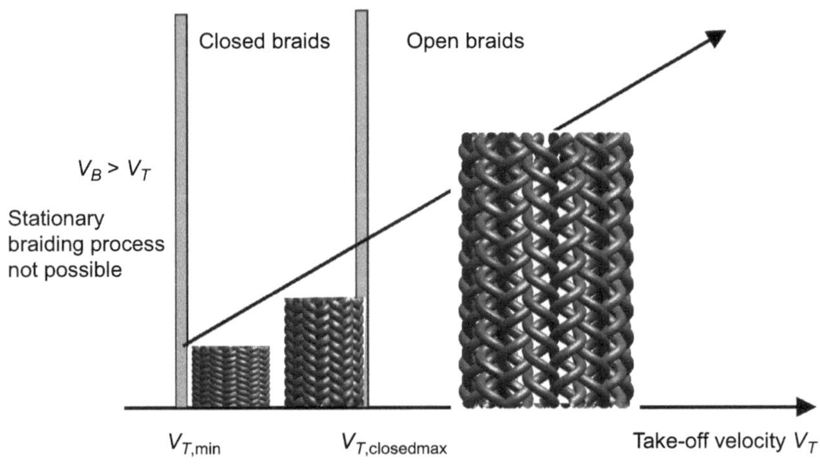

Figure 8.9 Open and closed braids for over-braiding of cylindrical core.

interlacement. This effect is used in several applications where, for instance, flat braids need to be deformed and mounted or sawed around curvilinear profiles. In such cases, they are produced slightly over the range for closed braids $V_{T,closed,max}$ so that there is sufficient flexibility, and they can be deformed at different curvatures without causing wrinkles.

If some middle ends (inlay yarns) are used, or part of the core is over-braided, then the friction between the braiding yarns and the core or inlay yarns determines the look of the structure. If the friction is sufficient, then the open braid is also open during the production process (Figs 8.9 and 8.10b).

Figure 8.10 Open and closed braids: (a) common case of closed braids are the products of multifilament yarns, whose filaments change their position and can fill the gaps to the neighbouring yarns; (b) typical examples for open braid can be found during the overbraiding of some profile with monofilaments with solid cross-section.

The categorization of open and closed braids not only simplifies the text and titles, but also the theoretical considerations. For open braids, there are plenty of well-developed, simple geometrical, or even mechanical, methods for computing the braid paths (Kessels and Akkerman, 2002; Rawal *et al.*, 2005; Rawal *et al.*, 2007). For closed braids, it is still problematic managing the huge computer power effectively to compute the entirety of contact interactions at the braiding zone properly.

8.3 Forces on the braid building yarn segment

The forces acting on the yarn segments involved in the braid building process differ significantly, depending on whether the braid is open or closed. In open braids, the yarn segment is not pressed laterally on the side by the yarn of the previous carrier,

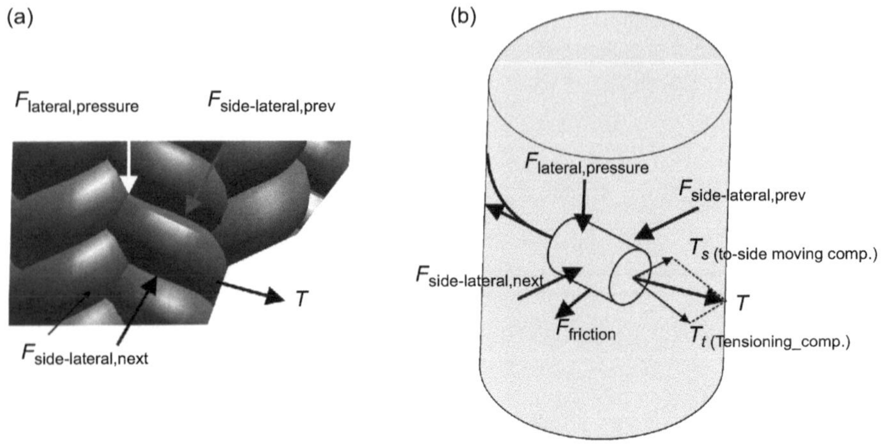

Figure 8.11 Forces over one yarn segment. presented as the segement is in the braid (a) and represented isolated for better visibility (b)

and then by the yarn of the following carrier, so the $F_{side-lateral,prev}$ and $F_{side-lateral,next}$ do not exist, or can be set to zero (Fig. 8.11). For closed braids, these two forces are acting and can cause deformation of the yarn cross-section and/or movement of the yarn segment to either side. Usually $F_{side-lateral,prev}$ starts working first, because the yarn segment from the previous carrier is already in its final position and the yarn of the next carrier is still on its previous braiding step, so $F_{side-lateral,next}$ starts acting only later.

The yarns from the opposite moving direction press the yarn segment with the force $F_{lateral,pressure}$. This force stabilizes the position of the yarn segment at its back end, causing a friction force on the upper side of the yarn segment (the friction force is not shown in the Figure). On the bottom side, the yarn segment is pressed into yarn(s) moving in the opposite direction, resulting in the friction force $F_{friction}$. Depending on the geometry around the analysed braiding segment, it can happen that the tension force T is not orientated parallel to the segment axis but at an angle with it. In such cases, depending on the angle between the actual yarn tension force and the yarn segment axis (Fig. 8.12), the side moving projection of the tension force T_S can become larger than the friction force $F_{friction}$ and the yarn segment changes its initial position.

A proper analysis of the forces on a single yarn segment requires a separate analysis of all the potential configurations – with and without inlay yarns, and with and without core (mandrel) – which exceeds the scope of this chapter. Such analysis is mainly important for over-braiding of parts with non-cylindrical cross-sections, where the yarn segments can lose their stability. As a result of geometry, the pressure of the yarn to the core becomes lower and the friction force becomes insufficient to hold it there. Such situations are thoroughly investigated in the theory of wound packages in some Russian books (Proshkov, 1986) where the stability of yarn over a surface is studied analytically (Fig. 8.13).

Figure 8.13 shows the forces and important angles for winding a single yarn segment on a surface without contact with other yarns, but these forces and angles are also valid for braiding open braids. The conditions for static equilibrium of the yarn

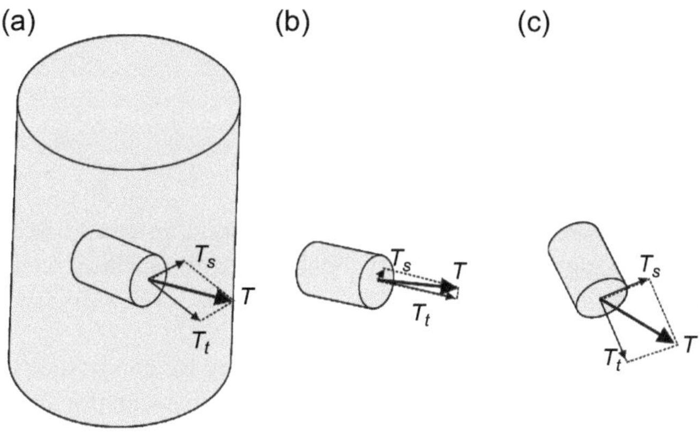

Figure 8.12 The carrier moves during the braiding and the yarn tension force changes its direction. If the yarn tension vector is not more tangential to the initial yarn segment axis, a side moving component tries to move the segment (a) yarn segment on the braid (b) yarn tension vector almost tangential to the yarn axis (c) yarn tension vector T changed the orientation and the side moving component Ts tries to move the segment.

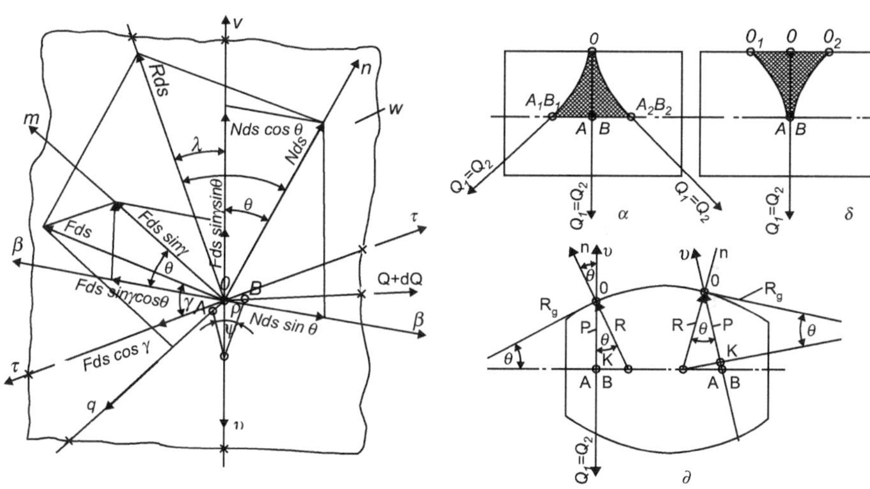

Figure 8.13 Angles and forces over a yarn segment on a surface.
Source: Picture from Proshkov, 1986.

segment are based on this Figure (Proshkov, 1986). The first condition is the relationship between:

1. the (helix) winding (braiding) angle α,
2. the geodesic angle of the contacting curve on the surface θ and the friction angle $\varepsilon_{\max}$, and
3. the friction of coefficient between the yarn and surface μ

$$\tan\alpha \leq \tan\theta \leq \tan\varepsilon_{\max} = \mu \qquad [8.3]$$

Additionally, the following condition has to be satisfied between the forces T_1 and T_2 on both sides of the yarn segment and the geodesic angle:

$$T_2 \le T_1 \cdot e^{\int_0^{\psi} \cos\theta \sqrt{\mu_{max}^2 - tg^2\theta}\, d\psi} \quad [8.4]$$

With these conditions, it can be established if braiding can be achieved over a conical surface with angle θ without yarn slippage over the surface. As Equation [8.3] shows, surfaces with higher friction allow over-braiding, even at higher braiding and geodesic angles.

There are also several papers about filament winding for composites whose analytical and numerical results can be directly applied to the investigation of over-braiding, such as the prediction of the yarn path for open braids (Koussios *et al.*, 2004) or to the winding of forms with concave profiles in (Hongya *et al.*, 2008). In most cases, these works mainly apply geometrical methods to calculate the yarn path, and so are unsuitable for closed braids where the forces between the yarns are important.

8.4 Relationship between take-off velocity and braiding angle

8.4.1 *Take-off velocity and the braiding angle for open tubular braids*

The relationship between the take-off velocity and braiding angle for open tubular braids (braiding velocity equal to take-off velocity) is analysed in several works (Du and Popper, 1994; Ko *et al.*, 1989; Rawal *et al.*, 2005). Here we present a simplified version.

If the main surface of one tubular braid is enveloped, the yarn path of one yarn in one repeat is represented by the line AC, and the braiding angle by the angle α. The line BC on the triangle ABC (Fig. 8.9) corresponds to the circumference of the braid with diameter D, so the relationship between them will be:

$$\tan\alpha = \frac{\pi \cdot D}{L_{\text{Repeat}}} = \frac{\pi \cdot D}{V_T \cdot t_{\text{Repeat}}} \quad [8.5]$$

where V_T is the take-off speed and t_{Repeat} is the time for one repeat.

As described in Chapter 14 (Productivity calculations in braiding), in a tubular braiding machine the slots in one track make up half of the total number of slots on the machine, so for a tubular machine:

$$\frac{2 \cdot N_{\text{Slots\_in\_track}}}{N_{\text{Slots\_per\_horngear}}} = N_{\text{HornGears}}$$

Considering that the relationship between the revolutions per minute of one horn gear and its angular velocity is

$$2\cdot\pi\cdot n/60=\omega_{\text{HornGears}}$$

the time for one repeat can be expressed as:

$$t_{\text{Repeat}[s]}=\frac{N_{\text{Slots\_in\_track}}\cdot 60}{N_{\text{Slots\_per\_horngear}}\cdot n_{[\text{rpm}]}}=\frac{N_{\text{HornGears}}\cdot 60}{2\cdot n_{[\text{rpm}]}} \qquad [8.6]$$

From here, the relationship between the braiding angle, horn gear angular velocity, braid diameter and take-off speed can be given as:

$$\tan\alpha=\frac{\pi\cdot D}{V_T\cdot\left(N_{\text{HornGears}}\cdot 60/2\cdot n_{[\text{rpm}]}\right)}=\frac{2\cdot n_{[\text{rpm}]}\cdot\pi\cdot D}{N_{\text{HornGears}}\cdot V_T\cdot 60}=\frac{\omega_{\text{HornGears}}\cdot D}{N_{\text{HornGears}}\cdot V_T} \qquad [8.7]$$

This equation is commonly used in another form where the radius of the braid instead of the diameter R is used, and the angular velocity of the yarn ω_{Yarn} is used instead for the horn gears:

$$\tan\alpha=\frac{2\cdot n_{[\text{rpm}]}\cdot\pi\cdot D}{N_{\text{HornGears}}\cdot V_T\cdot 60}=\frac{\omega_{\text{HornGears}}\cdot D}{N_{\text{HornGears}}\cdot V_T}=\frac{\omega_{\text{Yarn}}\cdot R}{V_T} \qquad [8.8]$$

Equations [8.7] and [8.8] show that, even if all the other conditions are the same, if the braid or its core is thicker, then the braiding angle also increases (Fig. 8.14).

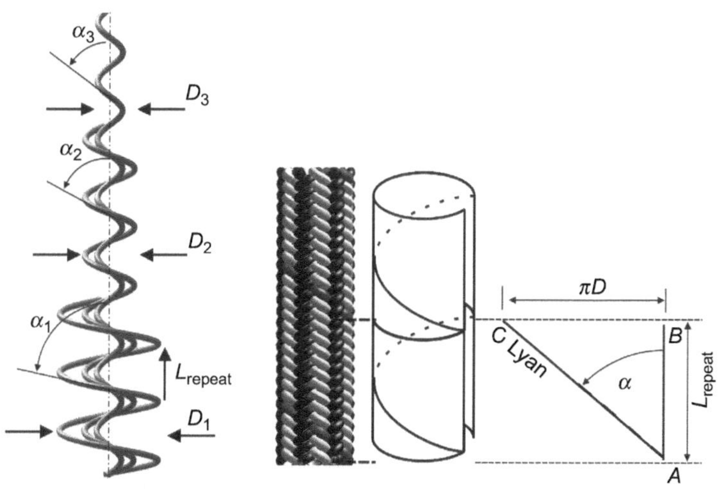

Figure 8.14 Braiding angle and braiding diameter for tubular braids.

It should be emphasized that if some consideration about the braiding angle and the size of the machine is required in understanding horn gears, Equation [8.7] is more suitable than the popular Equation [8.8], because it allows the reader to decide if the machine runs at the same speed ($\omega_{\text{HornGears}} = const$) or not, and what happens if more horn gears are available. If a large machine with more horn gears is used, then each carrier needs more time for one repeat and, to maintain the same angular yarn speed ω_{Yarn}, the horn gears have to rotate faster, which is not always possible. Increasing the number of horn gears makes the yarn rotate slowly, so the braiding angle increases as well. In order not to get confused by all the steps in reaching this equation and to re-emphasize its importance in the tubular braiding of open braids, the equation is repeated in its plain form here:

$$\tan\alpha = \frac{\omega_{\text{HornGears}} \cdot D}{N_{\text{HornGears}} \cdot V_T} \qquad [8.9]$$

8.4.2 Take-off velocity and the braiding angle for flat braids

For flat braids it can be proven (see Chapter 14 *Productivity calculations in braiding*) that most relationships are the same as for tubular if the circumference of the braid is replaced with the double braid width $\pi \cdot D = 2 \cdot w$, (Fig. 8.15)

During the production of flat braids, every carrier has to go along the entire track and has to "visit" all the slots of the horn gears, so the relationship between the total number of slots in one track and the slots per horn gear is

$$\frac{N_{\text{Slots\_in\_track}}}{N_{\text{Slots\_per\_horngear}}} = N_{\text{HornGears}} \qquad [8.10]$$

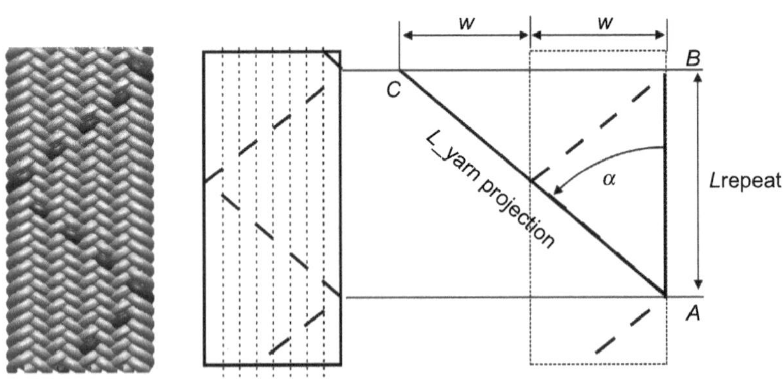

Figure 8.15 Braiding angle and braiding diameter for flat braids.

In Equation [8.10] and the following equations, the numbers of slots per horn gear $N_{\text{Slots\_per\_horngear}}$ are taken from the "normal" horn gears in the middle part of the track because in addition to their angular velocity, resp. revolutions per minutes are considered. In the most common case, there are four slots per horn gear $\left(N_{\text{Slots\_per\_horngear}} = 4\right)$ and only the two turning horn gears at both ends of the horn gear set have five slots. The five-slot horn gears have to keep the same circumferential velocity as the four-slot horn gears, so they do not need to be given special consideration. One normal horn gear makes $n_{[\text{rpm}]}/60$ revolutions per second, which means that per second one carrier can go through $N_{\text{Slots\_per\_horngear}} \cdot n_{[\text{rpm}]}/60$ slots, or the time to travel along the entire track is:

$$t_{\text{Repeat[s]}} = \frac{N_{\text{Slots\_in\_track}} \cdot 60}{N_{\text{Slots\_per\_horngear}} \cdot n_{[\text{rpm}]}} = \frac{N_{\text{HornGears}} \cdot 60}{n_{[\text{rpm}]}} \qquad [8.11]$$

Replacing this time in the equation about the braiding angle Fig. 8.15:

$$\tan\alpha = \frac{2 \cdot w}{L_{\text{Repeat}}} = \frac{2 \cdot w}{V_T \cdot t_{\text{Repeat}}} \qquad [8.12]$$

the following relation is valid:

$$\tan\alpha = \frac{2 \cdot w \cdot n_{[\text{rpm}]}}{V_T \cdot 60} \cdot \frac{N_{\text{Slots\_per\_(normal)\_horngear}}}{N_{\text{Slots\_in\_track}}} \qquad [8.13]$$

For flat braiding the width cannot be changed by altering the core, but inserting thicker inlay yarns (middle ends) can influence the width w. In such cases, if the width becomes larger at the same conditions, the braiding angle will increase.

8.4.3 Braiding density – picks per cm or inch for open braids

The density of braids is usually determined as the number of yarns (picks) per centimetre or per inch. If the braiding for time Δt is considered, the following length will be taken off:

$$L = V_T \cdot \Delta t \qquad [8.14]$$

For this time, one horn gear with $N_{\text{Slots\_per\_horngear}}$ slots will rotate

$$N_{\text{Slots\_per\_horngear}} \cdot n_{[\text{rpm}]} \cdot \Delta t / 60 \qquad [8.15]$$

slots through any observation point selecting somewhere on the track (Fig. 8.16).

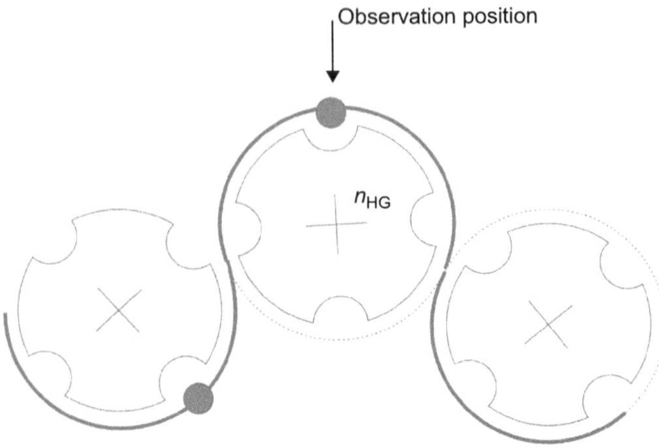

Figure 8.16 In order for the number of picks (yarn pieces) per cm to be calculated, the number of carriers going through one place during the time in which the required length of braid is taken off needs to be counted.

Depending on the carrier arrangement, the carrier will take half of these slots at most. If the carrier arrangement is "F Full – E Empty" (where $F \leq E$), only $K_{\text{Occupation}}$ part of the time slots with carriers resp. yarns will move around the observation point and create a yarn there:

$$K_{\text{Occupation}} = \frac{F}{F+E} \qquad [8.16]$$

The yarn pieces, resp. pics per unit length then will be:

$$\text{ppL} = \frac{\text{Pics}}{\text{Length}} = \frac{(F/F+E)\cdot N_{\text{Slots\_per\_horngear}} \cdot n_{[\text{rpm}]} \cdot \Delta t/60}{V_T \cdot \Delta t} \qquad [8.17]$$

or

$$\text{ppL}_{[\text{pics\_per\_meter}]} = \frac{F}{F+E} \cdot \frac{N_{\text{Slots\_per\_horngear}} \cdot n_{[\text{rpm}]}}{V_{T[\text{m/s}]} \cdot 60} \qquad [8.18]$$

The use of picks per centimetre (ppc) is common for braiding and the take-off speed is relatively slow, and so it is usually given as metre per minute:

$$\text{ppc} = \frac{1}{100} \cdot \frac{F}{F+E} \cdot \frac{N_{\text{Slots\_per\_horngear}} \cdot n_{[\text{rpm}]}}{V_{T[m/\text{min}]}} \qquad [8.19]$$

or if it has to be recalculated in ppi (picks per inch), then:

$$\text{ppi} = \frac{2.54}{100} \cdot \frac{F}{F+E} \cdot \frac{N_{\text{Slots\_per\_horngear}} \cdot n_{[\text{rpm}]}}{V_{T[\text{m/min}]}} \qquad [8.20]$$

For example, the common configuration of "1 Full 1 Empty" carrier arrangement for four slots per horn gear, rotating with 100 rpm and a take-off speed of 1 m/min, gives two picks per cm:

$$\text{ppc} = \frac{1}{100} \cdot \frac{1}{1+1} \cdot \frac{4 \cdot 100}{1} = 2 \qquad [8.21]$$

Reducing the take-off speed by four to 0.25 m/min increases the pick density four times:

$$\text{ppc} = \frac{1}{100} \cdot \frac{1}{1+1} \cdot \frac{4 \cdot 100}{0.25} = 8 \qquad [8.22]$$

If a carrier arrangement "2 Full 2 Empty" is used, the results remain the same, but in the case "1 Full 3 Empty," the picks per centimetre will be reduced twice, because only every fourth slot in the track has a carrier, not every second:

$$\text{ppc} = \frac{1}{100} \cdot \frac{1}{1+3} \cdot \frac{4 \cdot 100}{1} = 1 \qquad [8.23]$$

8.4.4 Take-off velocity and the braiding angle for closed braids

Both Equations [8.9] and [8.13] can be used in calculations for open braids if the diameter of the core or the braid width is known. They explain the relationship between the parameters if something is changed. These equations alone cannot be used to predict the braiding angle for an unknown structure because they do not take into account any information about the yarn properties. These factors should also be included and, although they can be ignored for open braids because the parallel yarns do not contact each other, for closed braids the situation changes significantly.

There are some works which try to analyse these relationships for closed braids, but the only accurate way is by using numerical methods found in computer software. The methods used by Damien Durville (Fig. 8.17) for the simulation of multifilament structures (Vu and Durville, 2012; Vu *et al.*, 2012) can be used successfully in such computations, but currently these methods still require highly qualified powerful personal computers. Over the next decade, this will change because the most recent research results demonstrate that computation of complex problems is possible by more effective use of available hardware, and especially using the power of a large number of processors on graphic cards.

Figure 8.17 Numerical simulation of braids from multifilament yarn. In this case, the interaction between each filament is considered and correct results as well for closed braids can be received, if the longer computational time is not a problem.
Source: Pictures courtesy of Damien Durville, Vu *et al.*, 2012.

8.5 Braid tension variances during tubular braiding and overbraiding

During braiding the carriers change their position, travelling around the horn gears. The angle β, which is closed between the machine axis and the yarns, changes during this motion in the range (Fig. 8.18):

$$\beta_2 \leq \beta \leq \beta_1 \qquad [8.24]$$

$$\tan \beta_{1,2} = \frac{R \pm R_{HG}}{H} \qquad [8.25]$$

For a tubular machine with

- a horn gear radius of 0.12 m,
- radius of the circle where the horn gear centres are located $R = 0.24$ m, and
- braiding point height $H = 0.3$ m

the angle is between $50° \geq \beta \geq 21°$, while for a larger machine with more horn gears placed around a circle with

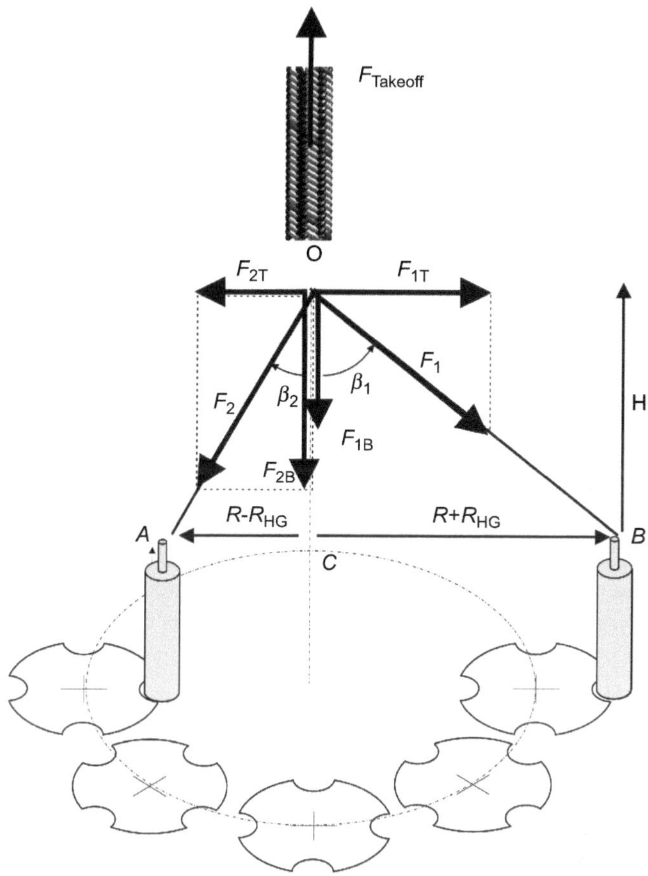

Figure 8.18 Angles and forces between the yarns and the braid.

- radius $R = 1$ m and
- braiding point $H = 1.0$ m

the variations will be lower, in the range $48° \geq \beta \geq 41°$.

It can also be concluded that where the braiding point is at a lower position, which is common for denser, close braids, the H will be lower and the differences between the angles will again become larger.

The tension of the yarn when the carrier is at the most external position F_1 can be equal to the yarn tension F_2 of the yarn moving in the opposite direction when at the innermost position, but because of the difference of the angles, the projections from these two forces will be different. For instance, the projection which causes twisting of the current cross-section will be:

$$F_{1T} \approx F_1 \cdot \sin\beta_1; \quad F_{2T} \approx F_2 \cdot \sin\beta_2 \qquad [8.26]$$

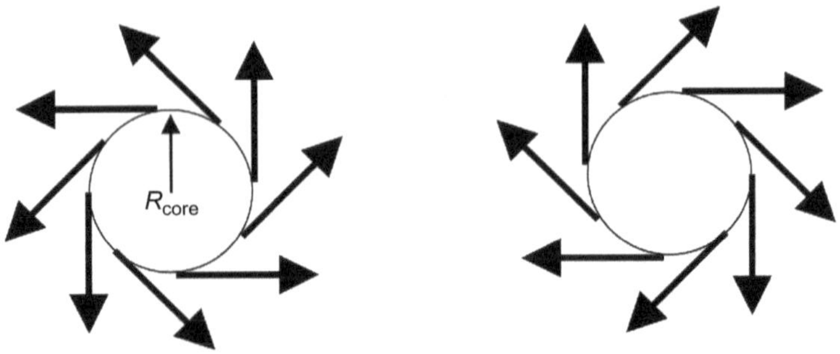

Figure 8.19 The resulting forces cause twisting, and the direction of the twisting moment changes during every motion of the carrier around some horn gear.

Here, a precise geometrical consideration would involve taking into account the radius of the rope or the core as well, but this will be ignored because the aim of the current section is to demonstrate the reason for an event and describe it. Because of this simplification in Equation [8.26], the nearly equal sign is used instead of the equal sign, because the equal sign would require additional terms.

As was shown, the angles β are not equal, so the resulting vector of common action of both yarns is a resulting twisting force, which is not zero:

$$F_{T\,\mathrm{Res}} = F_1 \cdot \sin\beta_1 - F_2 \cdot \sin\beta_2 \qquad [8.27]$$

This twisting force for the first example above with angles between $50° \geq \beta \geq 21°$ will be:

$$F_{T\,\mathrm{Res}} = F_1 \cdot \sin 50° - F_2 \cdot \sin 21° = F \cdot 0.4 \qquad [8.28]$$

which is 40% of the yarn tension. Multiplied by the number of yarn pairs which are equal to half the number of carriers and the radius of the rope or core (Fig. 8.19), the maximum twisting moment can be estimated as:

$$M_{\mathrm{Twisting}} = \frac{1}{2} N_{\mathrm{Carriers}} \cdot \left(F_1 \cdot \sin\beta_1 - F_2 \cdot \sin\beta_2\right) \cdot R_{\mathrm{Core}} \qquad [8.29]$$

Equations [8.27] and [8.29] show the maximum value of the twisting force and twisting moment, respectively. These values are reached for a moment. Shortly after that, the carriers change their position again and some carrier from the opposite track will be in the outer position, resulting in the twisting moment changing its direction. If the change in the angle β for one carrier at central angle φ starting from some reference point on the machine is represented to follow the sinus law

$$\beta_1 = \beta_1(\varphi) = \frac{1}{2}(\beta_1 + \beta_2) \cdot \sin(\omega t) \qquad [8.30]$$

then the law for the change of angle for a yarn in the opposite track will just be shifted at k sections,

$$\beta_2 = \frac{1}{2}(\beta_1 + \beta_2)\cdot \sin\left(\omega t + \frac{k\cdot\pi}{4}\right) \qquad [8.31]$$

and the entire angle change can be calculated.

The calculations of angle changes and twisting moment are required by machine builders because they are responsible for developing the take-off mechanism that fixes the core or rope and works despite these changes.

The carrier arrangement influences the twisting moment. At arrangement 1 Full 1 Empty, the fluctuations of the forces have half the amplitude and double the frequency compared to the 2 Full 2 Empty arrangement. By 2 Full 2 Empty, every two carriers are moving one after another as a group, which means that twice as many yarns are in the same position. Of interest is the situation with 1 Full 3 Empty, where three possibilities for the placement of the carriers of the second track are available – on the first, second or the third free place. If the second free place is used, again the amplitude of fluctuations will be higher compared to the situation where the first or third free place is used.

Changing the angle β also causes change in the axial tension of the yarn, so for the forces F_{B1} and F_{B2} as with Equation [8.27], the resulting force can be written:

$$F_B = F_1 \cdot \cos\beta_1 - F_2 \cdot \cos\beta_2 \qquad [8.32]$$

Using analogical considerations it can be proven that as well as the tensile force over the braid which determines the take-off force $F_{Take-Off}$, it also changes its amplitude periodically. This can also be detected optically during braiding, especially for carrier occupation 2 Full 2 Empty, and has to be taken into account during the construction of the take-off devices.

As the example at the beginning of this section shows, these variations are larger if the machines are smaller. For machines with large numbers of horn gears, or especially for the changed geometry of the radial braiding machines, the angle variations can become negligible, but in such cases the number of carriers usually increases rapidly so such effects can still be registered.

8.6 Influence of the braiding needles over the braiding process of flat braids

The braiding of flat braids differs mechanically from the braiding of tubular braids in the direction of the yarn tension forces. During tubular braiding, the carriers on each track move continuously in one direction. During the braiding of flat braids only one track is available, and each carrier changes direction after it reaches the end horn gear (Fig. 8.20)

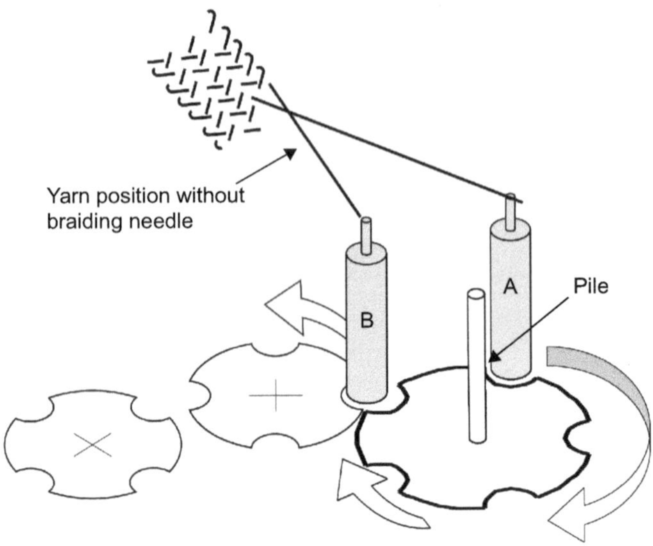

Figure 8.20 The carriers change their direction of motion at the end horn gear during flat braiding, and the tensile force of the yarns changes its acting direction at the braiding point.

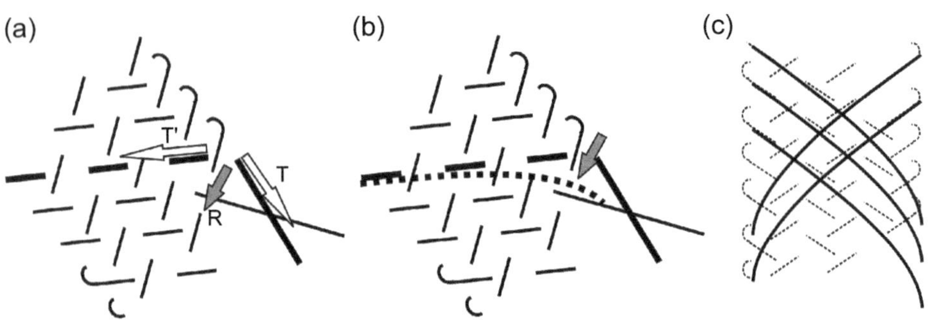

Figure 8.21 Because of the change of the yarn direction, a resulting force from the yarn tension moves the boundary segment to the middle of the braid and deforms the yarn path. (a) Ideal position of the yarn and the working forces after the direction change; (b) the yarn changes the position under the influence of the resulting force; (c) appearence of the braid.

After the carrier changes direction, its yarn is already incorporated with the opposite system of yarns and starts being placed under another angle. If it were a woven structure, this change would correspond to the change from weft to warp direction, for instance. In the case of braiding, the yarn, which was orientated at angle α will after the change be at angle$-\alpha$. The tensile force of the yarn then starts acting in the other direction (Fig. 8.21a), and the resulting force from the yarn tension over the yarn segment at the boundary starts moving to the centre of the braid (Fig. 8.21.b). After some time the yarn is actually already interlaced with enough other yarns, so this reconfiguration stops, but for certain materials with a low friction coefficient

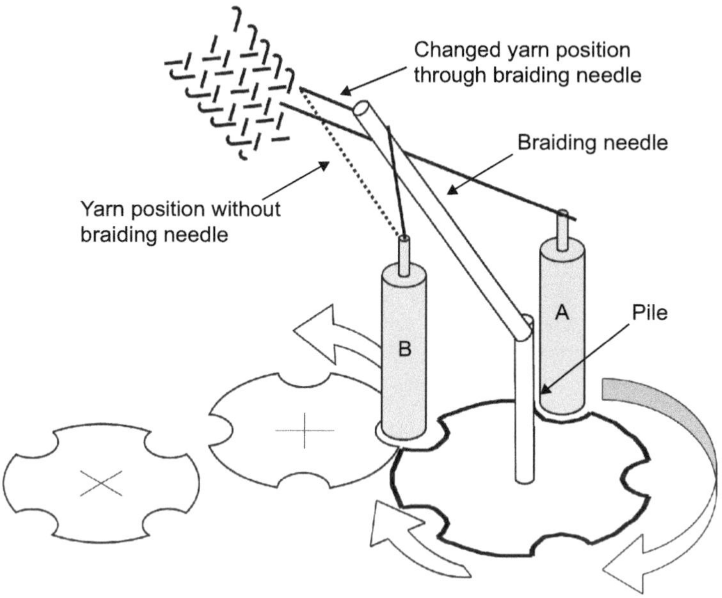

Figure 8.22 The yarn remains on the braiding needle for a while after its carrier changes direction of movement.

(such as glass, carbon, aramid) and especially in the case of not very dense braids, the yarn paths change from lines to curves, as shown in Fig. 8.21c.

To prevent this more than simply optical defect, so-called "braiding needles" are used (Fig. 8.22). Braiding needles are thin metal parts, like needles, which are fixed on the piles of the end horn gears so that the upper part of the yarn is kept at the same orientation for a while after changing its direction (Fig. 8.23).

This static orientation of the yarn piece between the braid and the braiding needle acts with tensile force in the correct direction and does not try to move the boundary segment (Fig. 8.24). After some time, the following yarns are already interlaced and, although the yarn will slip down from the needle, it will not be able to further deform the braid, because the other interlacements keep it stable. It is important that the braiding needles are adjusted to the proper position so that the yarn remains there some time, but once the yarn has started interlacing in the opposite direction it has to leave the needle surface.

Depending on the height of the braiding point, the needle orientation also has to be adjusted, so that the angle β_N is between the minimum and maximum angles which the yarns make with the horizontal surface (Fig. 8.25).

To adjust the angle, the needle can be rotated on the pile (Fig. 8.26). Small changes in the position of the needle's head can be achieved by shifting its mounting point. For larger corrections, different length needles are used. During the braiding, with or without incorrectly positioned needles, the boundary areas interlace faster than the middle area of the braid and build irregular structures (Fig. 8.27)

Figure 8.23 Braiding needles. The right needle holds one yarn and keeps its previous orientation in the area of the braiding zone.

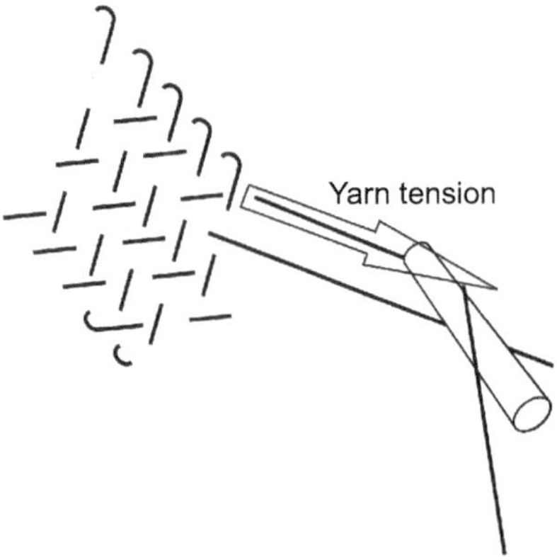

Figure 8.24 The yarn is held by the needles and the yarn tension of this yarn piece remains in the same direction as before the change of the moving direction. The boundary segment is not forced to move to the middle of the braid.

8.7 Bobbin winding tension and the braiding process

The braider rule is "good winding is half the braiding." The quality of the winding of the yarn bobbins is very important for the quality of the braided products, and one of the most important parameters that determine winding quality is the winding tension. Most issues about winding are included in Chapter 10 *Yarn*

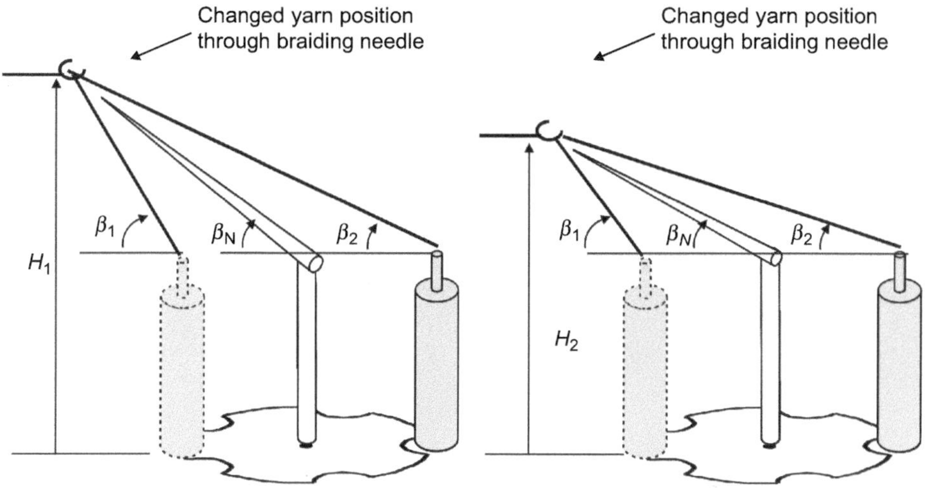

Figure 8.25 The orientation and position of the braiding needle always have to be adjusted if the height of the braiding point changes.

Figure 8.26 This machine has four braiding needles but only the two are clearly visible. It can braid two flat braids or one with double width.

winding operations in braiding but one of them relates to the forces during braiding and has to be mentioned here.

The materials used for braiding usually have linear elastic, or at least elastic, behaviour at the limits of the tension used during winding and braiding. Elastic means that if the material is deformed, after removing the deforming forces it relaxes back to its original state. For linear materials such as yarns and ropes, this means that it relaxes to its initial length.

Figure 8.27 Flat braid, produced without needles. The monofilaments use slide (no sticking, low friction coefficient) and the yarn segments move several millimetres from their initial location. The yarns at the selvages have different orientation and braiding density, compared to those in the middle of the product.

If tension T_W is used during winding, then the material is wound and fixed on the bobbins with elongation ε_W

$$\varepsilon_W = \frac{T_W}{EA} \qquad [8.33]$$

where E is the elasticity module, A is the cross-section of the yarn, and EA is the spring constant of the material.

If the yarn tension during the braiding is T_B, then the elongation of the yarns during the time they are in the braiding zone and later in the braid will be:

$$\varepsilon_B = \frac{T_B}{EA} \qquad [8.34]$$

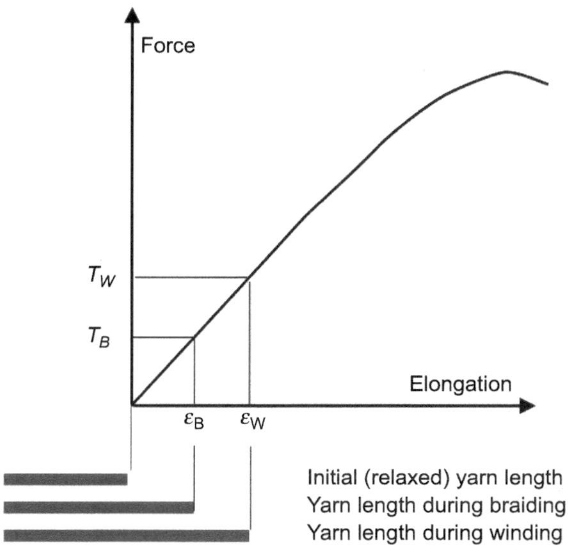

Figure 8.28 Yarn tension and elongation in relaxed state during winding and during braiding.

It is an accepted rule that the tension during winding has to be equal to or higher than the braiding tension:

$$T_W \geq T_B \tag{8.35}$$

so that the wound package is kept stable and can resist the braiding tension without deformations and displacements. If the braiding tension is higher than the winding tension, the forces during the unwinding will be higher compared to those during the winding, and this could lead to inclination of the yarn between other windings, which can lead later to significant irregularity of yarn tension, quality problems in the braid or, in the worst case, broken yarns or carriers.

In the normal case, if the braiding tension is lower than the winding tension, the yarn relaxes in the braiding zone and is braided with lower elongation compared to that on the bobbin (Fig. 8.28).

Problems can occur if some of the bobbins are wound with different winding tension from the others. This can happen if, for instance, some yarn springs out of the tension brake or becomes wound around some machine part, which then increases the friction and the winding tension. In such cases, most of the yarns will relax to one length, but the differently wound yarn will relax to another length. Usually, the effect of this difference is immediately visible after the braiding point as some kind of spiral or zigzag effect. If the differences were not high, the effect can remain invisible but will be detected by a lower breaking force of the rope, because the differently wound yarn(s) will break earlier or later and will not contribute to the rope strength simultaneously with the other yarns.

8.8 Braiding tension influences over the braided product

The yarn tension during braiding has to be kept equal for all carriers, resp. all yarns. If some carriers have more hard springs, their yarns will have higher braiding tension, which will lead to spiral effects during tubular braiding (Fig. 8.29). The tenser yarns remain in a tighter position, and the yarns with lower tension position themselves around the others. If only one carrier, or only a carrier in one direction, has a different tension, the spiral effect will be more regular and more visible.

The different yarn tension during braiding is used in flat braiding for producing fancy braids with zigzag effects (named "rick rack" or "ric rac") (Fig. 8.30). As demonstrated in the book by Professor Lepperhoff (Lepperhoff, 1953), whose original drawings are reproduced here, the yarns with higher tension (50 g – the first version of the book is from 1914, when grammes and not Newtons were used) will stay almost straight and tight. The yarns with lower tension then interlace and build braids around the straight yarn, creating the decorative effect. The exact distribution given in (Lepperhoff, 1953) is 50, 30,25, 22, 20, 17, 15, 12, 12, 12, 15, 17, 20, 22, 25, 30, 50 g force.

The amplitude of the rick rack effect and the type of corners depend on the distribution of the yarn tensions across the springs in the carriers. Normally, between the carriers with the highest and lowest tension there are almost equal numbers of carriers in both directions with increasing and decreasing tension respectively (Fig. 8.31). A smooth curve in the tension is difficult to achieve, because usually only about five to ten different springs are available.

Using a symmetrical distribution, symmetrical effects with 3D relief are also possible, because the middle area will be pressed by the side yarns and more to the front (Lepperhoff, 1953) (Fig. 8.32). The weights for the yarn tension according to

Figure 8.29 Spiral structure as a result of higher tension of some of the yarns.

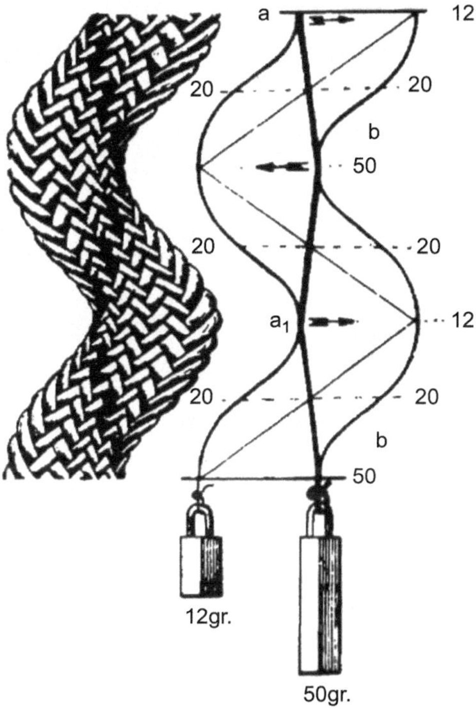

Figure 8.30 Rick rack braid as a result of the different yarn tensions (Lepperhoff, 1953).

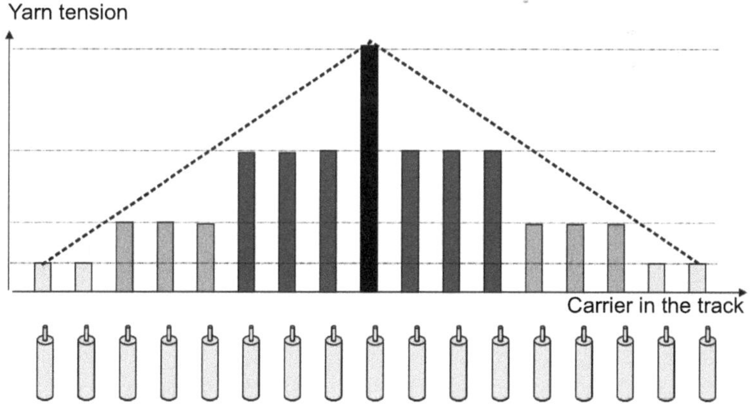

Figure 8.31 Yarn tension distribution on the carriers in order to get rick rack effect.

(Lepperhoff, 1953) arrange two pics and between these the yarns have lower tension (Fig. 8.33):

from 1 to 17 = 50, 20, 15, 12, 10, 10, 8, 8, 8, 8, 8, 10, 10, 12, 15, 20, 50,
from 18 to 33= 50, 20, 15, 12, 10, 10, 8, 8, 8, 8, 10, 10, 12, 15, 20, 50.

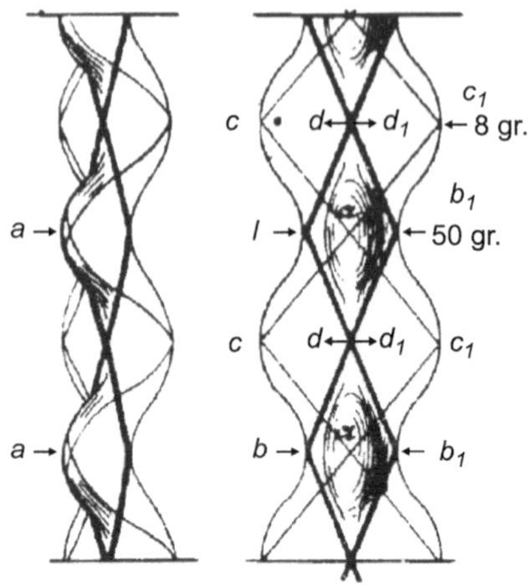

Figure 8.32 Flat rick rack braids with symmetric effect in side and front view (Lepperhoff, 1953).

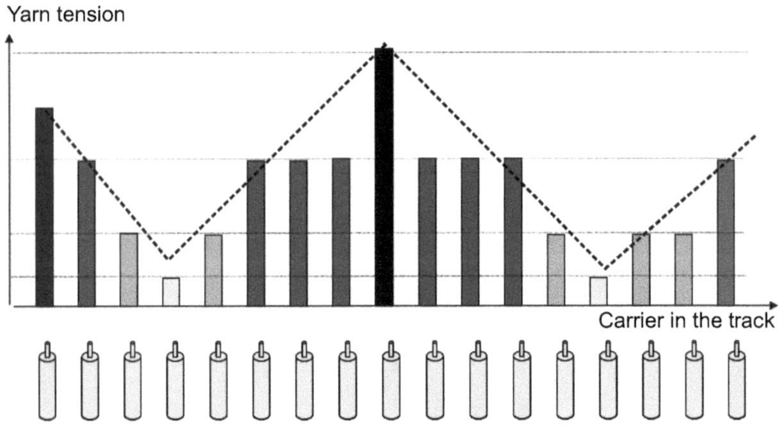

Figure 8.33 Yarn tension distribution for double zigzag effect, similar to that presented in Fig. 8.34.

Using modified distribution of yarn tension (Lepperhoff, 1953):

35, 30, 25, 20, 15, 10, 10, 6, 6, 6, 4, 6, 10, 10, 12, 15, 22, 27, 30, 50, 50, 50, 50 followed by 25, 20, 15, 15, 12, 12, 12, 15, 15, 20.

and, additionally, if one of the braiding needles is placed almost up to the former and the other – at some distance from it (Fig. 8.35)(Lepperhoff, 1953), more decorative effects can be created (Fig. 8.34). For example, if one of the needles keeps the yarns there longer, an asymmetric effect appears.

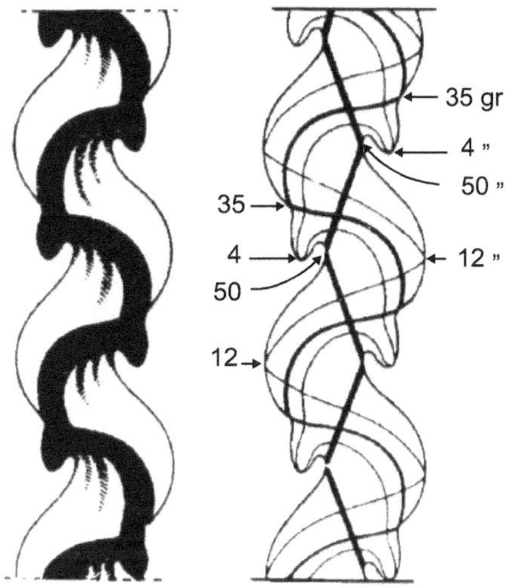

Figure 8.34 Double rick rack braid using more sets of yarn tensions (Lepperhoff, 1953).

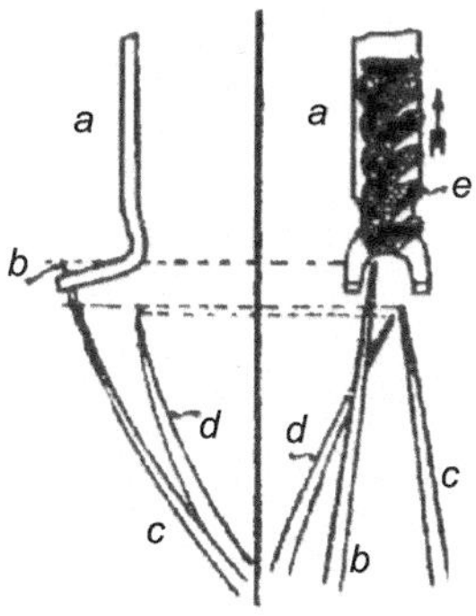

Figure 8.35 Using differently positioned braiding needles and an additional one leads to building of larger asymmetric effects (Lepperhoff, 1953) – left part is the side view and the right part the bottom view of the braiding area.

8.9 Control of the yarn tension in the braid former

For the braiding of ropes, yarn tension and regular configuration of the yarns can be used to control the braiding process. If the yarns are contacting a ring shortly before the braiding point, it has to stay orientated perpendicularly to the take-off velocity (Fig. 8.36) (Schneider, 1991). All yarns contact with the ring, and the contact forces are the same in all directions, so the ring stays in its normal state (Fig. 8.36a and 8.36c). If one yarn breaks, its contact force no longer acts over the ring (Fig. 8.36b and 8.36d); the consequence is a moment of resulting force and the ring rotates. This change in position can be detected and the machine stops, halting production of any more defective braid. The same effect happens if one of the yarns increases in tension, for instance because of a knot failing to pass through the yarn guides. Then its contact force is higher and the ring rotates. For thicker ropes, a special former in a spherical bearing can follow the orientation of the completed rope and detect defects. The rope orientation changes as well if one yarn is missing or the tension becomes too high.

8.10 Sources of further information and advice

There are more papers on the prediction of the yarn path of open braids with exact geometrical considerations (Rawal *et al.*, 2007). The changes of the braiding angle

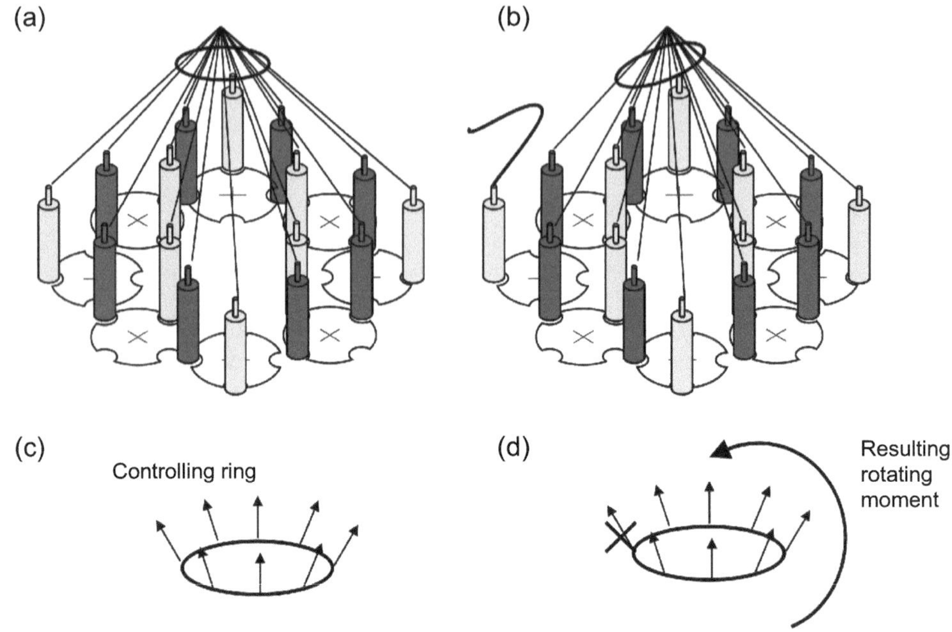

Figure 8.36 Controlling ring and its working principle. (a) and (c) all yarns have constant tension, the ring is in normal state; (b) and (d) one yarn is broken, its contact force is no longer acting on the ring, the ring rotates. This change of position can be detected, and the machine stops. The same effect, but in the opposite direction, appears if the yarn tension of one yarn increases because of some problem.

at different conditions – with or without ring and around non-cylindrical forms, etc. are investigated and published by (Nishimoto *et al.*, 2010) and some more papers in– the same research group. Several equations for braided products can be found in (Ko *et al.*, 1989) but the book is extremely difficult to find.

More models and predictions using FEM and other analytical and numerical methods will be provided in a separate book because they are beyond the scope of this book.

Details about the patterning of fancy braids can be found in (Lepperhoff, 1953) (Goseberg *et al.*, 1981) and (Engels, 1994).

References

Du, G.W. and Popper, P. (1994). Analysis of a circular braiding process for complex shapes. *Journal of the Textile Institute*, 85 (3), 316–337.

Engels, H. (1994). *Handbuch der Schmaltextilien. Die Flechttechnologie. Teil I: Maschinen und Verfahren zur Erzeugung konventioneller Geflechte*. Mönchengladbach.

Geo. Gleistein & Sohn GmbH. *Brief rope lore* [online]. Available from: http://www.gleistein.com/en-brief-rope-lore/ (Accessed 2013).

Goseberg, F., Brabender, K. and Moeller, P. (1981). *Textiltechnik Maschinengeflechte: Ausbildungsmittel Unterrichtshilfen*. Frankfurt am Mai: Arbeitgeberkreis Gesamttextil.

Hongya, Fu., Xianfent, W. and Zhenyu, H. (2008). Abnormal form patch winding. *Journal of Advanced Materials*, **40** (2) 72–79.

Kessels, J. and Akkerman, R. (2002). Prediction of the yarn trajectories on complex braided preforms. *Composites Part A: Applied Science and Manufacturing*, 33 (8), 1073–1081.

Ko, F.K., Pastore, C.M. and Head, A.A. (1989). *Handbook of Industrial Braiding*: Atkins & Peirce.

Koussios, S., Bergsma, O. and Beukers, A. (2004). Filament winding. Part 1: determination of the wound body related parameters. *Composites Part A: Applied Science and Manufacturing*, 35 (2), 181–195.

Lepperhoff, B. (1953). *Die Flechterei.* 3rd edn.: Eugen G. Leuze Verlag Saulgau/WTTBG.

Nishimoto, H. *et al.* (2010). Prediction method for temporal change in fiber orientation on cylindrical braided preforms. *Textile Research Journal* [online], 80 (9), 814–821. Available from: http://trj.sagepub.com/cgi/content/abstract/80/9/814.

Proshkov, A. (1986). *Mechanisms for Yarn Winding (Topics of Design): In Russian, Original Title: Прошков, А.Ф. Механизмы раскладки нити (вопросы проектирования), Москва, Легпромбытиздат.* Moscow: Legpromizdad.

Rawal, A., Potluri, P. and Steele, C. (2005). Geometrical modeling of the yarn paths in three-dimensional braided structures. *Journal of Industrial Textiles* [online], 35 (2), 115–135. Available from: http://jit.sagepub.com/cgi/content/abstract/35/2/115.

Rawal, A., Potluri, P. and Steele, C. (2007). Prediction of yarn paths in braided structures formed on a square pyramid. *Journal of Industrial Textiles* [online], 36 (3), 221–226. Available from: http://jit.sagepub.com/cgi/content/abstract/36/3/221.

Schneider, H. (1991). *Flechtschole.* EP 0 498 944 A2.

Vu, T.D. and Durville, D. (2012). Modelling of contact-friction interactions in fibrous assemblies for the simulation of the mechanical behaviour of braided ropes. *In:* M. Raous and P. Wriggers, eds. *Presentations.*

Vu, T.D., Durville, D. and Davies, P. (2012). Simulation of the bend-over-sheave behaviour of braided synthetic ropes. *In:* G.A. Holzapfel and R.W. Ogden, eds. *ESMC 2012: Book of Abstracts.* Graz: Technischen Universität Graz.

Carrier mechanics in braiding operations

9

9.1 Introduction

The forces in the yarns and in the braid around the braiding point were discussed in Chapter 8 (The mechanics of the braiding process.) The present chapter continues the mechanical analysis of the braiding process in the area around the carriers and the machine. These equations are also important because they determine the forces acting in the braiding zone and the machine dynamics. At the end, some thoughts about future developments are proposed as most of the problems discussed in this chapter are connected with current carrier construction based on pure mechanical compensation without any electrical regulation.

9.2 Bobbin mass and yarn tension

One of the most important functions of the carriers during the braiding process is to keep the yarn tension as constant as possible. The basic principles behind the realization of this aim are presented in Chapter 7 "Carriers for braiding machines"; here some mechanical relations are presented. In most examples of commonly used carriers today there are still fluctuations in the yarn tension connected with the carrier construction and nature of the unwinding process. These fluctuations cannot be removed completely, but an understanding of their reasons can help reduce their magnitude.

One of the most "non-constant" factors is the weight of the bobbin – the carrier starts with a full bobbin and runs until the bobbin becomes empty. The full bobbin has a larger radius R than the empty bobbin, so the moment $M_{\text{yarn\_tension}}$ of the tension force, F, rotating the bobbin to start the unwinding process, changes linearly during the braiding process (see Fig. 9.1):

$$M_{\text{yarn\_tension}} = F \times R \qquad [9.1]$$

This moment is the only one which tries to move the bobbin; working against it are two other moments – the bobbin's friction forces and its inertia forces.

The moment of the friction forces between the bobbin and the carrier

$$M_{fr} = m \cdot r_{fr} \qquad [9.2]$$

Braiding Technology for Textiles. http://dx.doi.org/10.1533/9780857099211.2.211

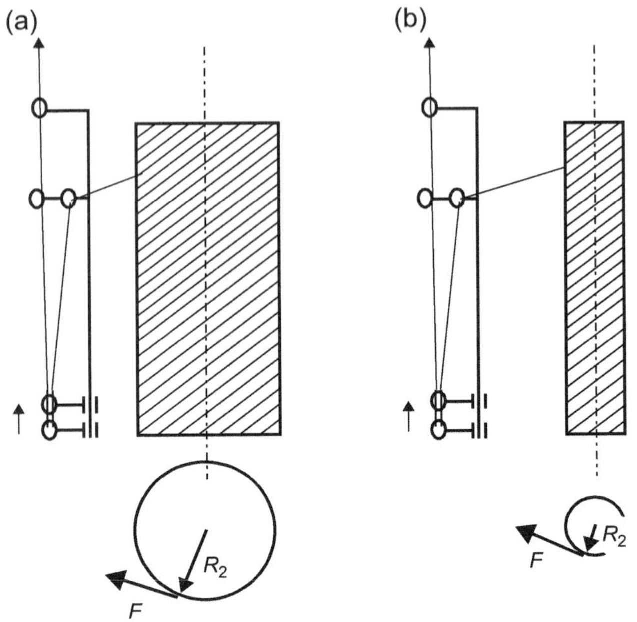

Figure 9.1 The bobbin radius changes the magnitude of the moment, created by the yarn tension, which has to rotate the bobbin in order to unwind the yarn.

where m is the mass of the bobbin, including the empty bobbin and the yarn material, and r_{fr} is the friction radius.

The friction radius for rotational motion has a meaning analogous to the friction coefficient for translational motion, and has a length dimension. For more details, the reader should refer to books on basic mechanics, machine theory or machine elements.

The mass of the bobbin changes during braiding according to the quadratic equation on the radius R

$$m(R) = (R^2 - R_h^2) \cdot \pi \cdot h \cdot \rho + m_{\text{bobbin\_empty}} \tag{9.3}$$

where R_h is the radius of the empty bobbin, h is the height of the winding, ρ is the wound package density (not the yarn density or material density – the wound package has gaps between the yarns, and the yarns have gaps between the fibres), and $m_{\text{bobbin\_empty}}$ is the mass of the empty bobbin.

The bobbin does not rotate during the braiding process because the brake has to hold it in a fixed position, so the angular acceleration of the bobbin $\ddot{\varphi}$ at the moment the brake is released is zero. If the dynamics of the starting process have to be investigated, the differential equation of motion has to be integrated:

$$J \cdot \ddot{\varphi} = F \cdot R - m \cdot r_{fr} \tag{9.4}$$

A thorough investigation of the dynamics of the process requires additional information about the release of the bobbin brakes – some brakes are released smoothly (based on friction), while for others the bobbin release is a process with impulse type dynamics (brakes based on a pin that holds the bobbin geometrically) and correct modelling of the yarn tension variations. This involves building a complete model for the yarn and bobbin motion, which could form the subject matter for an entire chapter and is not described here. Some works about this kind of modelling can be found in Ma *et al.* (2012), Büsgen (1993), and much analogical investigation has been carried out in the area of simulating warp yarn tension on weaving machines (Milkov and Kyosev, 2000; Cremer, 1996), etc. Investigation of the process dynamics is especially important for fast-running machines and for bobbins with a larger mass where the transient processes during unwinding can significantly affect the yarn tension.

In the current case the yarn tension force is important, from a practical point of view, to get the bobbin rotating and releasing yarn. Due to the sticking friction force being higher than the slipping one, the maximum value of the required yarn tension will be the initial one, so in the normal case there is no need to do any further investigation of the process dynamics. Based on the initial static equilibrium from Equation [9.4]:

$$0 = F \cdot R - \left(R^2 - R_h^2\right) \cdot \pi \cdot h \cdot \rho \cdot g \cdot r_{fr} - m_{\text{bobbin\_empty}} \cdot g \cdot r_{fr} \qquad [9.5]$$

the value of the yarn tension required to overcome the friction moment can be defined as

$$F \cdot R > \left(R^2 - R_h^2\right) \cdot \pi \cdot h \cdot \rho \cdot g \cdot r_{fr} + m_{\text{bobbin\_empty}} \cdot g \cdot r_{fr} \qquad [9.6]$$

or

$$F > \left[R \cdot \pi \cdot h \cdot \rho + \frac{1}{R} \cdot \left(m_{\text{bobbin\_empty}} - \pi \cdot h \cdot \rho \cdot R_h^2 \right) \right] \cdot g \cdot r_{fi} \qquad [9.7]$$

Equation [9.7] consists of two terms. The former is proportional to the current bobbin radius and is connected to the mass of the yarn material in the bobbin; the latter is inversely proportional to the current bobbin radius and is connected to the constant mass of the empty bobbin. Some example calculations using simple MATLAB script using Equation [9.7] are given to demonstrate the changes in yarn tension during braiding until one bobbin becomes empty. The computational details and code are provided in the appendix; here, only the graphical results are presented. Figure 9.2 demonstrates the change in yarn tension required to unwind the bobbin, in a standard case chosen for this example and in the case of double reduced friction between the bobbin and the carrier. In both cases the required tension decreases in the time because the weight of the bobbin becomes less, and any reduction in the friction has a directly proportional effect on the yarn tension.

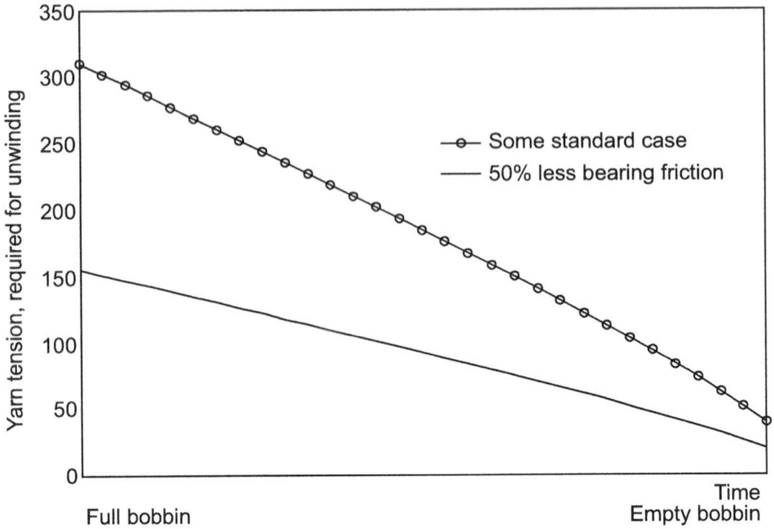

Figure 9.2 The yarn tension, required to cause rotation of the bobbin during the unwinding changes, because the bobbin radius and the bobbin weight are both decreasing. The friction between the bobbin and the carrier has a significant influence on that tension.

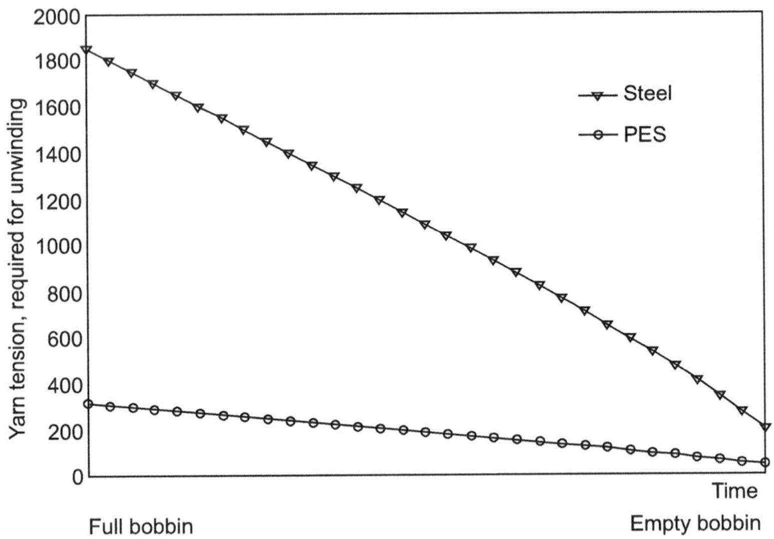

Figure 9.3 The bobbin radius changes the magnitude of the moment, created by the yarn tension, which has to rotate the bobbin in order to unwind the yarn.

Figure 9.3 compares the same standard case with one where, instead of polyester yarn, steel is used and the force required to move the bobbin increases significantly because of the increased weight. The comparison in Fig. 9.3 is valid only if the same carrier and bobbins are used for both materials, which is not the case in normal

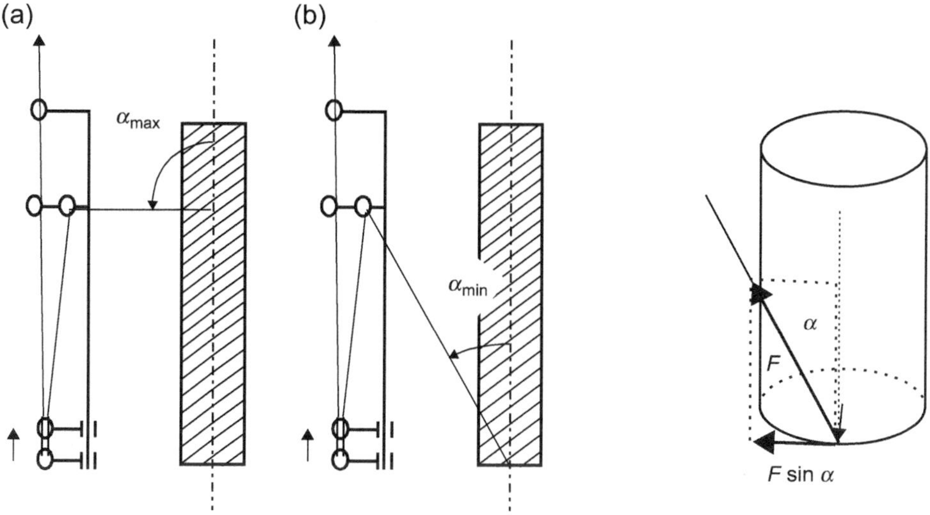

Figure 9.4 The unwinding angle influences the magnitude of the unwinding force. Only the projection of the tension participates in the bobbin motion. For yarn guides, which have fixed positions, this leads to fluctuations in the yarn tension.

conditions. Due to the higher weight of the metal wires, they are braided using special carriers with bobbins which are normally fixed using ball-bearings or needle axial bearings to reduce friction.

9.3 Unwinding angle and yarn tension

In the above examples, only the change in weight and radius of the unwinding (bobbin) were considered. There is one additional parameter which also affects the yarn tension – this is the angle α between the yarn and the bobbin at the point of unwinding. This angle influences the magnitude of the force, created by the yarn tension rotating the bobbin – only the projection $F \cdot \sin\alpha$ of the tensile force F rotates the bobbin (Fig. 9.4), and so the dynamic equation of the bobbin motion becomes:

$$J \cdot \ddot{\varphi} = F \cdot R \cdot \sin\alpha - \left(R^2 - R_h^2\right) \cdot \pi \cdot h \cdot \rho \cdot g \cdot r_{fr} - m_{\text{bobbin\_empty}} \cdot g \cdot r_{fr} \qquad [9.8]$$

The ideal case is when the angle $\alpha = 90°$, because then the entire yarn tension is used to rotate the bobbin. Each additional angle less than 90° means that more yarn tension will be required to overcome the friction of the bobbin and start the unwinding process.

During unwinding, the angle changes continuously between the maximum value of 90° and the minimum value $\alpha_{\min}$, and consequently the yarn tension varies according to the equation:

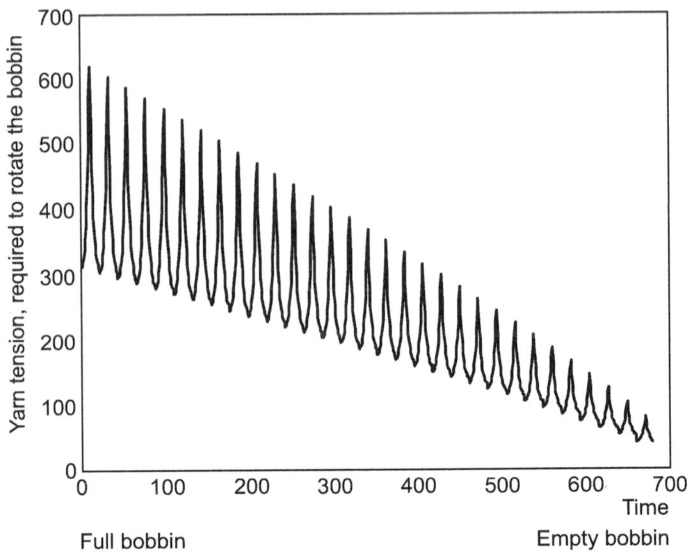

Figure 9.5 The unwinding angle changes depending on the winding type and cause periodic change of the yarn tension, required to start unwinding from the bobbin.

$$F(t) > \frac{1}{\sin\alpha(t)} \cdot \left[R(t)\cdot\pi\cdot h\cdot\rho + \frac{1}{R(t)}\cdot\left(m_{\text{bobbin\_empty}} - \pi\cdot h\cdot\rho\cdot R_h^2\right)\right]\cdot g\cdot r_{fr} \quad [9.9]$$

In the above equation, the angle of radius R and the angle α are both functions of time. The radius becomes smaller and smaller until the yarn ends, but the angle α changes its value depending on the type of winding. For parallel windings, the change is slower, so the fluctuations have a larger period. For cross-winding (usually precision winding in braiding) the angle changes very quickly and the fluctuations have a very short period. An example time curve of the yarn tension required to move the bobbin from static equilibrium is presented in Fig. 9.5, and the code is also explained in Chapter 15 (Using MATLAB® for calculations in braiding).

The yarn tension during braiding follows the curve from Fig. 9.5 only at the points where the bobbin brake is released and new yarn length is let out from the bobbin. Once the brake closes, the yarn tension is determined by the compensation mechanism of the carrier. This causes some peaks in the yarn tension during unwinding (Fig. 9.6), which are reduced with well-chosen carrier springs and intelligent carrier construction but can be significant for machines built by inexperienced machinery producers unfamiliar with carrier mechanics.

For weight-balanced carriers it is kept almost constant, while for spring-balanced carriers there are also small or large fluctuations between the inner and outer positions around the horn gears, which are not shown in Fig. 9.6. The small fluctuations are

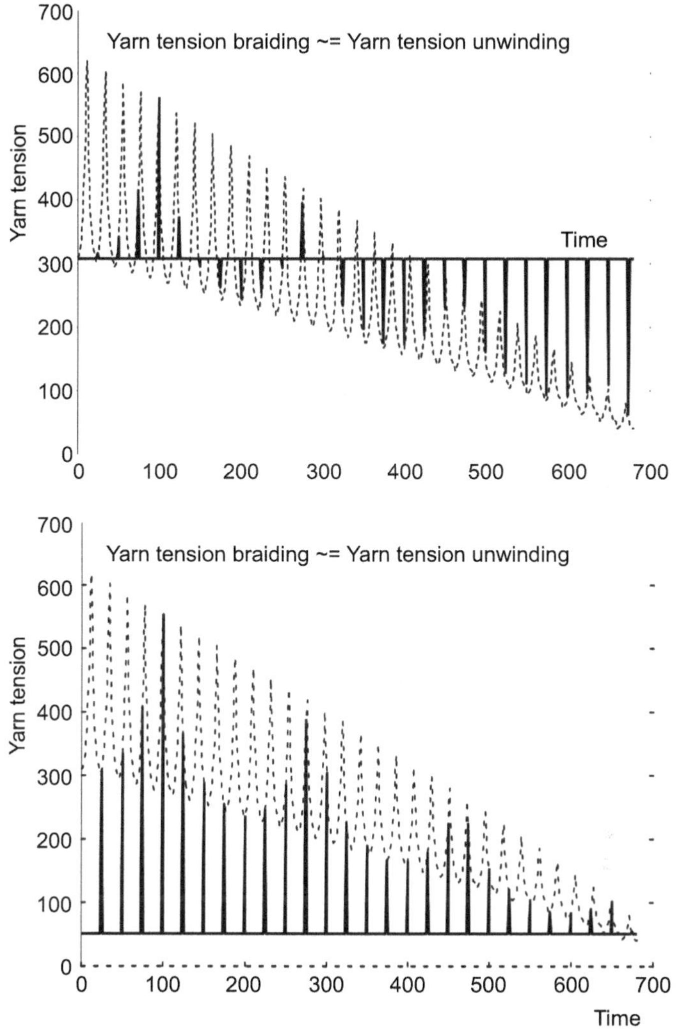

Figure 9.6 The yarn tension during the braiding follows the curve from Fig. 9.5 only at times where the bobbin brake is released and new yarn length is taken out from the bobbin. Once the brake closes, the yarn tension remains with less fluctuations, depending only oncompensation processes.

simple to calculate from the linear spring motion, and are left to interested readers to do for themselves.

One way used to reduce the fluctuations caused by the changing unwinding angle is to change the first yarn guide from fixed to movable, or replace it with yarn guiding bars. In this way, the unwinding geometry changes and the variations in the unwinding angle are significantly reduced (Fig. 9.7).

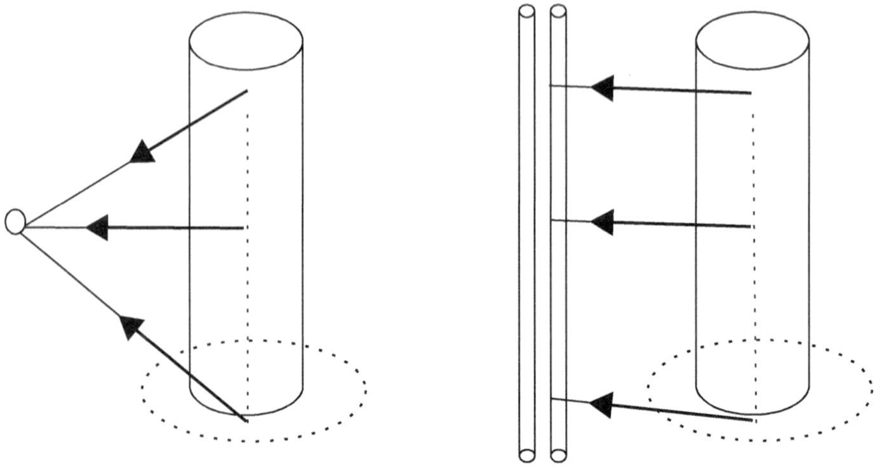

Figure 9.7 Making the yarn guide movable, or replacing it with guiding bares on the modern intelligent designed carriers, reduces the fluctuations in the yarn tension significantly.

9.4 Yarn velocity during length compensation

During braiding, the distance between the braiding point and the last yarn guide on the carrier changes (Fig. 9.8) between the two values:

$$L_{\max} = \sqrt{H^2 + (R + R_{HG})^2} \qquad [9.10]$$

$$L_{\min} = \sqrt{H^2 + (R - R_{HG})^2} \qquad [9.11]$$

where H is the height of the braiding point, measured from the carrier top,
R is the radius of the circle, on which the horn gears are mounted, and
R_{HG} is the radius of each horn gear.

The differences in yarn length are compensated for by a special mechanism (see Chapter 7 "Carriers for braiding machines") so that the yarn tension remains constant. If the carrier is on the inner side of the machine, the distance between the carrier and the braiding point is shorter than the middle value. There is some excess yarn which has to be retained in the carrier for later when the carrier moves to the outer side of the horn gears – then the distance is longer, and more yarn length is required. The compensation mechanism applies some tension to the yarn so that it comes back into the carrier, and this "giving" and "taking" of the yarn length works well. The critical question nowadays, where machines are already working significantly faster than years ago, is: up to what velocity will this simple mechanism work? Let the position of one point (Fig. 9.8) on the last yarn guide of the carrier be considered. The velocity of this point is the (vector) sum of two velocities:

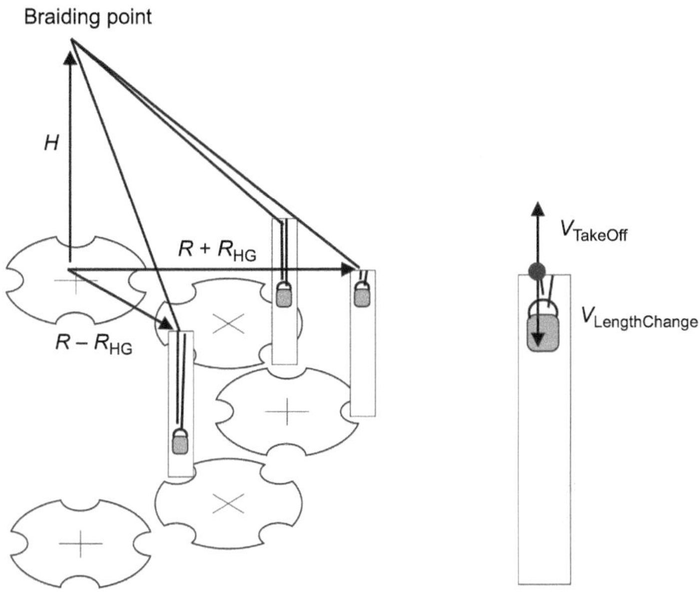

Figure 9.8 Yarn length differences on tubular braiding machine (or flat braiding machine with more horn gears).

$$V_{\text{point}} = V_{\text{TakeOff}} \pm V_{\text{LengthChange}} \tag{9.12}$$

where V_{TakeOff} is the take-off velocity, and

$V_{\text{LengthChange}}$ is the velocity required for that point to compensate the changes in length between the carrier and braiding point.

One carrier moves from its innermost position to the outer position in the same time as its horn gear completes half a revolution $\Delta\varphi_{HG} = \pi$ (the carrier moves along two arcs, each equal to one quarter of the horn gear revolution). If the horn gear makes n revolutions per minute, the time for this motion is

$$\Delta t_{[s]} = \frac{\Delta\varphi_{HG}}{\omega_{HG}} = \frac{\pi}{\left(\pi \cdot n_{HG}/30\right)} = \frac{30}{n_{HG[\text{min}^{-1}]}} \tag{9.13}$$

The *mean* velocity of the yarn point *P*, based only on the change in the length differences, is then

$$v_{\text{LengthChange}[\text{m/s}]} \approx \frac{\Delta L_{[\text{m}]}}{\Delta t_{[\text{s}]}} = \frac{n_{HG[\text{min}^{-1}]}}{30} \cdot \left(\sqrt{H^2 + \left(R + R_{HG}\right)^2} - \sqrt{H^2 + \left(R - R_{HG}\right)^2} \right) \tag{9.14}$$

For the compensation in yarn length differences to work, the yarn available in the compensation device should not be completely used during the half cycle between the two positions of the carriers (inner and outer). Let us assume that the compensation device can take only ΔL length, as required for the compensation; then the take-off velocity should always be lower than the velocity of the length difference change $v_{\text{LengthChange}}$

$$\Delta L = V_{\text{LengthChange}} \cdot dt > V_{\text{TakeOff}} \cdot dt \qquad [9.15]$$

or the critical take-off velocity would be

$$v_{\text{critical[m/s]}} \approx \frac{n_{HG[\text{min}^{-1}]}}{30} \cdot \left(\sqrt{H^2 + \left(R + R_{HG}\right)^2} - \sqrt{H^2 + \left(R - R_{HG}\right)^2} \right) \qquad [9.16]$$

The take-off velocity has to be lower than this velocity for the compensation to be able to work. If the take-off velocity is equal to the critical speed

$$v_{\text{critical}} = v_{\text{TakeOff}} \qquad [9.17]$$

then during the half cycle, where the carrier goes from the outer to inner position, the required yarn length for braiding will be taken from the difference in distance between the braiding point and the carrier. During the next half cycle, where the carrier moves from the inner to the outer side, the required yarn length increases because of the geometry of the machine and also due to the take-off; so it will be taken directly from the bobbin and the compensation will not have a chance to work. If the take-off velocity is higher than the critical velocity:

$$v_{\text{critical}} < v_{\text{TakeOff}} \qquad [9.18]$$

then during the entire braiding process the yarn will be taken directly from the bobbin and the compensation will not have a chance to work. Any differences because of the unwinding angle, position of the yarn, friction of the bobbin, defects during winding, etc. will also directly affect the braided product. High performance fibres such as glass or carbon can at such speed create forces (for instance, if the unwinding is disturbed) which lead to the carriers breaking.

Figure 9.9 shows the amplitude in the critical velocity for a particular machine. These results could look wrong initially – if the braiding point is higher, then the length differences are smaller and the critical compensation velocity also becomes smaller.

Smaller length differences are better from one point of view – less compensation is needed. However, this can become a challenge ensuring the compensation device works – if there are no fluctuations in velocity, a completely different type of carrier construction should be selected – which can only "give" yarn maintaining a constant tension.

One way to overcome this problem, and reduce the risk of the compensation not working, is to increase the compensation length using a larger number of levers and multiple compensation. In such cases, the yarn reserve is not used completely during

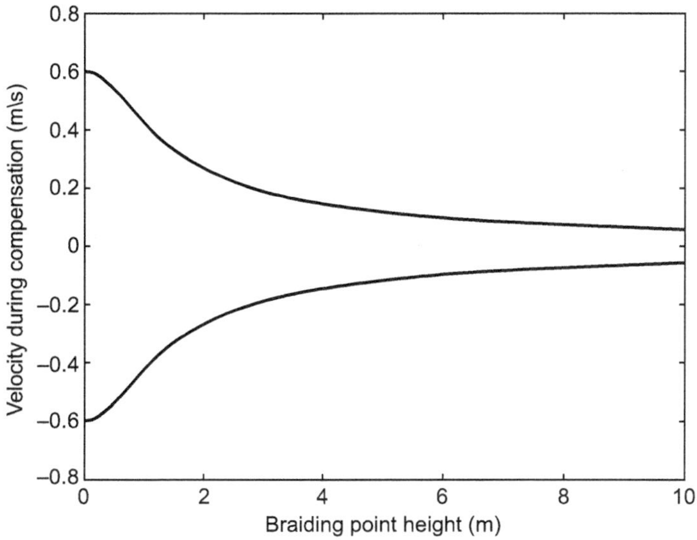

Figure 9.9 Velocity of the yarn at end of the compensation device (last eye of the carrier), depending on the height of the braiding point. As higher the braiding point is, as less the differences in the length are, the "dancing" of the compensation lever will be less. Calculations for tubular braiding machine with radius of circle where the horn gear centres are located $R = 1$ m, each horn gear with radius m and horn gear speed 150 min^{-1} at a height of the braiding point from 0 to 10 m over the surface of the highest carrier eyes.

one cycle, and once it is used and a new unwinding step starts, more length can be stored in the compensation area.

This specific problem is even more important for radial braiding machines – where length differences are almost not present. Nevertheless, having some yarn buffer significantly improves the regularity of the braided products because it reduces fluctuations in the yarn tension during unwinding from the bobbin and allows large variations in the take-off speed, which are common during complex over-braiding.

9.5 Yarn tension fluctuations for spring-balanced carriers

Spring-balanced carriers have an additional source of tension fluctuation – the tensioning springs (Fig. 9.10). The springs become an acting force due to their deformation. The deformation of the spring depends on the available yarn length – at the inner position on the machine the carriers are close to the braiding point, so the excess yarn has to be stored in the carrier and the spring relaxes. Less deformation Δxmeans less tension force on the yarn F_{yarn}:

$$F_{\text{yarn}} = \frac{1}{2} K \cdot \Delta x \qquad [9.19]$$

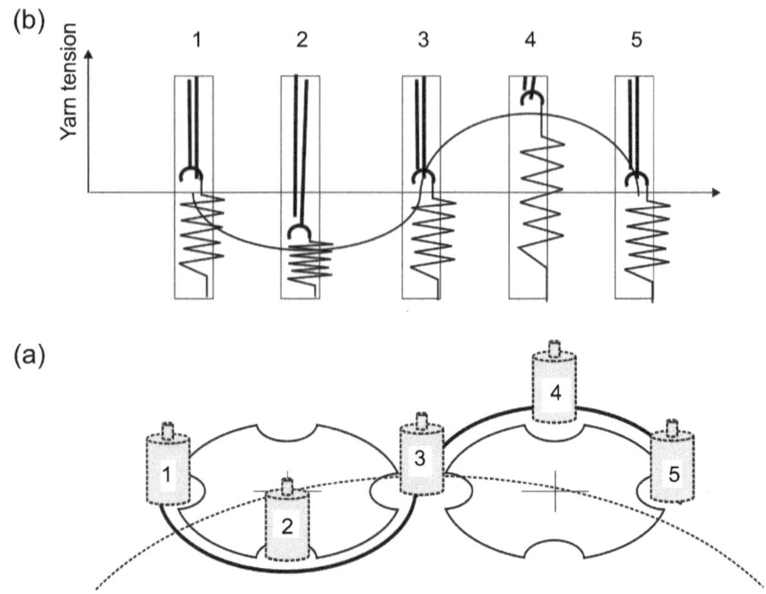

Figure 9.10 The spring changes its force depending on the deformation (B), which depends on the available yarn length. The yarn length changes, so the yarn tension varies as well during the braiding process.

If the carrier is on the outer side of the horn gears, the distance to the braiding point is larger; the whole reserved length is used and the spring deforms more, increasing the yarn tension. Normally, the springs are made as long as possible so that the deformation Δx during the compensation is relatively small, as well as the yarn tension fluctuations.

9.6 Maximum velocity of weight-balanced carriers

Weight-balanced carriers are used nowadays only in special cases, because of several disadvantages, as mentioned in Chapter 7 "Carriers for braiding machines." As a result of the weight remaining constant during the entire braiding process, the yarn tension is also kept almost constant, and so weight balancing is preferred in cases where even small yarn tension fluctuations are not allowed. This section highlights some limitations of the process where constant tension is no longer guaranteed, for example if the weight sticks on the carrier or starts vibrating.

Figure 9.11 presents a projection of the carrier path on the plate's plane. The carrier is moving along a path of a circular arc and its weight is pressed to part of the carrier (tube, leading bar) by centrifugal force.

In the radial direction there is no motion – the weight contacts with the wall or a guiding bar of the carrier, and static equilibrium is satisfied. According to Newton's third law "action = reaction"

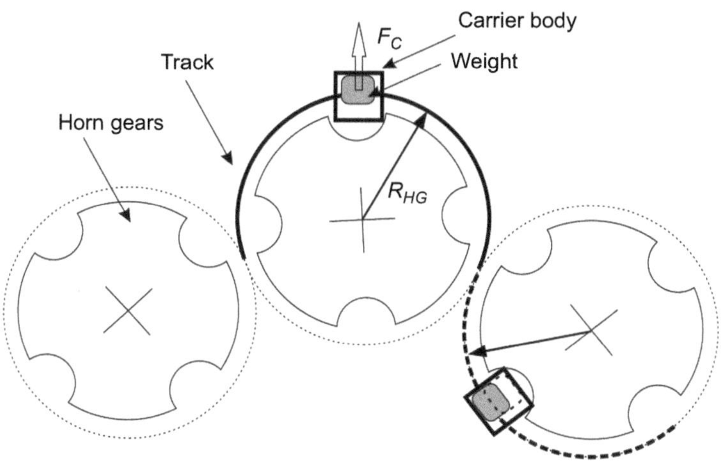

Figure 9.11 Tracks position and the horn gears for determining the limit velocity of the weight-balanced carriers.

$$R = F_C \quad [9.20]$$

where R is the reaction force of the carrier wall and
F_C is the centrifugal force of the weight.

The friction force F_{Fr} between the weight and the carrier wall is:

$$F_{Fr} = \mu \cdot R \quad [9.21]$$

where μ is the friction coefficient between these parts.

The centrifugal force F_C is

$$F_C = m \cdot r \cdot \omega^2 \quad [9.22]$$

where r is the radius of curvature of the current carrier path, and
ω is the angular velocity of the carrier.

Thus, the friction force, would be

$$F_{Fr} = \mu \cdot m \cdot r \cdot \omega^2 \quad [9.23]$$

In the vertical direction, free motion is required for the yarn length to be compensated, so the dynamics equilibrium according to Newton's second law of motion will be:

$$m \cdot \ddot{y} = 2 \cdot T + F_{Fr} - G \quad [9.24]$$

where m is the mass of the compensating weight,

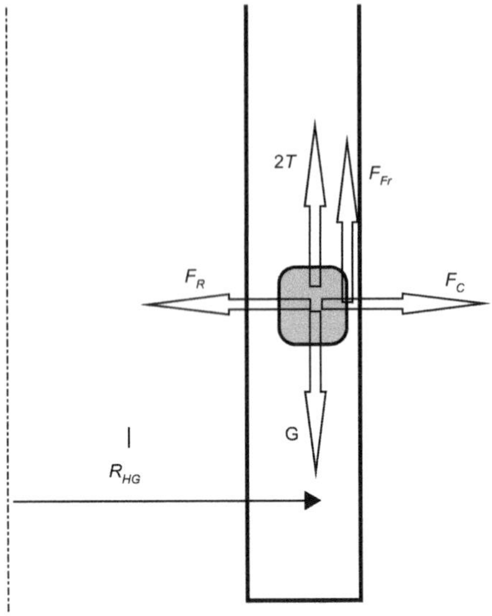

Figure 9.12 Acting forces on the compensation weight.

T is the yarn tension force, G the gravitational force of this weight, and
$\ddot{y}$ is the acceleration of the weight in the vertical direction (Fig. 9.12).

Here the direction of the friction force is chosen for the example where the weight moves down. This is its working direction if the yarn length has to be compensated when the carrier is travelling from the outer to the inner part of the track

$$m \cdot \ddot{y} = 2 \cdot T + \mu \cdot m \cdot r \cdot \omega^2 - G \quad [9.25]$$

In this case, compensation is needed, the yarn tension is reduced to zero for a very short time, and the task of the weight is to keep it constant, moving the weight so that the entire length of yarn is tensioned again.

If $T = s$

$$m \cdot \ddot{y} = \mu \cdot m \cdot r \cdot \omega^2 - m \cdot g \quad [9.26]$$

or

$$\ddot{y} = \mu \cdot r \cdot \omega^2 - g \quad [9.27]$$

To have an acceleration greater than zero and for the weight to start moving down, the right-hand side of Equation [9.27] has to be negative so that the gravitational force is higher than the friction force between the weight and the carrier wall.

Thus, the condition:

$$\mu \cdot r \cdot \omega^2 < g \qquad [9.28]$$

has to be satisfied to prevent the weight sticking on the carrier wall and enable it to work as a compensator for the yarn length and tension.

Example:

Normally, the horn gear radius for a machine with 80 stitches is 0.04 m,

$g = 9.8$ N/ms^2,

$\mu = 0.3$

the maximum angular velocity of the horn gears has to be:

$$\omega < \sqrt{\frac{g}{\mu \cdot r}} = \sqrt{\frac{9.8}{0.3 \cdot 0.04}} = 28.57\,\mathrm{s}^{-1} \qquad [9.29]$$

This angular velocity corresponds to 272.89 min^{-1} of the horn gear per minute, which is actually the normal velocity for several machines today, but with spring-carriers.

Equation [9.28] gives the angular velocity limit of the horn gears from which no more compensation is possible because the compensating weight will be unable to move. The actual maximum possible angular velocity can be lower than this limit because the weight will be unable to compensate as well if it starts swinging (oscillating). To observe this oscillation limit, the dynamics of the process need to be considered.

The carrier travels along the half circular arc of the track and then moves to the next horn gear, where it travels along the opposite track (Fig. 9.11)

The time for completing the half arc is:

$$t_{\mathrm{Half\_horn\_gear}} = \pi / \alpha \qquad [9.30]$$

Which, for angular velocity at the above example of 28 s^{-1}, will be 0.11 s. After that time, the weight will jump to the opposite carrier wall and be pressed there with the same force, but on the other side. For that time $t_{\mathrm{Half\_horn\_gear}}$ the weight has to travel from its middle position to the upper one (if the carrier is on the outer side, maximum yarn length is required) and then fall down to the middle position as well (Figs 9.13 and 9.14).

The motion law of the weight for a machine with horn gears arranged around a circle (the usual case for tubular and flat braiding machines) is a sinus-like curve, as shown in Fig. 9.13. The exact function of the yarn length differences can be derived using Pythagoras' theorem and the circular arc for the motion path, but it is not important for the current discussion.

This change of motion direction causes forced vibrations of the weight with the same frequency. Equation [9.25] describes the motion law in the vertical direction

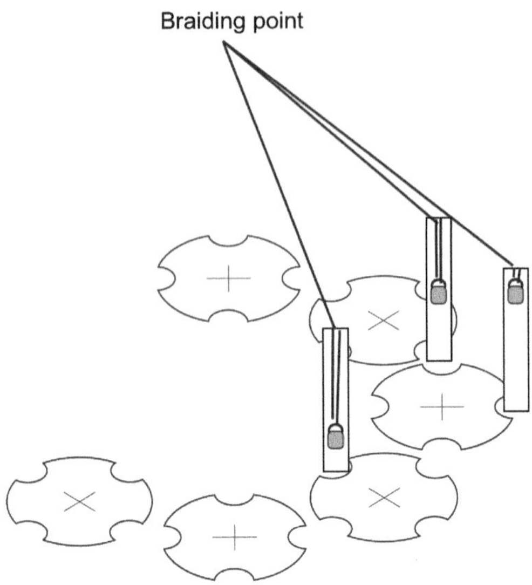

Figure 9.13 Position of the weight depending on the carrier position resp. the current yarn length between the carrier and braiding point.

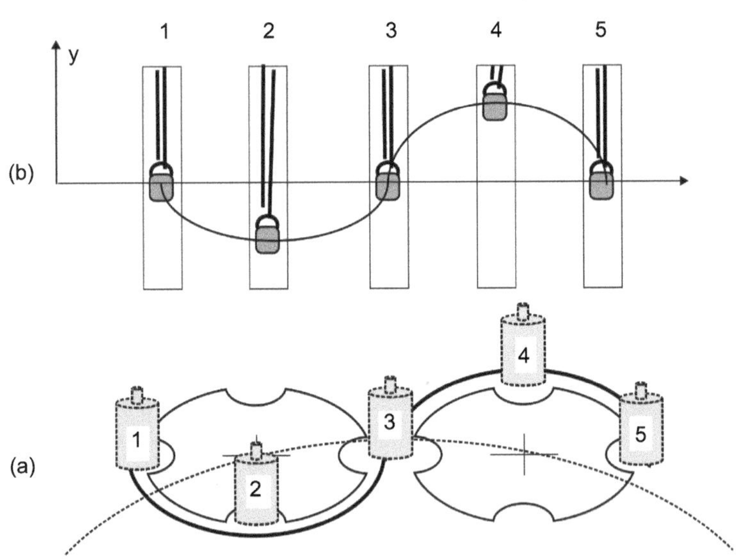

Figure 9.14 Motion law of the weight (B), presented for the time, where one carrier moves once from the inner side of the horn gear and once outside. At position 2 the distance between the carrier and the braiding point is minimal and at position 4 it is maximal.

of the weight in this case. The forced vibration will be caused here by the yarn tension T. If linear elastic behaviour of the yarn during the braiding process is assumed, the stress in the yarn σ is linearly dependent on the elongation ε and the elasticity modulus E:

$$\sigma = \varepsilon \cdot E \quad [9.31]$$

In the textile industry the determination of the yarn cross-section A is a challenging problem, and the tension $T = \sigma \cdot A$ in the yarn is used so that Hooke's law [9.31] can be rewritten in the form:

$$T = \varepsilon \cdot E \cdot A \quad [9.32]$$

Normally the constant EA means the elasticity modulus, and since it is usually determined directly by experimental data it will be written as one parameter.

The elongation of the yarn here depends on the available length:

$$T = \varepsilon \cdot EA = \frac{L - L_0}{L_0} \cdot EA = \Delta L \cdot \frac{EA}{L_0} \quad [9.33]$$

Actually the elongation of the yarn depends on the position of the weight and carrier:

$$\Delta L \approx C_L \cdot \sin(\omega \cdot t) - 2 \cdot y \quad [9.34]$$

where C_L is half the size of the length variations depending on the braiding geometry. If the yarn length compensation is working properly, the coordinate y will change constantly so that the elongation of the yarn will be kept constant, $\Delta L = const$.

Combining Equations [9.34] and [9.33] in [9.25], the motion law becomes:

$$m \cdot \ddot{y} = 2 \cdot (C_L \cdot \sin(\omega \cdot t) - 2 \cdot y) \cdot \frac{EA}{L_0} + \mu \cdot m \cdot r \cdot \omega^2 - G \quad [9.35]$$

$$m \cdot \ddot{y} + 2 \cdot 2 \cdot y \cdot \frac{EA}{L_0} - \mu \cdot m \cdot r \cdot \omega^2 + G = 2 \cdot C_L \cdot \sin(\omega \cdot t) \cdot \frac{EA}{L_0} \quad [9.36]$$

$$m \cdot \ddot{y} + \frac{4 \cdot EA}{L_0} \cdot y + G - \mu \cdot m \cdot r \cdot \omega^2 = 2 \cdot C_L \cdot \frac{EA}{L_0} \cdot \sin(\omega \cdot t) \quad [9.37]$$

$$\ddot{y} + \frac{4 \cdot EA}{m \cdot L_0} \cdot y + g - \mu \cdot r \cdot \omega^2 = 2 \cdot C_L \cdot \frac{EA}{m \cdot L_0} \cdot \sin(\omega \cdot t) \quad [9.38]$$

Equation [9.38] is the classical equation of a one-mass system with forced vibrations with natural frequency:

$$\omega_0 = \sqrt{\frac{4 \cdot EA}{m \cdot L_0}} \qquad [9.39]$$

If the angular velocity of the horn gears is the same as the natural frequency of the system, then it will resonate and the compensation will not work properly.

For a precise calculation of the frequencies which are unsuitable for such carriers, it also has to be considered that the length L_0 used in the equation is actually not constant. One can take the length of the yarn in the carrier between the bobbin and the yarn guide for this measurement. Depending on the friction between the yarn and the yarn guides, the vibrations can be transmitted to the next part, or can be partially damped in small areas, so determining the oscillating yarn length is not trivial.

9.7 Future trends

The issues discussed in this chapter are connected with the construction of current carriers, which are purely mechanical assemblies. Several ideas exist, some patents, etc., about the introduction of electrical regulation of the yarn tension and general electro-mechanical realization of all the carriers' functions. If each carrier had a motor drive moving the bobbin, it could "give" and "take" the required yarn length just by rotating the bobbin. If the bobbin were too heavy, an additional compensation device could be controlled and moved so that only the necessary yarn length was in the braiding zone. With even less mechanical intelligence and more automation, putting in a yarn tension sensor and some compensation device should work to keep the yarn tension constant independent of the type of machine, the current radius of the bobbin, its winding radius, etc. All these are good ideas, but with one problem – the carrier moves from horn gear to horn gear, and it is impossible to mount an electricity cable between the cable and other parts of the machine.

Putting a battery on each carrier could help in some way, and could be one solution. Using some kind of wireless energy transfer could also be a partial solution. Controllers, electro-mechanical devices and batteries are becoming even smaller, cheaper and more powerful; so in the very near future some electric carriers can be expected to appear and start being used in industrial applications, probably initially only for very sensitive applications. If that time comes, this chapter would no longer be relevant, and then it should be extended or exchanged with a consideration of the control and regulation of automated devices. But the mechanics are simple and stable and so mechanical carriers will surely continue to be used in a large number of applications in future decades.

References

Büsgen, A. (1993). *Neue Verfahren zur Herstellung von dreidimensionalen Textilien für den Einsatz in Faserverbundwerkstoffen: Dissertation.*

Cremer, E. (1996). *Prozesssimulation der Kettfadenzugkraefte in Webmaschinen.* Dissertation. RWTH Aachen, Germany.

Ma, G., Branscomb, D.J. and Beale, D.G. (2012). Modeling of the tensioning system on a braiding machine carrier. *Mechanism and Machine Theory* (47), 46–61.

Milkov, M. and Kyosev, Y. (2000). Mechano-techological modeling of system "warp-fabric" on the weaving looms. *In:* M. Todorov, ed. *Summer School "Applications of Mathematics in Engineering and Economics"*: Proceedings, pp. 137–140.

Yarn winding operations in braiding

10

10.1 Introduction

The winding of yarns on bobbins is an essential part of the preparation for braiding. Incorrect settings applied to the winding can lead to a braid with visible irregularities, significantly reduced elongation, and greatly reduced strength, or it can result in the breakage of yarn carriers or other machine parts. This is why the braiding experts say: “Good winding is half the braiding.”

Usually the material to be braided is stored on large bobbins and has to be transferred to the yarn carriers. The first step in preparing the yarn is therefore to unwind it from the storage bobbins (Fig. 10.1). This initial procedure may seem trivial, but special equipment and techniques are often required to ensure its stability and regularity. If the process of unwinding is so unstable that the yarn breaks, then the operators have an immediate problem. If instability in the unwinding causes changes in the winding tension on the carriers, then it may only be discovered later that the optical and mechanical properties of the braids have been affected, which will be a costly mistake for the manufacturer.

To regulate tension, several control devices are used between the unwinding and winding areas. Such devices can apply a range of solutions – from simple weights and friction brakes to very complex electronic regulation. Additional devices can identify specific defects in the yarn, such as knots and varying thickness, and may be able to apply some finishing to correct them.

Lastly, a package of wound yarn suitable for the carriers on the braiding machine has to be formed. The parameters of the winding process determine the density and quality of the bobbin used in the braiding machine and ultimately the quality of the braided product.

10.2 Unwinding

10.2.1 Axial unwinding

Yarn or wire for braiding usually originates from different producers, countries or even continents from those of the manufacturers who will be using it. So, to simplify the logistics of distributing and storing this material, it is held on large bobbins weighing from 3 to 15 kg. Such bobbins cannot be used on the braiding machine, because their weight would limit the speed and other functionalities of the carriers, and they would simply be too large. The yarn is therefore unwound from the large,

Braiding Technology for Textiles. http://dx.doi.org/10.1533/9780857099211.2.231

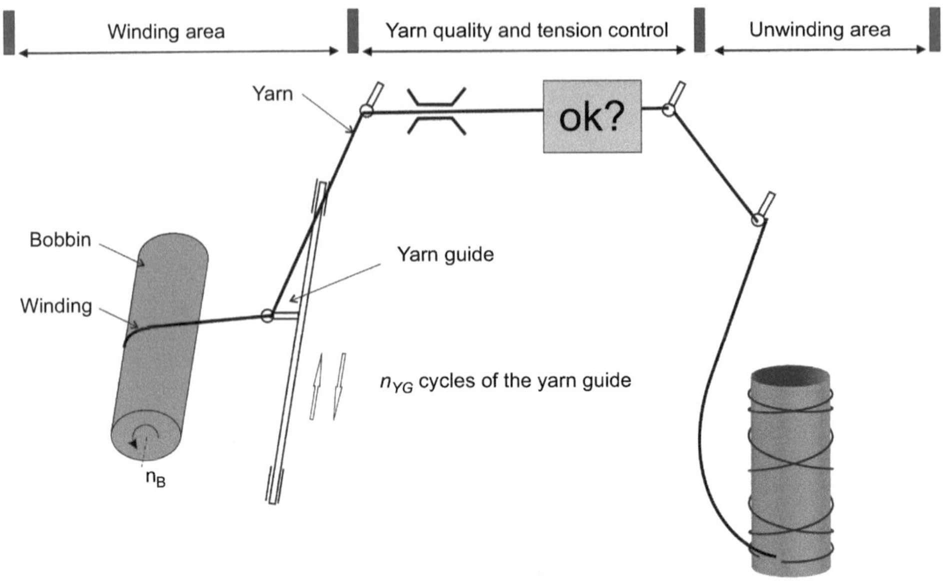

Figure 10.1 Basic zones during the winding of the yarn to bobbin for braiding.

storage bobbins and wound on to specially designed bobbins suitable for the braiding machine.

Different forms of bobbin are used for storing materials between the producer and the final user. Usually these are without flanges and are cylindrical or conical in form (Figs 10.2 and 10.3), with the winding of one half-layer crossing over the winding of the next half-layer. Such bobbins can be unwound axially (over the head) – the yarn guide is mounted on the bobbin axis and the yarn builds a "balloon," as it is called, around the bobbin during the process of unwinding. There must be sufficient clearance between the yarn guide and the bobbin, and sufficient distance from other bobbins and machine components, so that the yarn balloon remains stable and the yarn is unimpeded. The higher the yarn guide, the greater the width of the balloon, which then requires more space per bobbin. Rings or plates can be used to limit the size of the balloon, should the yarn guide be set too high, but these will lead to variations in the tension of the yarn.

Most materials with a high enough coefficient of friction can be wound on a stable, random cross-winding package (Fig. 10.3a). Smooth filament yarns, glass filaments, and wire require flanges on both sides of the bobbin to keep the package stable (Fig. 10.3b). Bobbins with flanges can be unwound axially in certain cases, as for instance if the upper flange is smaller than the lower one.

Some materials, such as those with a glass or PES core coated with PVC or polyurethane, or monofilaments, come wound not on bobbins with flanges but on normal, cylindrical bobbins, with precise cross-winding (Fig. 10.4). Because of the lower coefficient of friction and the stiffness of the yarn, the winding will collapse immediately if the yarn does not remain in tension (Fig. 10.4a). During the

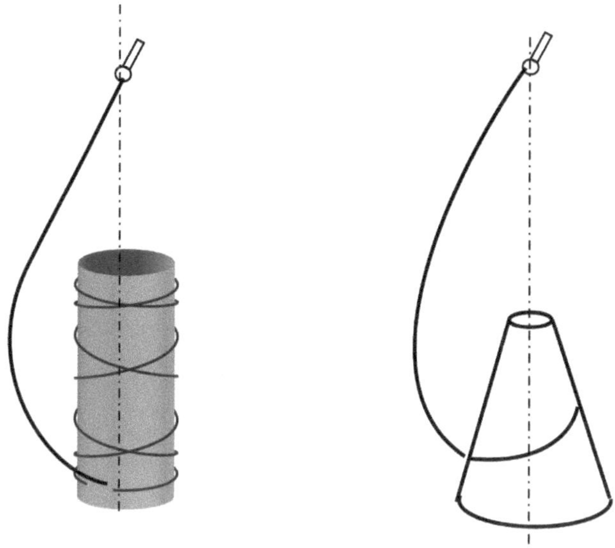

Figure 10.2 Cylindrical and conical bobbins for axial unwinding.

Figure 10.3 Random cross-winding and parallel winding on body with flanges for axial unwinding.

unwinding and winding process this need not be a problem, if the winding speed is high enough, but once the bobbin to be used by the braiding machine is full and has to be changed, the yarn will fall down. Other factors can also cause the winding to slip down and terminate the unwinding process, such as small variations in yarn tension, vibrations in the machine, or anything that affects the unwinding balloon.

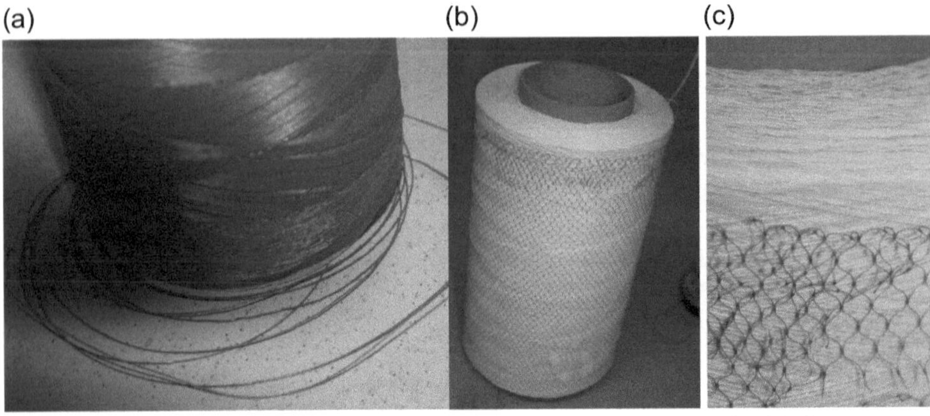

Figure 10.4 Monofilament or coated yarn on precise cross-winding bobbin. (a) the outer windings are falling down once the yarn tension is removed; (b) and (c) fixing this problem using fine mesh.

One solution is to use tangential unwinding (discussed in the next section), but this requires special and more expensive equipment. Another, simple solution for such bobbins is to use a thin, fine mesh (Fig. 10.4b and 10.4c), which keeps the windings stable on the surface of the bobbin but still allows the unwinding process to continue.

Some high performance materials, such as carbon- and glass-fibre from rovings, should not be bent over a small radius during production, because up to 30% of their fibres can be broken in that way. In addition, the very high linear density of such rovings (>1000 tex) makes winding on a bobbin with a smaller radius problematic, because the thickness increases very fast and changes the winding geometry significantly. Such materials are therefore usually wound on bobbins of larger radius. These bobbins are unwound from inside to out (Fig. 10.5), and cannot be held in a creel, but have to stay on a horizontal surface during the unwinding process.

10.2.2 Tangential unwinding

When the yarn is wound on a bobbin with flanges, or when the bobbin is heavy (for example with metal wire) or the material to be unwound is delicate, tangential unwinding is appropriate. In this case the yarn is pulled out in a direction tangential to the bobbin surface (Fig. 10.6). There are two variants of tangential unwinding, for yarn tension-driven bobbins (passive) and motor-driven bobbins (active).

The yarn tension-driven bobbin rotates as a result of the force applied by the yarn tension (Fig. 10.6a). As the bobbin becomes smaller, the winding radius decreases, so if friction and inertial forces remain constant, the yarn tension will need to be increased to overcome these forces. In practice, this is not normally a problem, because the bobbin weight also decreases. A more detailed analysis of these forces is given in Chapter 8 (The mechanics of the braiding process), where the influence of

Figure 10.5 Negative bobbin of glass/polypropylene yarn 1780 tex. The unwinding starts from the innermost layer.

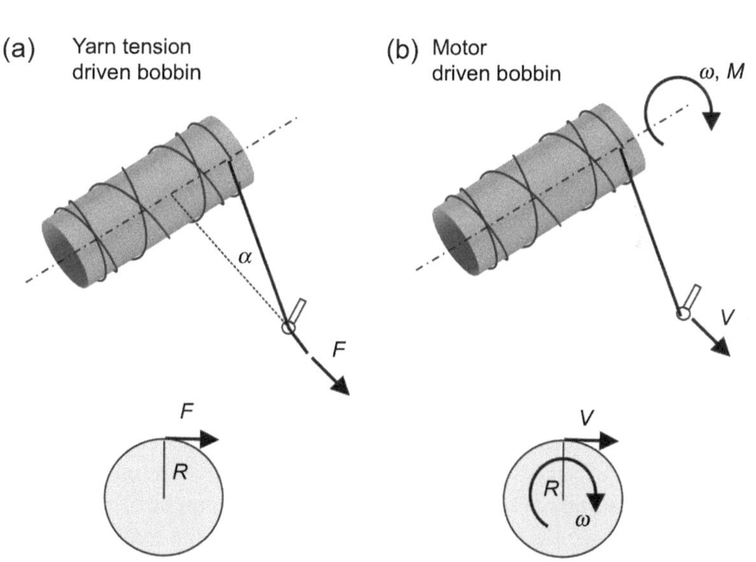

Figure 10.6 Tangential unwinding (a) yarn tension-driven (b) motor-driven.

carrier and bobbin geometry on yarn tension during braiding is presented. Identical relations apply here.

If the material of the yarn is unsuitable for transferring force to the bobbin, either because the bobbin is too heavy or the yarn is too elastic (e.g., elastodien, elasthan), then a separate motor drive is used to rotate the bobbin during the unwinding process

Figure 10.7 Bobbin with copper wire prepared for motor-driven unwinding on the universal winding machine of the company Jakob Müller AG, Frick, Switzerland.

(Figs 10.6b and 10.7). The speed of the yarn is then determined by the speed of the motor, and the yarn tension has to be controlled precisely by managing the differences between the speeds of the unwinding and winding bodies.

The synchronization of winding and unwinding speeds is not straightforward, especially when one of the bodies (the bobbin to be unwound) is heavy, which significantly reduces the reaction speed of the control system. A compensating device is therefore required between the unwinding and winding bodies. A compensating roll is commonly used, known as a "dancer," because it dances up and down when in operation (Fig. 10.8). If excess yarn is unwound as a result of the motion of the bobbin, the dancer moves down and activates a switch that stops the motor. The dancer is then pulled back up by the excess yarn as it is taken in and the motor switches on again. The more intelligent systems have an additional switch and sensor – if the dancer moves higher, this means that the current speed cannot keep up with the amount of yarn required, and then the additional sensor, having detected the new position of the dancer, switches the motor to a higher speed. Such systems are used mainly for wire and other strong materials and not for fine yarns, as the weight of the dancer and the friction of its motion would otherwise significantly affect the tension of the yarn.

10.2.3 Multiple yarns

In modern yarn preparation departments, the bobbins are usually fixed on a creel. Multiple bobbins can then be unwound at the same time and under the same conditions. The large number of spaces on a creel brings several advantages:

- if the creel can accommodate them, then different types of bobbins can be unwound without any change of equipment;

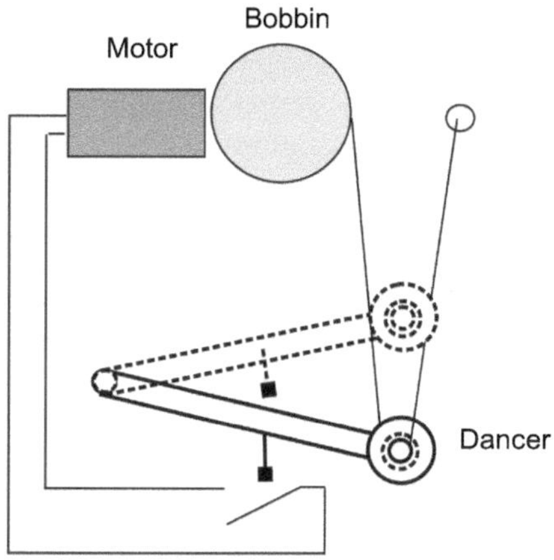

Figure 10.8 Device for unwinding of wire with active bobbin motion and dancer for length compensation and regulation.

- if a winding machine with several heads is used, then several bobbins can be unwound simultaneously;
- for the winding of ply yarns, for instance for rope or candlewick, yarns from several bobbins can be unwound under the same constant conditions and then wound on a single bobbin to form one high-quality package.

Figure 10.9 shows a creel with active bobbin motion, able to deal with different types of bobbin. Such creels can have 4, 8, 16 or 32 bobbin positions and anything up to the same number of control devices (August Herzog, 2013). Creels for axial unwinding (pay-off creels) are also available. Modern equipment is designed for managing yarn control and yarn tension electronically; users can save the unwinding parameters for a particular product and reload them on a later occasion.

10.3 Control and regulation devices

Control and regulation devices are very important for the consistent operation of the winding body and therefore ultimately for the quality of the final braided product. Only a few of the principles are presented here; control devices exist in many variants that work on mechanical, optical, or capacitive principles. The type of control device chosen in a particular case will depend on need, costs and required accuracy.

10.3.1 Detection of yarn breaks

One of the simplest mechanical devices for detecting yarn breakage is shown in Fig. 10.10. It consists of a yarn guide made of porcelain, mounted on a swinging

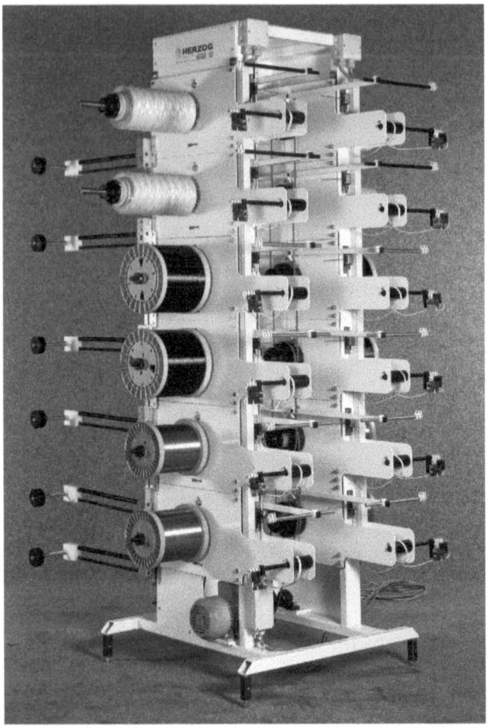

Figure 10.9 Creel with driven bobbins and several bobbin places, photo courtesy of (August Herzog Textilmaschinenfabric GmbH & Ko KG, Oldenburg, 2013).

metal part. As long as the yarn is going through the yarn guide (Fig. 10.10b), the tension of the yarn forces the guide to remain in a vertical position. If the yarn breaks, then the guide swings down under its own weight (Fig. 10.10c) and switches off the winding unit. Several such controlling guides can be mounted as a packet, so that in the case of winding ply yarn the machine will stop if any of the yarns break. Guides not in use (Fig. 10.10a, position A) are fixed in the upper position behind a special wire (Fig. 10.10a, position C), so that they do not interfere with the process.

10.3.2 Detection of knots

One common technique to check for knots and other points of excessive thickness is to use a plate with an opening that determines the maximal acceptable diameter of the yarn at any point. Such a plate (Fig. 10.11) is held in its normal position by a spring. If a point of excessive thickness in the yarn, such as a knot, reaches the plate, it rotates, which operates a switch to stop the machine. Such points of thickness in the yarn are not always checked on braiding machines, but they can cause breakage of carrier parts, if the eye openings are smaller than the yarn diameter at those points.

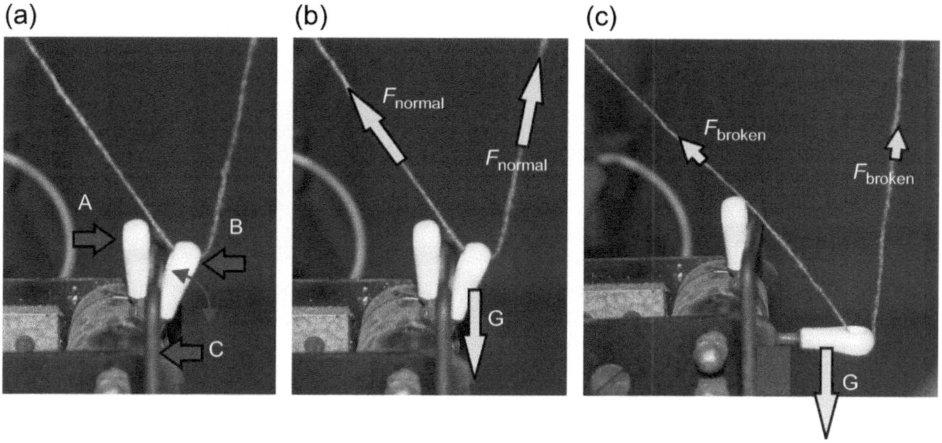

Figure 10.10 Yarn breaks detection. If the yarn is broken, the yarn tension can no longer support the weight of the yarn guide; the guide falls down and stops the machine.

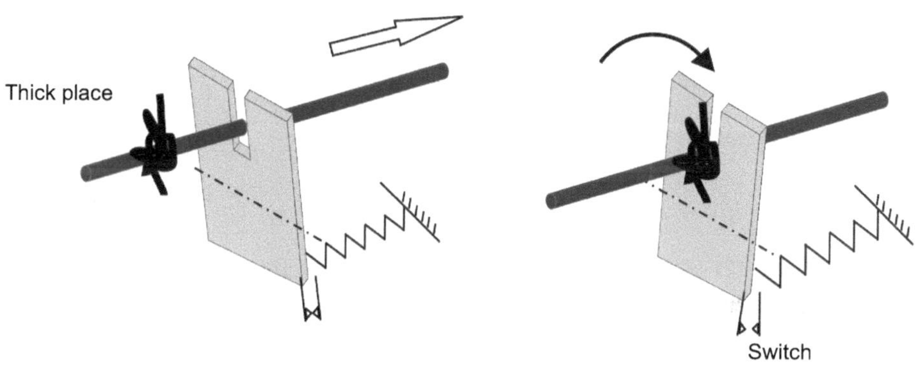

Figure 10.11 Checking for thick places or thick knots.

10.3.3 Passive adjustment of yarn tension

Yarn tension during the winding process determines the density and stability of the bobbin package. Furthermore, variations in tension during winding variously cause elongations and relaxations of the yarn on the bobbin. During and after the braiding, those parts of the yarn that were more deformed, or that tend more to relax (depending on the material type), will shrink more than other parts of the yarn, which can cause optical defects in the braid and significant reduction of rope strength. This is because those yarns that shrink more are therefore shorter and will be the first to start carrying the load. To prevent such defects, the yarns have to be wound with constant tension, as far as possible. Yarn tension is commonly maintained at a particular value by the application of slipping friction forces, based on normal friction, cylinder friction or combined friction.

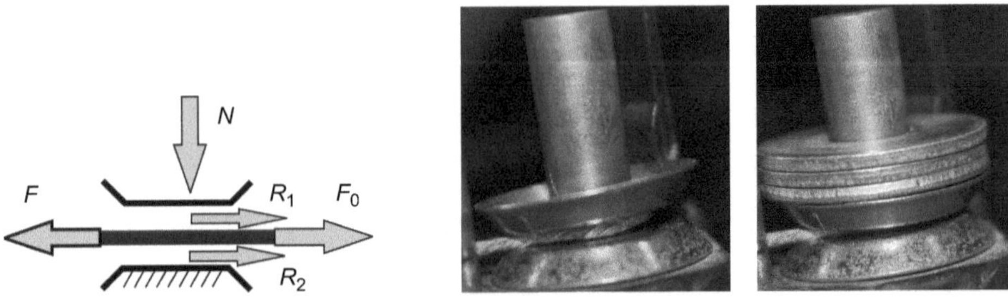

Figure 10.12 Disc brake working on the additive principle (a) and practical realization as combined brake (a) diagram of the forces (b) brake with one disk (c) with additional disks for increasing the friction and thus the yarn tension.

10.3.3.1 Normal friction (additive principle)

Normal friction (Fig. 10.12) is based on contact between the yarn and two disc brakes (down and up). The friction force added to the yarn tension then depends on the normal force N and the friction coefficient between the yarn and the disc:

$$R_{1,2} = \mu \cdot N \qquad [10.1]$$

so that the final yarn tension after passing one brake will be:

$$F = 2 \cdot \mu \cdot N + F_0 \qquad [10.2]$$

These kinds of brakes follow the additive principle, in that the friction forces are added to the previous yarn tension F_0. If the friction forces are significantly higher than the yarn tension before the brake, then one such brake can reduce variations in yarn tension during the unwinding process and thus be sufficient to guarantee the consistency of the winding process. The simplest way to regulate the tension is by adding or removing weights (Fig. 10.12c). Modern creels can use an electromagnetic field or pneumatic pressure to apply the normal force N, so that, in the case of multiple brakes, all can be adjusted to keep the same yarn tension.

10.3.3.2 Cylinder friction (capstan friction, multiplicative principle)

Another method for applying friction force to the yarn is to keep the yarn sliding around a cylinder (Fig. 10.13). According to the capstan equation or belt friction equation, also known as the Euler–Eytelwein formula (after Leonhard Euler (1707–1783) and Johann Albert Eytelwein (1764–1848)), the force after the cylinder will be:

$$F = F_0 \cdot e^{\mu \cdot \alpha} \qquad [10.3]$$

where α is the angle swept by the yarn around the cylinder in radians, and $\mu$is the coefficient of sliding friction between the yarn and the cylinder. As Equation [10.3]

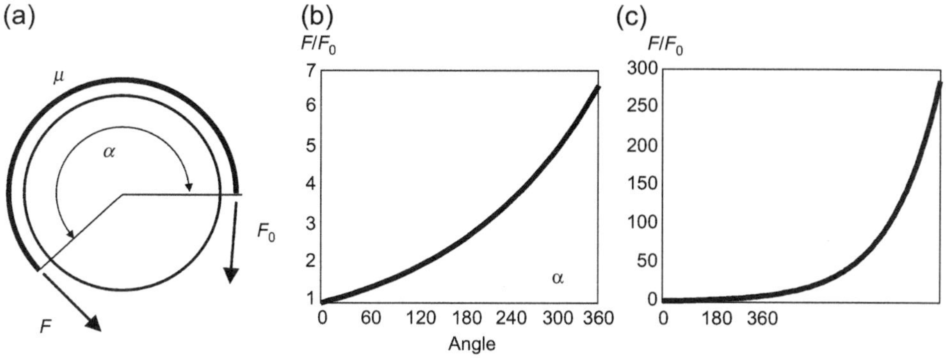

Figure 10.13 (a) Cylinder friction brake (b) and (c) relation between the yarn tension and the initial yarn tension depending on the angle (for) friction coefficient μ = 0.3.

shows, the yarn tension after the cylinder depends on the initial yarn friction – the final value is the initial force multiplied by some factor. This means that the friction force added around the cylinder is not independent; not only the initial tension but also variations in the initial tension will be multiplied. For an angle greater than three whole revolutions ($3 \cdot 2 \cdot \pi$ radians), the friction is so high that even a very small initial yarn tension is enough to keep the yarn stuck on the cylinder (Fig. 10.13c shows the case of yarn tension after the cylinder being up to 300 times greater than the initial tension). A tendency to stick to the cylinder is not good for regulating yarn tension, so smaller angles are commonly used (Fig. 10.12b). The use of cylinder friction brakes has the advantage of enabling the yarn tension to be changed only by changing the geometry, without the need for external forces.

One application of cylinder brakes is when several winding bodies are involved. It can be proved that the force then depends on the total angle by which the yarn goes around the cylinders (Fig. 10.14a):

$$F = F_0 \cdot e^{\mu \sum \alpha} = F_0 \cdot e^{\mu(\alpha_1 + \alpha_2 + \alpha_3 + \alpha_4 + \alpha_5)} \qquad [10.4]$$

Half of the cylinders can be mounted on a fixed part and the other half on a mobile part, so that the friction force can then be adjusted by simple displacement of the mobile part (Fig. 10.14b).

The classical disc brake (Fig. 10.12b and 10.12c) is, in practical terms, a combined application of normal and cylinder friction, because the simplest solution for bearing and fixing the upper disc is to use a central shaft.

10.3.4 Active yarn regulation

Brakes create tension in the yarn up to a certain level, adding a friction force to the initial yarn tension or multiplying it by the friction component. In both cases, the final yarn tension is still dependent on the initial force and, where there are large fluctuations

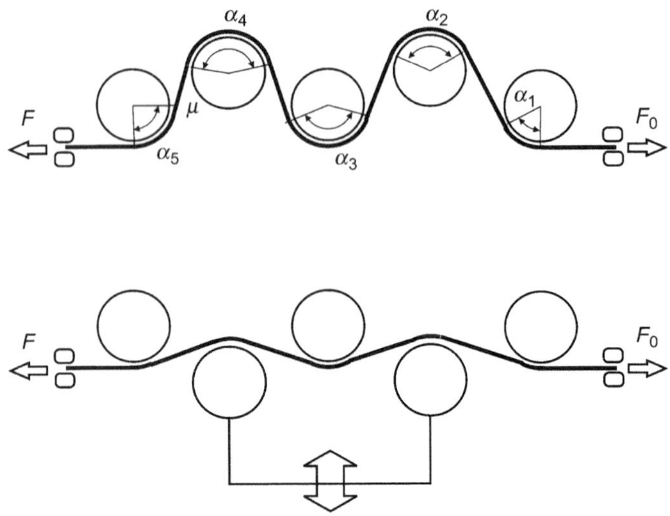

Figure 10.14 Yarn tension with several cylindrical bodies.

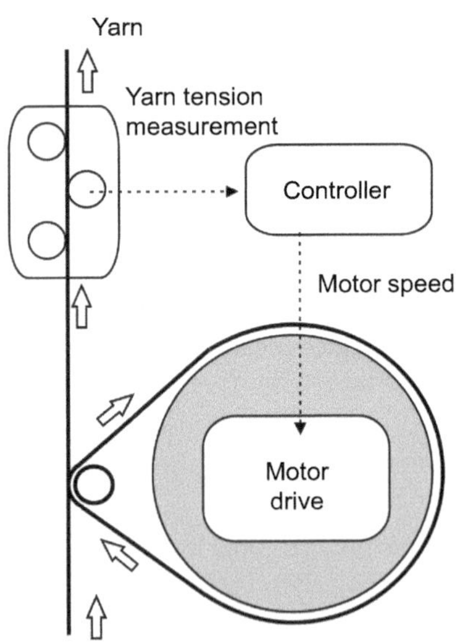

Figure 10.15 Working principle of yarn feeding control device.

in tension during unwinding, these will remain to some extent in the winding tension. Only a system with automatic regulation can completely eliminate variations in the yarn tension. Such a system uses a tension measuring unit (Fig. 10.15) to give a signal to the controller. The controller calculates the required reaction and adjusts the speed of the motor drive. This drive rotates a special roller, around which the yarn makes at least three revolutions; the minimum number of revolutions is again determined by Equation [10.3]. However, in this case the friction needs to be sufficient to keep the yarn sticking to the surface of the roller, because the level of control will be reduced if there is slippage. If the yarn tension is lower than the required value, the motor starts rotating slowly, so that the required peripheral velocity is lower than the yarn velocity and the motor drive works as a brake. If the yarn tension is higher than the required value, the motor starts rotating faster and feeds the yarn, so that the tension can be reduced. Units of such systems are available, for instance, from the company BTSR (BTSR Yarn feeding control, 2013).

10.3.5 Length measurement

For one braiding machine with three carriers, working with cotton or polyester, it is not very important whether the length of the yarn on each bobbin is exactly the same, but for one machine with 46 or 288 carriers, differing yarn lengths on the bobbins is a major problem. This is because after the first bobbin is empty, all other bobbins have to be unwound or "cleaned" of their remaining yarn length in order to be empty and ready for the next bobbin winding process. In mid-sized and large braiding companies, there are workers who attend to this task daily, as well as machines that simplify the process and wind the remaining material into a form suitable for applications other than braiding. If the material is expensive (there are, for example, yarns and wires priced at more than €100–200 per kg) then each wasted piece of the material is effectively a waste of money for the company.

There are relatively simple devices that can be used on the winding machine to measure and register the yarn length and thus reduce the problems described above. Most of them again rely on the sticking effect of cylinder friction. They use a roller, which rotates with the same peripheral velocity *V* as the yarn:

$$V = r \cdot \omega = r \cdot \frac{\pi \cdot n}{30} \qquad [10.5]$$

If there are sensors for velocity then the yarn length can be calculated by integration of [10.5] against time, but the simplest method is to use a device that counts the revolutions of the roller. A roller with radius r has circumference:

$$U = 2 \cdot \pi \cdot r \qquad [10.6]$$

and so, after making *n* revolutions, a length of:

$$L = n \cdot 2 \cdot \pi \cdot r \qquad [10.7]$$

will be wound. Modern winding machines are usually equipped with a length-measuring device and are able to stop automatically after the required length is wound.

10.4 Winding

10.4.1 Winding types

Winding for braiding imposes different requirements on bobbins from winding for weaving or knitting. The bobbins have to be as small as possible to reduce the size of the machine and the inertial forces that have to be countered during the process, but they still have to collect as great a length of yarn as possible to minimize the frequency with which bobbins have to be changed (which takes time). These two contrary conditions can be satisfied simultaneously only if the density of the wound package is high: a greater length of yarn length stored in a smaller volume on the bobbin. Two types of winding structure give the maximal density: parallel winding and precise cross-winding.

Parallel winding requires the use of bobbins with flanges to keep the sides of the bobbin stable (Fig. 10.16). It is the most common type of winding to be used in the braiding process because it is simple to produce. Precision cross-winding creates a very high-density bobbin on which greater yarn length can be stored in the same volume (Fig. 10.17): each winding is placed, with great precision, very near the previous one, leaving minimal gaps in the bobbins. Precision cross-winding bobbins do not need flanges on the sides, but they can be prepared only with special machines, and adjustments to the winding process are more complex than for parallel winding.

10.4.2 Winding kinematics

To perform a winding process, two independent motions are required – rotation of the bobbin and reversible motion of the yarn guide (Fig. 10.18). If at the beginning of

Figure 10.16 Bobbin with flanges with yarns in parallel winding.

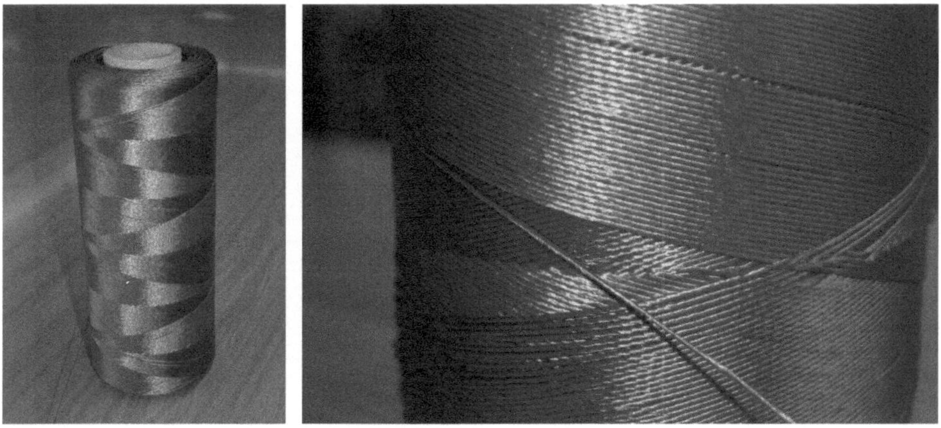

Figure 10.17 Precision winding package.

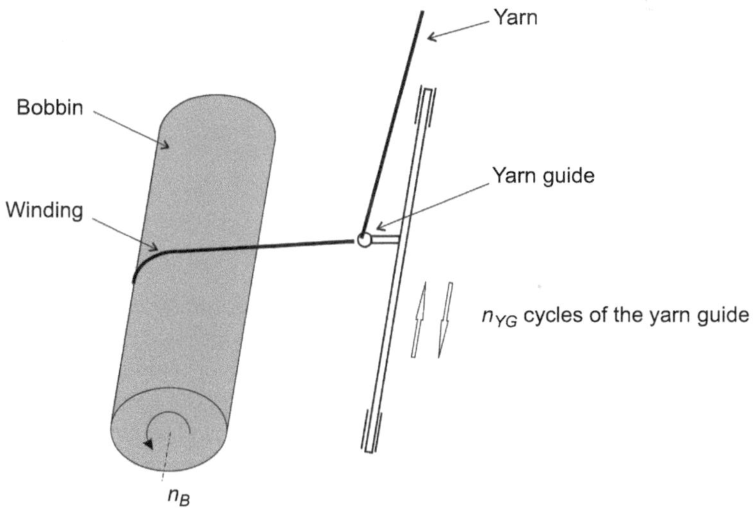

Figure 10.18 Basics of yarn winding.

the process the end of the yarn is fixed on the bobbin, then during the rotation of the bobbin the next piece of yarn will be wound over it. The ratio:

$$\kappa = \frac{n_{YG}}{n_B} \qquad [10.8]$$

where n_B and n_{YG} are the number of the revolutions of the bobbin and number of the cycles of the yarn guide per unit time (usually a minute), respectively, determines the number of yarn guide cycles per bobbin revolution.

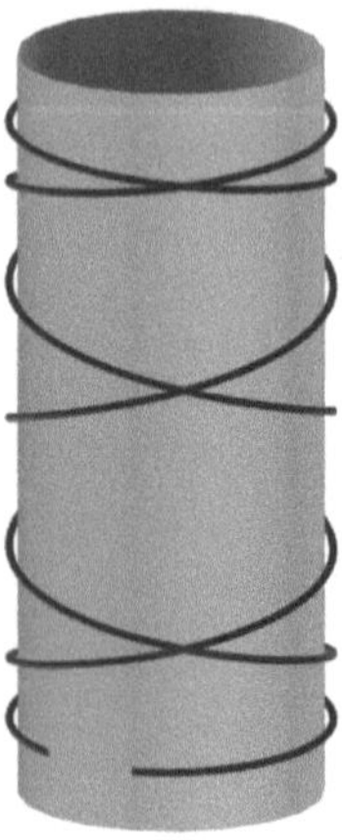

κ = 1/7.1 = 0.14085

Figure 10.19 Winding ratio and the number of windings per layer.

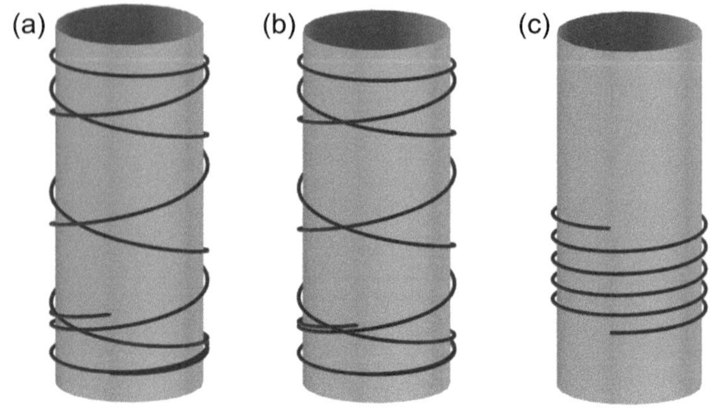

Figure 10.20 Winding types (a) cross-winding, wild; (b) cross-winding, precision; (c) parallel winding.

The inversely proportional relation:

$$K = \frac{1}{\kappa} = \frac{n_B}{n_{YG}} \qquad [10.9]$$

is not so common in the winding literature, but it more naturally explains the winding ratio, because it gives the number of bobbin revolutions per cycle of the yarn guide. The number of bobbin revolutions is equal to the number of windings in one winding layer. The ratio is illustrated in Fig. 10.19 for $\kappa = 0.14085 = 1/7.1$. This layer consists of seven complete windings and 0.1 of the eighth, though the last part is not drawn.

Depending on the value of κ, four types of wound package can be classified:

- parallel winding
- cross-winding, wild
- cross-winding, precision
- cross-winding with defects ("tapes" or "images").

Parallel winding is one of the most commonly used windings in bobbins for braiding, in those cases where the tubes have flanges. The bobbin makes a large number of revolutions for one cycle of the yarn guide, and so K is large and accordingly κ is small (Fig. 10.20c). The limiting value for the winding ratio is:

$$\kappa < \kappa_{\text{parallel,lim}} \qquad [10.10]$$

where $\kappa_{\text{parallel,lim}}$ represents the case in which one winding lies exactly on the next with no space between them. If the yarn guide starts to move more slowly than the speed corresponding to the $\kappa_{\text{parallel,lim}}$ value, then new windings will be placed over previous windings in some places. These positions might not be stable, and if a winding slips to an adjacent location this can lead to problems later during braiding, because one part of the yarn, when it needs to be unwound, might be trapped by another.

If $\kappa > \kappa_{\text{parallel,lim}}$ then the wound package has cross-windings. As the name suggests, the windings in the first half-layer cross the windings of the second half-layer at an angle not near 0 (as it is for parallel windings).

Depending on the exact value of the winding ratio κ, three sub-types of winding can be produced. As mentioned above, if:

$$K = \frac{1}{\kappa} \qquad [10.11]$$

is a whole (and small) number, the windings will be placed successively over the same path, usually leading to defects, and an unstably wound package will be produced. (In some special cases, this is done deliberately, for producing effect packages with very low packing density). If the value of $\kappa$is selected so that the windings of one layer start very close to the windings of the previous layer, precision winding is the result (Figs 10.20b and 10.17). In the other cases, however, a so-called "wild" cross-package is created, in which the crossing points of the winding have a more stochastic character, rather than an exact form (Figs 10.20a and 10.3a). Wild windings are commonly used in preparation for weaving and knitting, because they can be produced with simple machinery (drum winding). But they cannot be applied to the braiding process, because their packing density is up to three times less than it is for parallel and precision winding bodies. The well-ordered and compact windings of the latter two types accordingly allow up to three times more yarn length to be held in a package. More details about the calculation of the winding for all kinds of package can be found in Simon and Hübner (2012).

10.4.3 Practicalities of parallel winding

During parallel winding the motion of the yarn guide has to be adjusted so that each winding lies directly next to the previous one (Fig. 10.22b). If the yarn guide moves too slowly, some windings, rather than being merely very close to the previous ones, will actually lie on top of them, thus causing an irregular layer to be formed (Fig. 10.21). Effectively the same result is obtained if the yarn guide moves too fast, because initially there will be gaps left between the windings which subsequent windings fill in irregularly. The amplitude of the yarn guide also has to correspond exactly to the bobbin length. If the yarn guide turns back too early, then the winding layer at the end will be thin and unstable. If the yarn guide has a larger amplitude than required, then the winding body will end with excessively thick layers (Figs 10.21a and 10.22a).

In any case, the yarn tension during winding has to be higher than the tension during braiding, so that the wound package remains stable during the braiding process and there is no motion or deformation of the single windings. Figure 10.23shows

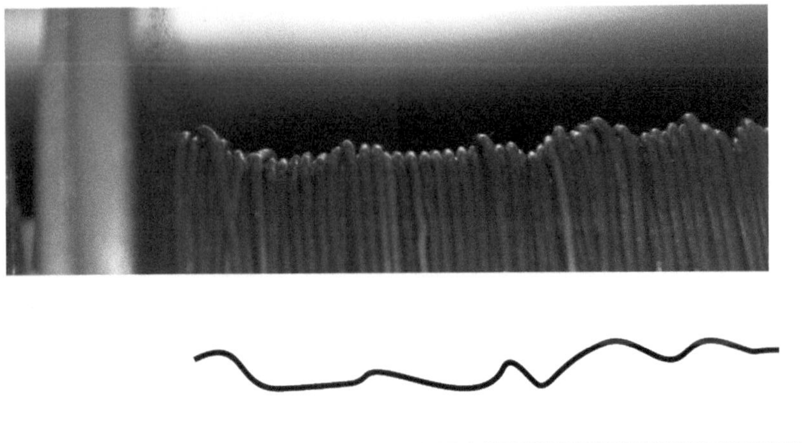

Figure 10.21 Irregular layer during the winding of monofilament yarn. The yarn guide velocity has to be adjusted and the yarn guide should not go so far to the left.

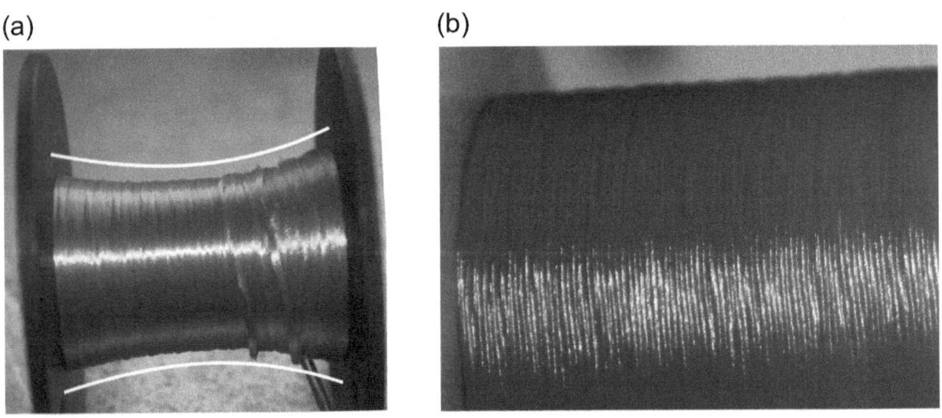

Figure 10.22 Bobbin with incorrect yarn guide amplitude (a) and correct parameters of the winding process (b).

different possible deformations of the yarn cross-section depending on the radial pressure, which from other side depends on the yarn tension. Different yarns react differently on the radial pressure – textured and staple yarns can compress (Fig. 10.23b); multifilament rowings hardly compress at all – their filaments redistribute and their cross-section becomes wider (10.23c).

10.5 Machines

Machines that wind bobbins for braiding commonly have several (for instance, two or four) locations for winding and a semi- or fully automated bobbin-changing device. Automated machines (Fig. 10.24) carry a bobbin magazine with an automatic feeder,

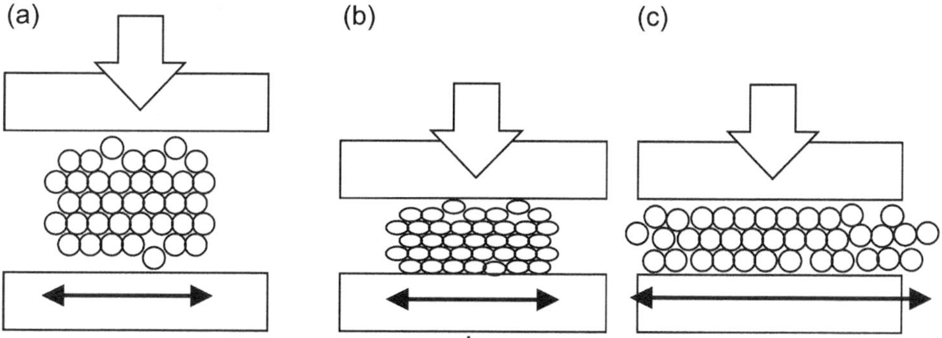

Figure 10.23 Deformation of the yarn cross-section under lateral loading, for instance during the winding. (a) initial, unloaded state (b) deformed shape of compressible yarns, for instance texturized or staple yarn (c) deformed shape of incompressible yarns, for instance multifilament rowing without twist.

Figure 10.24 Fully automated winding machine with four heads.
Source: Photo courtesy of August Herzog Textilmaschinenfabric GmbH & Ko KG, Oldenburg, 2013.

so that once the yarn is prepared and the machine is adjusted, no more human interaction is required during the winding process. This arrangement is especially convenient when there are a large number of bobbins. In most cases the machines are adjustable for different bobbin sizes. Semi-automated machines (Fig. 10.25) stop when the bobbins

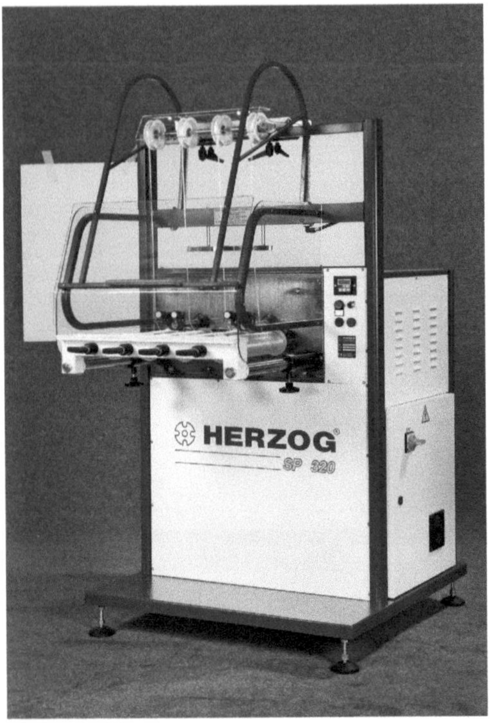

Figure 10.25 Semi-automated winding machine with four heads.
Source: Photo courtesy of August Herzog Textilmaschinenfabric GmbH & Ko KG, Oldenburg, 2013.

are filled, the filled bobbins having to be taken off and new, empty bobbins having to be fixed by the operators. Because they do not have the additional, automatic components, the investment costs for such semi-automatic machines are lower. Whether it is more efficient to use an automatic or semi-automatic machine depends on the number of the bobbins per day or week which have to be prepared and on personnel costs. Of course, there are smaller machines with one or two bobbins and fewer automatic components for simple cases, and also specially designed winding machines with larger carriers, and to make the appropriate choice, contacting the machine manufacturers, some of which are listed at the end of the chapter, is recommended.

10.6 Winding calculations

The easiest and most accurate way to estimate the winding time and length of the material on one bobbin is with practice. Only on the real machine can the complete set of winding conditions be considered and a good feeling achieved for how the material moves and how the unwinding works. In some cases, however, the production time

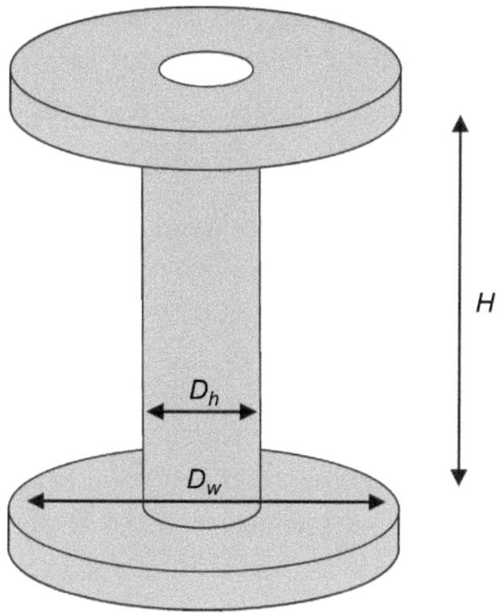

Figure 10.26 Dimensions of a bobbin with flanges.

and costs for some future products have to be estimated before obtaining the required machine. In these cases, yarn length can be approximately calculated using the following relations.

The volume of one winding body depends on the geometry of the bobbin (Fig. 10.26):

$$V = \frac{\pi \cdot (D_W^2 - D_H^2) \cdot H}{4} \quad [10.12]$$

where D_W is the maximum diameter of the wound packages, D_H is the diameter of the empty bobbin (the hulse) and H is the height of the bobbin (the inner distance between the flanges). The mass of the wound package is then:

$$M = \frac{\pi \cdot (D_W^2 - D_H^2) \cdot H}{4} \cdot \rho_{\text{wound\_package}} \quad [10.13]$$

where $\rho_{\text{wound\_package}}$ is the density of the wound package. The length L of the yarn on the bobbin in kilometres is:

$$L_{[\text{km}]} = \frac{1}{Tt_{[\text{tex}]}} \cdot \frac{\pi \cdot \left(D_{W[\text{cm}]}^2 - D_{H[\text{cm}]}^2\right) \cdot H_{[\text{cm}]}}{4} \cdot \rho_{\text{wound\_package}[\text{g/cm}^3]} \quad [10.14]$$

The most challenging part of using Equation [10.14] is to estimate the density $\rho_{\text{wound\_package}}$ of the wound package, which depends on the density $\rho_{\text{fiber\_material}[\text{g/cm}^3]}$ of the fibre material, the packing density μ_{yarn} of the fibres in the yarn, and the packing density μ_{windings} of the windings in the bobbin:

$$\rho_{\text{wound\_package}[\text{g/cm}^3]} = \rho_{\text{fiber\_material}[\text{g/cm}^3]} \cdot \mu_{\text{yarn}} \cdot \mu_{\text{windings}} \tag{10.15}$$

The packing density of the yarn is the relation between the total fibre volume in one yarn and the total yarn volume:

$$\mu_{\text{yarn}} = \frac{V_{\text{fibers}}}{V_{\text{yarn}}} = \frac{S_{\text{fibers}}}{S_{\text{yarn}}} = \frac{N_{\text{fibers}} \cdot S_{\text{single\_fiber}}}{S_{\text{yarn}}} < 1 \tag{10.16}$$

and the packing density of the windings is the relation between the (complete) yarn volume in one layer of the wound package and the total volume of this layer or package:

$$\mu_{\text{windings}} = \frac{V_{\text{yarns}}}{V_{\text{wound\_pakage}}} = \frac{S_{\text{yarns}}}{S_{\text{wound\_pakage}}} = \frac{N_{\text{yarns}} \cdot S_{\text{single\_yarn}}}{S_{\text{wound\_pakage}}} < 1 \tag{10.17}$$

For metal yarn, monofilaments and other incompressible materials, the packing densities are easy to determine geometrically, assuming that the yarn has a circular cross-section with diameter *d* and that this yarn requires space with a square form of side *d*:

$$\mu_{\text{windings}} = \frac{V_{\text{yarns}}}{V_{\text{wound\_pakage}}} = \frac{S_{\text{yarns}}}{S_{\text{wound\_pakage}}} = \frac{\pi \cdot d^2 / 4}{d^2} = \frac{\pi}{4} = 0.7853 \tag{10.18}$$

For textile yarns and multifilament yarns, however, this determination is difficult, because the packing density of both yarn and windings also depends on the winding type, winding tension and the exact location on the bobbin. Some layers are more compressed than others, depending on the tension and properties of the yarn. If all these parameters have to be exactly determined, then the bobbin has to be made, embedded in resin, cut, measured, and so on; but then, once the bobbin is produced, the yarn length on it can be measured practically, so there is no real need for more theoretical calculations from that point.

The usual range for the packing density of yarns, which depends on their twist, is around 0.5–0.8, with the more twisted yarns having the higher packing density. For winding, the packing density varies between 0.3 and 0.9, and so the final mass density can vary between 0.3 and 1.0 g/cm^3 or more for textile yarns, and up to 5 g/cm^3 for steel or other metal yarns.

Once the mass of the yarn on the bobbin is known, the yarn length can be determined using the definition for linear density in tex:

$$L_{\text{Yarn}[\text{km}]} \cdot \text{tex} = M_{[g]} \tag{10.19}$$

from which

$$L_{\text{Yarn[km]}} = \frac{M_{[g]}}{\text{tex}} \qquad [10.20]$$

The yarn length is required for the braiders to be able to estimate the length of the braid without knots or the time available for braiding until the bobbins are emptied.

Example: A bobbin with diameter 42 × 110 mm is used for polyester yarn with fineness 110 dtex (100 den.). The empty bobbin has diameter 17 mm.

The volume of the wound package is:

$$V = \frac{\pi\left(D_W^2 - D_H^2\right)H}{4} = \frac{\pi\left(4.2^2 - 1.7^2\right)\cdot 11}{4} = 127\,\text{cm}^3 \qquad [10.21]$$

If a wound package density of 0.8 g/cm³ is assumed (polyester has density 1.38 g/cm³, but considering the packing density, it will be reduced), then the mass of the material on the bobbin will be:

$$M = 127\ \text{cm}^3 \cdot 0.8\ \text{g/cm}^3 = 101.6\ \text{g} \qquad [10.22]$$

The approximate yarn length on this bobbin will then be:

$$L_{\text{Yarn[km]}} = \frac{M_{[g]}}{\text{tex}} = \frac{101.6}{11} = 9.236\ \text{km} \qquad [10.23]$$

To obtain reasonable values from such calculations, it is recommended that a table with bobbin densities for common materials is prepared experimentally on the machines to be used.

10.7 Sources of further information and advice

This chapter does not cover the chemical treatment aspects of yarn preparation, because, for modern materials these treatments are usually commercial secrets.

One of the most detailed explanations of winding, with theoretical background and calculations, can be found in Simon and Hübner (2012), although it is written in German.

Details of the winding machines, creels and other equipment can be obtained from the following manufacturers:

Germany

August Herzog Maschinenfabrik GmbH & Co. KG, http://www.herzog-online.com
Körting Nachfolger Wilhelm Steeger GmbH & Co. KG http://www.steeger-online.com

Spain

Talleres Ratera, http://www.ratera.com

England

Cobra, http://www.cobrabraids.com/

Winding machines and creels are also manufactured by companies that do not make braiding machines, such as:

SSM Schärer Schweiter Mettler AG, Horgen, http://www.ssm.ch/
Jakob Müller AG, Frick, www.mueller-frick.com
both in Switzerland, and many others.
Specialists in control devices include:
Hans Schmidt & Co GmbH, http://www.hans-schmidt.com/
BTSR International S.p.A http://www.btsr.com/

All these references are given for orientation only and do not pretend to be an exhaustive list. The listed companies have been represented for a long time on the market, and have provided enough information on their web pages. For serious research, the reader is advised to visit the International Textile Machinery Exhibition (ITMA).

References

August Herzog (2013). *Product Catalogue* [online], August Herzog Textilmaschinenfabric GmbH & Ko KG, Oldenburg. Available from: www.herzog-online.com (Accessed 01.2013).

BTSR Yarn feeding control (2013). *Constant Tension Feeders* [online], www.btsr.com. Available from: http://www.texgilles.de/fournisseure_btsr.html (Accessed 2013).

Simon, L. and Hübner (2012). *Vorbereitungstechnik fur die weberei, wirkerei und strickerei.* [S.l.]: Springer.

Part Three

Specialist braided structures using maypole braiding technology

Spiral braiding

11

11.1 Introduction

Spiral braids are a special group of tubular (occasionally flat) braids with a type of interlacing between the yarns different from that of "normal" braids. They require specialized technology: spiral braids can be produced by machines configured with carriers that move continuously, on the principle of maypole braiding, or with carriers designed to stop and wait in special parking positions. As both these types of machines differ from the classical maypole braiding machines, and the spiral braided products themselves have specific properties and applications, this chapter focuses on spiral braiding technology.

11.2 Terminology

A spiral can be a planar or three-dimensional curve (The American Heritage® *Dictionary of the English Language*, 2013). The three-dimensional case, defined by the same source as "a three-dimensional curve that turns around an axis at a constant or continuously varying distance while moving parallel to the axis; a helix," is relevant here. Note that the term "helix," which is used in other contexts, for instance to describe the configuration of protein macromolecules, is not commonly used to describe braided structures.

A spiral braid normally consists of a single system of yarns, with axes that, for the purposes of description, can be assumed to have an idealized spiral form (Fig. 11.1a). Each yarn is interlaced with the neighbouring yarn(s), so that one winding of one yarn goes around one or more other yarns. This type of interlacing is, in its simplest form, identical to that of a chain-link fence (also known as wire netting, or wire-mesh fencing) (Fig. 11.1b). In tubular spiral braids, the yarns have the form of a double spiral, because in addition to the local windings required for neighbouring yarns to interlace with each other, each yarn also rotates around the axis of the rope, as shown in Fig. 11.1c.

Spiral braiding is commonly used for tubular ropes. As can be seen from Fig. 11.2, removing one or more yarns from a tubular braid creates an open structure, which will take the form of a flat tape. This is achieved in production by removing one or more carriers from the machine, depending on the pattern type.

Braiding Technology for Textiles. http://dx.doi.org/10.1533/9780857099211.3.257

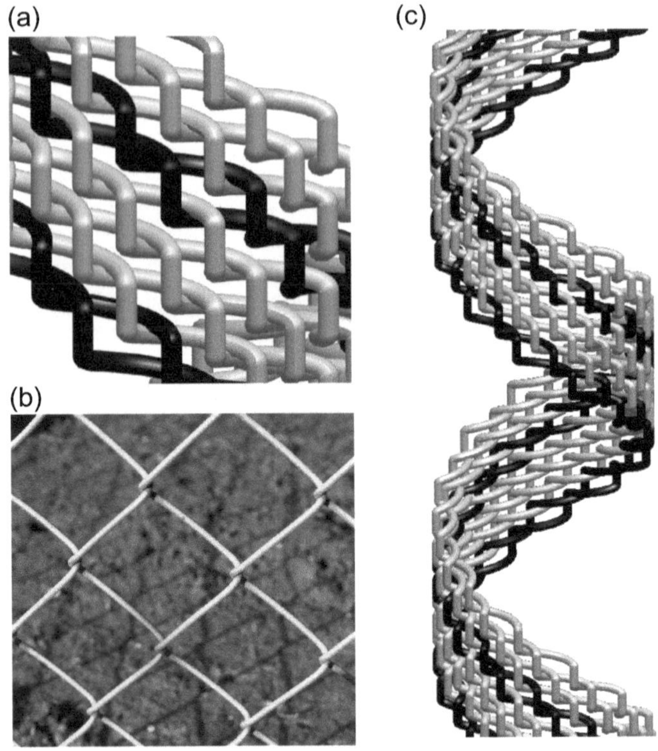

Figure 11.1 Yarn interlacement in a part of spiral braid (a), chain-link fence as analogy (b) and part of the yarns from tubular spiral braid (c) (Kyosev, 2012).

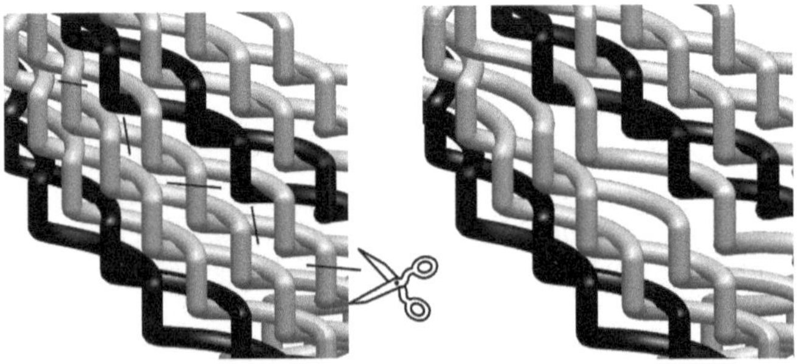

Figure 11.2 Removing one (or more, depending on the pattern type) yarn of the tubular spiral braid leads to flat spiral product.

11.3 Machine types for producing spiral braids

Machines can produce spiral braids with a variety of different working principles. According to Goseberg *et al.* (1981), two types of machine are used in industrial applications – continuous working maypole braiding machines with a special track, or specialized spiral braiding machines. It is also possible to produce spiral structures on pillow lace machines using a jacquard controller, which is discussed elsewhere.

11.3.1 Maypole braiding machines

The interlacing of two yarns with a single twist, to form a spiral, requires one carrier to move around another carrier. This motion can be achieved with a single horn gear attached to the side of a track, as shown in Fig. 11.3. This principle is applied in practice to every second horn gear in a circular arrangement of horn gears, as shown in Fig. 11.4, where the carrier track is represented by a solid line. The areas around the horn gears where carriers are not moving are marked with a dashed line. As can be seen in Fig. 11.4, only one track holds the carriers and all carriers move in one direction (here counter-clockwise). If all horn gears from the inner circle are excluded, the machine will only twist the yarns. During each rotation of two carriers around one inner horn gear, their yarns receive one twist and become interconnected.

The type of interlacing depends on the number of slots in the inner horn gears, though the scope for variation is very limited. Figure 11.4 shows three-slot horn gears, with each yarn floating over two and under two other yarns. If a floating length of 4 is required (where one yarn floats over four and under four others), then five-slot horn gears are used, as shown in Fig. 11.5. However, the space for inner horn gears is limited and the extension of the floating length corresponds to a reduction of interlacements, with only half the number of inner horn gears being available for this purpose. The limited space for inner horn gears leads to another problem for this type of machine – it is not possible to produce ropes with a low number of yarns. Such cases require the use of spiral braiding machines, which work on a slightly different principle.

11.3.2 Spiral braiding machines with carriers in hypocycloid motion

The spiral braiding machine is derived from the normal maypole braiding machine. As explained in the previous section, the interlacing of the yarns results from the

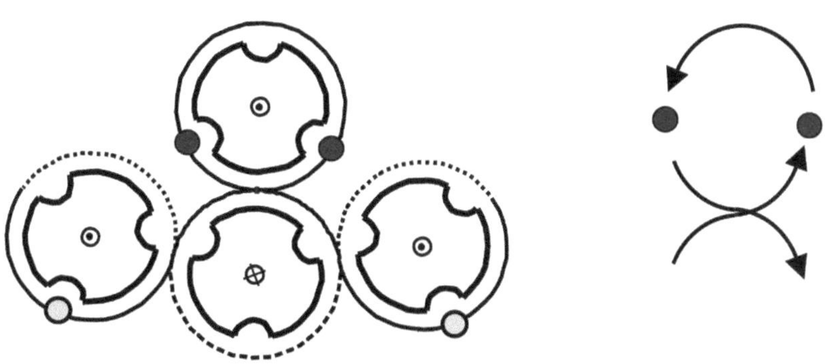

Figure 11.3 Placing additional horn gear on the side of some track causes twisting of the yarns of the two carriers, which pass this horn gear at the same time.

Figure 11.4 Tubular braiding machine for production of spiral braids (author's own presentation, following Goseberg *et al.*, 1981)

motion of the carriers around the inner horn gears. The outer horn gears are required only to move the carriers forward; they are not engaged in the interlacing itself and can be removed altogether from the machine, if some other mechanism is able to move the carriers forward between the inner horn gears.

As shown in Fig. 11.6, all external horn gears can be replaced by one large external horn gear with negative sections that rotate and move the carriers along the track from one inner horn gear to the next. This is the basic arrangement of elements now commonly used for producing spiral braids, with one small modification: the external horn gear is fixed and so, to generate the same relative motion, the axes of the inner horn gears are mounted on a rotating plate. This is a simpler and cheaper construction, because the large gear does not then have to be supported and driven. Spiral braiding machines are described in detail in the following sections.

Figure 11.5 Horn gear configuration and track for tubular spiral braiding with five-slot inner horn gears for producing of ropes with floating length of 4.

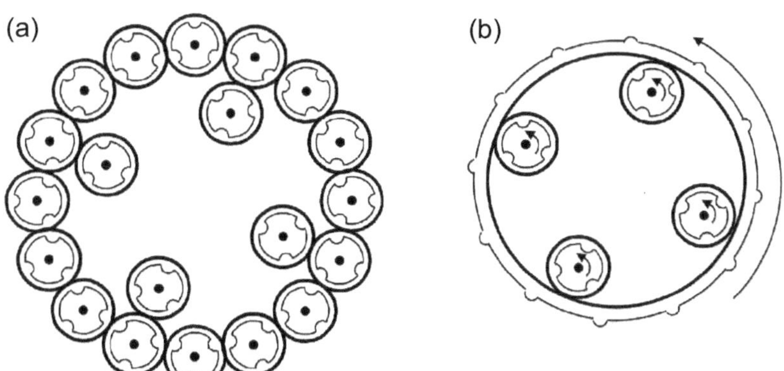

Figure 11.6 From (a) configuration with external horn gears to (b) configuration with one external horn gear.

11.4 Spiral braiding machines

11.4.1 Machine construction

The basic elements of a spiral braiding machine are presented in Fig. 11.7. The inner (small) horn gear 1 is supported on shaft 6, which is fixed on rotating plate 4. Channel T of the track is milled in track plate 3, which is also fixed on rotating plate 4. Track plate 3 thus rotates, together with rotating plate 4, around the main machine shaft 1. Inner horn gear 1 is connected with gear wheel 5. The cogs of wheel 5 are geared with the cogs of external gear 2 (at C). Because of this gearing, wheel 5 rotates around its axis with the motion of the rotating plate and in turn rotates inner horn gear 1. The carriers follow the track, being held from the slots S of the small and the large horn gears. External wheel 2 also has slots S.

The track for the carriers is shown in Fig. 11.8. The ideal configuration of the track consists of a channel in the form of circles around the small horn gears connected as smoothly as possible by circular arcs (Fig. 11.8a). This ideal configuration has one technical problem – the carrier "cannot decide" where to go when coming from the large arc to the next circle (Fig. 11.8b). The carrier should normally move forward to the next arc without making any changes in motion that involve small curvatures. To overcome this problem, the track is slightly modified – the large arcs are located at the outer part of the motion, which creates a more distinct crossing point (Fig. 11.8c): the carrier will direct its forward motion from the large arc into the small circle and vice versa (Fig. 11.8d). This arrangement leads of course to some dynamic disadvantages – any change in direction of the carriers creates inertial forces and thus greater abrasion at those points, more noise, and potentially vibrations in the carriers. For simplicity in the illustrations in this chapter, only the idealized form of the track is shown.

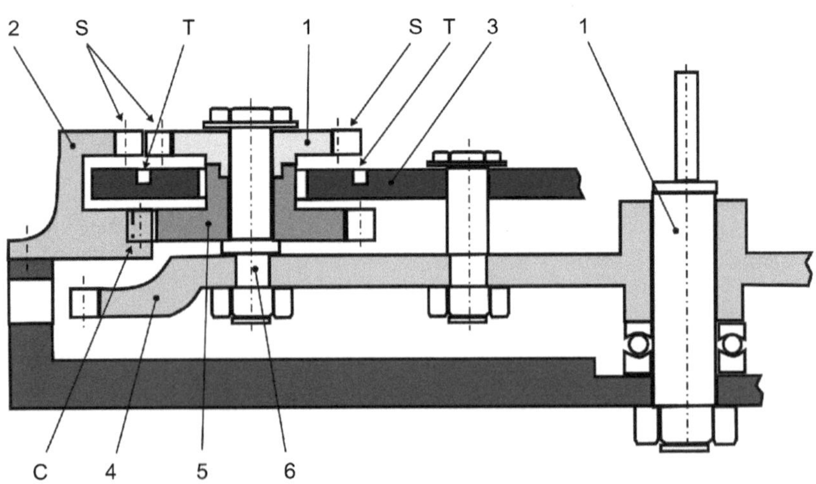

Figure 11.7 Basic configuration of a spiral braiding machine (following Goseberg *et al.*, 1981).

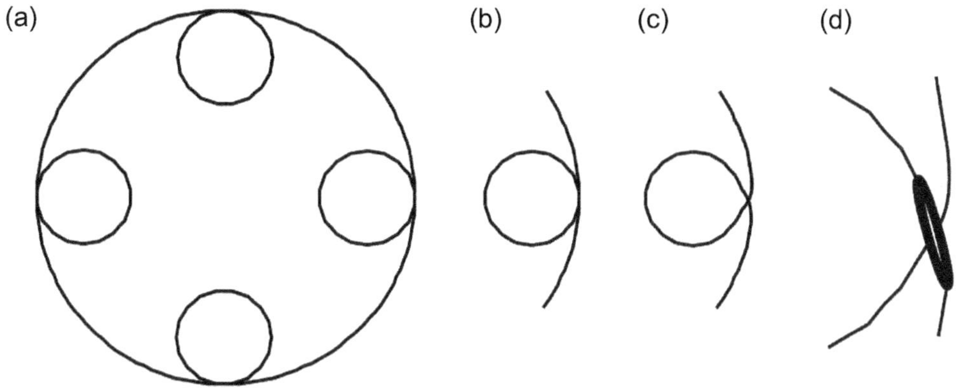

Figure 11.8 Track for a spiral braiding machine. (a) ideal track configuration for four horn gears (b) the part of the track around one horn gear – ideal configuration (c) real configuration – the track is slightly changed, so that the leading slider of the carrier can definitely select the only proper way (d) magnified configuration of (c) with the carrier slider.

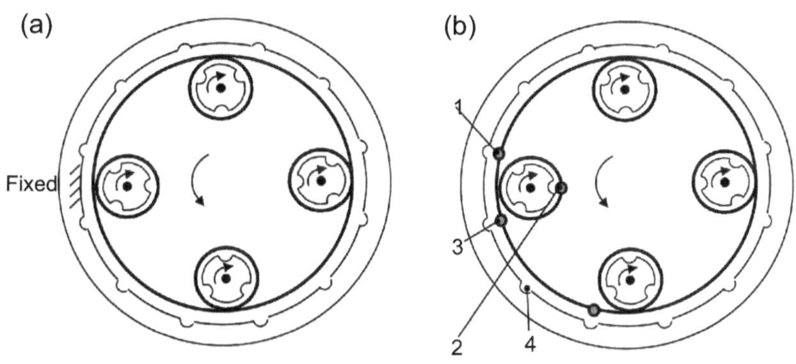

Figure 11.9 Track and carrier motion for spiral braids (a) schematic view of the horn gears with their slots and the track of the spiral braiding machine (b) carriers around one horn gear.

Figure 11.9a is a schematic view of the machine from its upper side. The track is represented by a thick, solid line. The shafts of the horn gears and the track plate rotate counter-clockwise; as result of the gearing between the wheels (Fig. 11.7c), the horn gears rotate clockwise. In the figure, the outer horn gear is fixed and does not move. The carriers are held in the slots of the outer horn gear and therefore also do not move; they stay there in "parking" positions until the track leads them to the working area of the next inner horn gear. In Fig. 11.9b, carrier 1 is already in parking position. Carrier 2 is turning around the inner horn gear, interlacing its yarn with that of carrier 3, which is also in parking position. Carrier 3 will next move on to the track around the horn gear and start turning around it. Position 4 in the figure is empty, without any carrier.

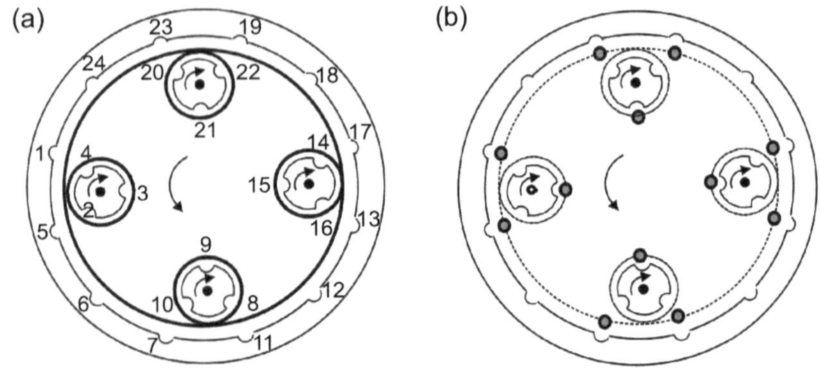

Figure 11.10 Track and carrier positions on a spiral braiding machine. (a) numbering of the positions; (b) carrier arrangement 1 Full 1 Empty.

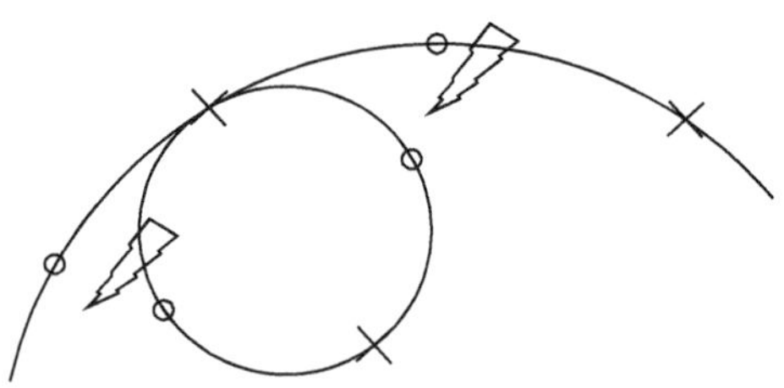

Figure 11.11 Internal horn gear with four slots is not suitable for spiral braiding machine because the carriers will be transferred to "full" position, which is not possible.

11.4.2 Carrier arrangement

The spiral braiding machine is derived from the normal maypole braiding machine and therefore obeys most of the same rules concerning carrier arrangement. Figure 11.10a demonstrates how positions are counted on the track of a spiral braiding machine with four three-slot horn gears and 12 slots on the outer horn gear. Note that this counting of positions is more difficult on a real machine, because the slots of the inner and outer gears meet and there is a carrier at most of these points (for example, it could be between positions 1 and 4, or 5 and 2, in this figure). The schematic drawing in Fig. 11.10a shows better the case of two positions on the track that have to be counted separately. After the carriers are arranged on the track, it has to be checked whether any one carrier position will meet any other carrier position with carriers at both of them, because in that case the machine would crash. Figure 11.10b shows a "1 Full 1 Empty" carrier arrangement, which is the commonly used configuration for spiral braids.

Spiral braiding machines are known to have three, five, and seven slots in the internal horn gears. Horn gears with four slots are not usable for the 1 Full 1 Empty carrier arrangement, because the carriers will collide, as shown in Fig. 11.11. An example of the 1 Full 1 Empty arrangement on a machine with seven-slot horn gears is presented in Fig. 11.12. Horn gears with six slots can be used for spiral braiding with a 2 Full 2 Empty carrier arrangement, using the same principles that apply to the 1 Full

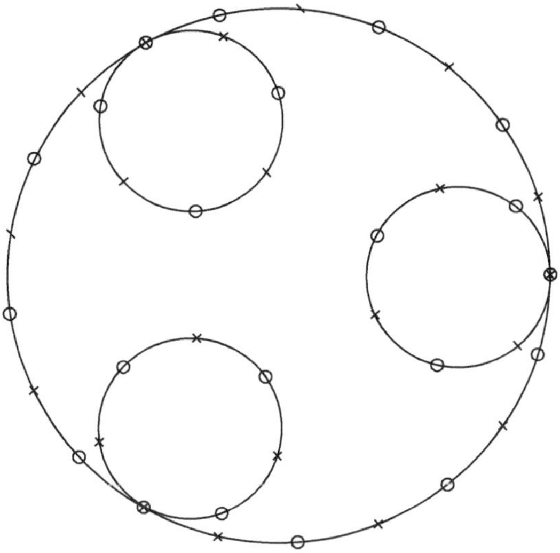

Figure 11.12 Carrier arrangement "1 Full 1 Empty" on a machine with seven-slot horn gears.

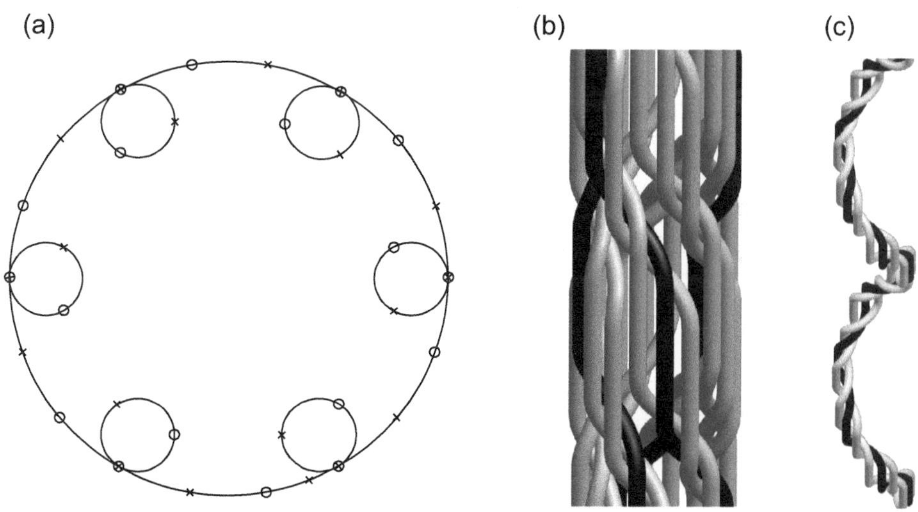

Figure 11.13 Arrangement of the spiral braiding machine where the yarns will be not interconnected (a) carrier arrangement "3 Full 3 Empty" (b) 3D presentation of the product (c) only one group of three yarns from this product.

1 Empty arrangement with three-slot horn gears but with the number of carriers and slots doubled.

There are some carrier arrangements which the machine will support but which do not lead to the production of rope. If the 3 Full 3 Empty arrangement is used in a spiral braiding machine with three-slot horn gears, then the carriers work in groups of three. Each yarn group is twisted once at each revolution around the inner horn gears, but the groups themselves are not connected. The arrangement is illustrated in Fig. 11.13. The absence of a connection between the triple groups of yarns can be seen in Fig. 11.13b, and the path of the yarns within a single group is shown (with the other yarns hidden) in Fig. 11.13c.

11.5 Equation of motion of the carriers

In classical spiral braiding machines, the large horn gear is fixed and does not move. The smaller, inner horn gears rotate inside it. This section will focus on the smaller, inner horn gears and will refer to them simply as the "horn gears."

The mechanical connection between these horn gears and the large horn gear (Fig. 11.7c) prevents any kind of slippage during motion. Figure 11.14 represents the track around one horn gear as a circle $k_1(C,r)$ with centre C and radius r, and the track inside the large horn gear as a circle $k_2(O,R)$ with centre O and radius R. During the motion of the horn gear, the small circle rolls along the inner side of the large circle, and because there is no slippage, the length of the arc AB of the large circle is equal to the length of the arc $A'B'$ of the small circle (Fig. 11.14).

After one revolution of the (small) horn gear, point A has to coincide with a carrier parking position, and after k revolutions point A has to return to its original starting place, which means that the circumference of the large circle has to be k times larger than the circumference of the smaller circle:

$$2\cdot\pi\cdot R = k\cdot 2\cdot\pi\cdot r \qquad [11.1]$$

The radii of the larger and smaller circles, or equivalently the large horn gear (external plate) and the horn gear, are thus related as

$$R = k\cdot r \qquad [11.2]$$

The whole number k is the number of revolutions that the rotating gear makes for one revolution of the central plate, and this determines the type of braided structure that will result. The motion of a point on the small circle, for instance A, will trace out a type of curve named a hypocycloid (described below).

The position of the centre of the horn gear, C', is described by the angle φ (Fig. 11.15):

$$\varphi = \omega\cdot t \qquad [11.3]$$

where ω is the angular velocity of the rotating plate. While the small circle is in motion, it also rotates around its own centre C' with angle α. Because there is no slippage, the arcs AB of the outer circle and $A'B'$ of the rotating circle have the same length, which can be expressed in terms of their radii:

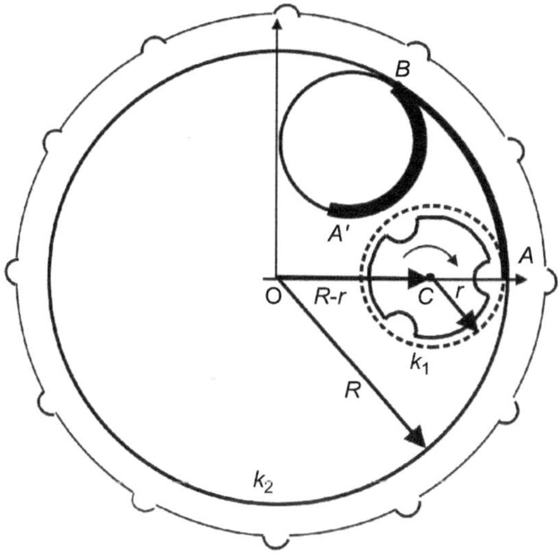

Figure 11.14 Geometrical relations for the horn gears in a spiral braiding machine.

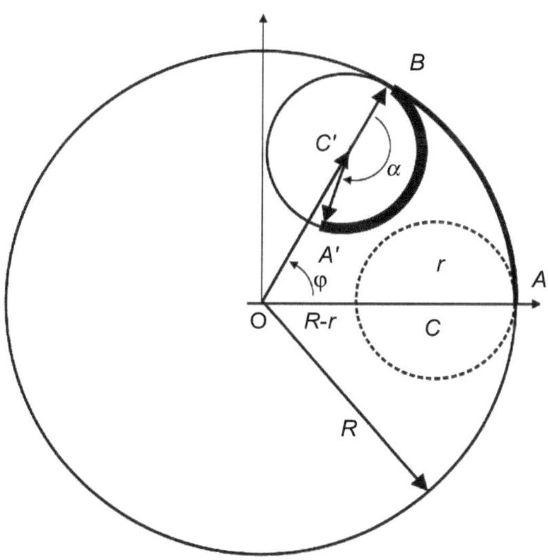

Figure 11.15 Geometrical relations for deriving the equation of motion of one carrier.

$$R \cdot \varphi = r \cdot \alpha \qquad [11.4]$$

from which, considering [11.2], it follows that:

$$\alpha = \frac{R}{r} \cdot \varphi = k \cdot \varphi \qquad [11.5]$$

The equation of motion of a point *P* can be derived from the position of the vector *OA*, presented as a vector sum of both *OC* and *CA*:

$$\overrightarrow{OA} = \overrightarrow{OC} + \overrightarrow{CA} \quad [11.6]$$

The projection (parametric equations) in a rectangular coordinate system will be:

$$x_A = (R - r) \cdot \cos\varphi + r \cdot \cos(\varphi - \alpha) \quad [11.7]$$

$$y_A = (R - r) \cdot \sin\varphi + r \cdot \sin(\varphi - \alpha) \quad [11.8]$$

or, again considering [11.2]:

$$x_A(\varphi) = (k-1) \cdot r \cdot \cos\varphi + r \cdot \cos((1-k) \cdot \varphi) \quad [11.9]$$

$$y_A(\varphi) = (k-1) \cdot r \cdot \sin\varphi + r \cdot \sin((1-k) \cdot \varphi) \quad [11.10]$$

or finally:

$$x_A(\varphi) = (k-1) \cdot r \cdot \cos\varphi + r \cdot \cos((k-1) \cdot \varphi) \quad [11.11]$$

$$y_A(\varphi) = (k-1) \cdot r \cdot \sin\varphi - r \cdot \sin((k-1) \cdot \varphi) \quad [11.12]$$

As Equations [11.11] and [11.12] demonstrate, the ratio k is directly connected with several mathematical and mechanical characteristics of carrier motion. If k is an integer (whole number), then the curve is closed and has k sharp corners (cusps), where the carriers have their parking positions. For spiral braiding machines it is usually the case that $k = 3$; 4; or 6. For the special case of $k = 2$ the curve becomes a straight line and the circles are called Cardano circles (Wikipedia, 2013). If $k = M / N$ (a rational number), where M and N are integers with no common factors (that is, the smallest integers that can form the ratio), then the curve has M sharp corners; such cases are again common in spiral braiding machines. The case in which k is an irrational number has no practical application for spiral braiding, however, because then the curve never closes, and so a carrier will never return to its initial parking position. For all cases where $k > 0$, the resulting curve is called a hypocycloid (Toponogov, 2006, p. 4). Figure 11.16 shows the curves for $k = 2$, 3, 4, 5 and 6.

Figure 11.16 shows three variant curves, all with the same number of parking positions or cusps (11), but formed on the basis of different ratios between the parameters. The braided structures produced by machines with the corresponding configurations will have the same number of visible ridges on the outside, because of the parking positions, but a slightly different colour pattern and significantly different strength. In the first variant, where $k = 5.5$, the yarns are interlaced only near the surface. A

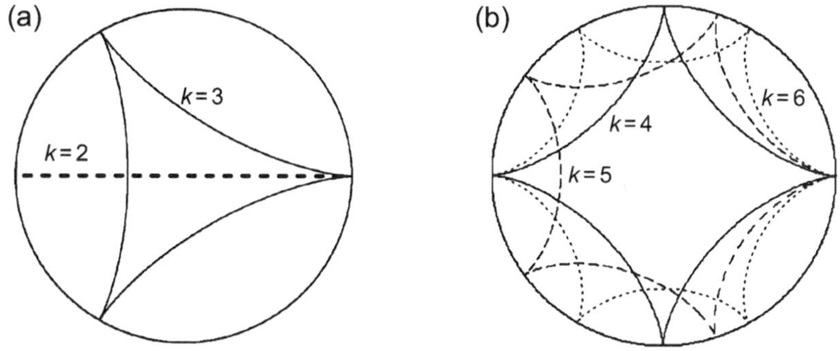

Figure 11.16 Hypocycloids as carriers paths (a) for $k = 2$ and 3 (b) $k = 4$, 5 and 6.

decrease in the ratio k corresponds to an increase in the number of yarns over which one yarn is floated, thus making the structure more compact and connected within its core.

11.6 Patterning basics for spiral braids

The braiding type for spiral braids depends on the carrier arrangement and on the number of the slots in the horn gear. The ratio k, described in the previous section, is defined in textile production as the ratio between the number of slots in the outer horn gear (that is, the fixed parking positions) and the number of slots in each small horn gear. Equation [11.2] can then be re-expressed as:

$$k = \frac{N_{\text{Slots\_outer\_horn\_gear}}}{N_{\text{Slots\_horn\_gear}}} \quad [11.13]$$

because the ratio of the radii of the gears is equal to the ratio of their circumferences. Equivalently, the lengths of the arcs between adjacent slots on both the outer and the small horn gears have to be equal. As shown in Fig. 11.17, the ratio k determines the "hardness" or "strength" of the interconnection between the yarns in a spiral braid rope. The whole (integer) part of k determines how many different yarns appear on the surface of the rope for one revolution around it, or equivalently how many interlacing points one yarn will have with the other yarns. Values of $k \approx 2$ mean that the yarn intersects the rope almost diagonally between two interlacements. Values of k around 5 mean that the yarn makes five interlacements in one revolution of the rope. The higher the value of k, the greater the number of interlacing points that one yarn will have around the surface of the rope and the less compact the rope will be. Figure 11.18 illustrates the influence of the ratio k on the compactness of the rope with two simulated examples of spiral braids, one where $k = 3$ (nine outer slots, three slots per horn gear, three horn gears) and one where $k = 8$ (24 outer slots, three slots per horn gear, eight horn gears).

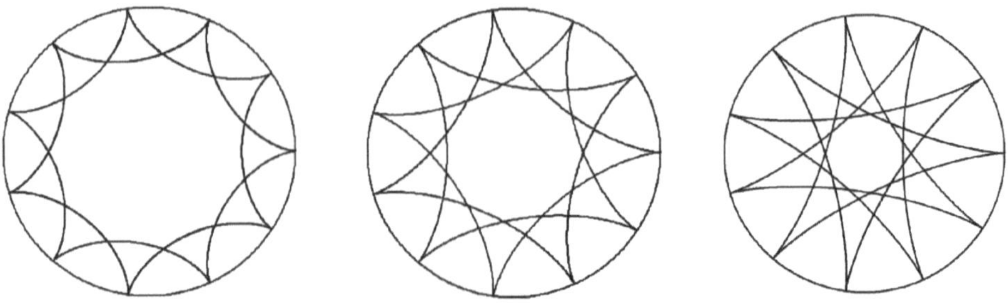

$k = 5.5 = 11/2$ $k = 3.667 = 11/3$ $k = 2.75 = 11/4$

Figure 11.17 Hypocycloids as carrier paths with 11 parking positions, but with different paths.

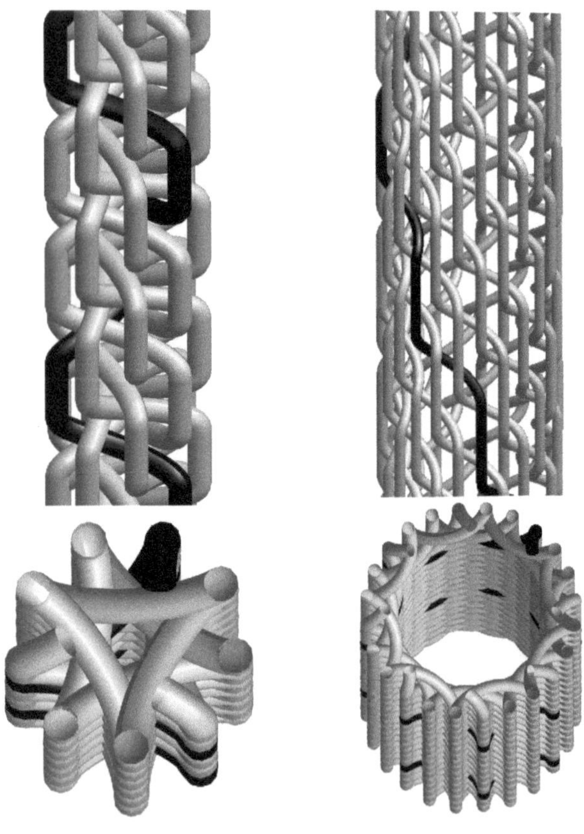

$k = 3$ $k = 8$

Figure 11.18 Spiral braids produced on a machine with $k = 3$ and $k = 8$ (Kyosev, 2012). The number of horn gears in the both cases is set to be equal to k.

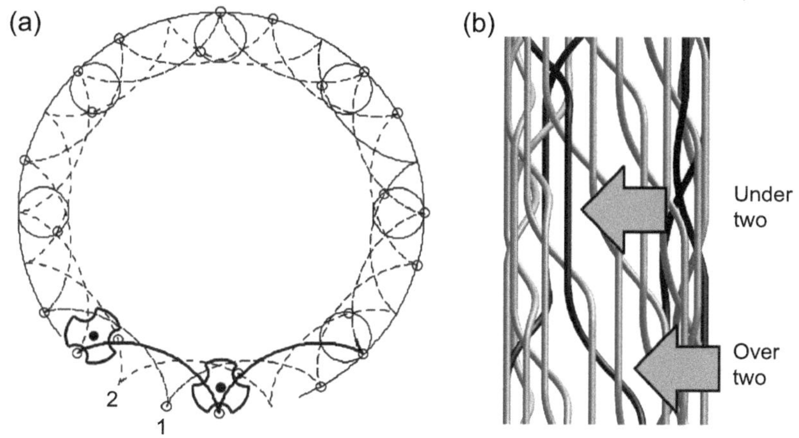

Figure 11.19 Spiral braiding machine with three-slot horn gears for producing ropes with floating length of 2. (a) One carrier moves over two carriers between the two parking positions. (b) The black yarn floats under two and then over two yarns.

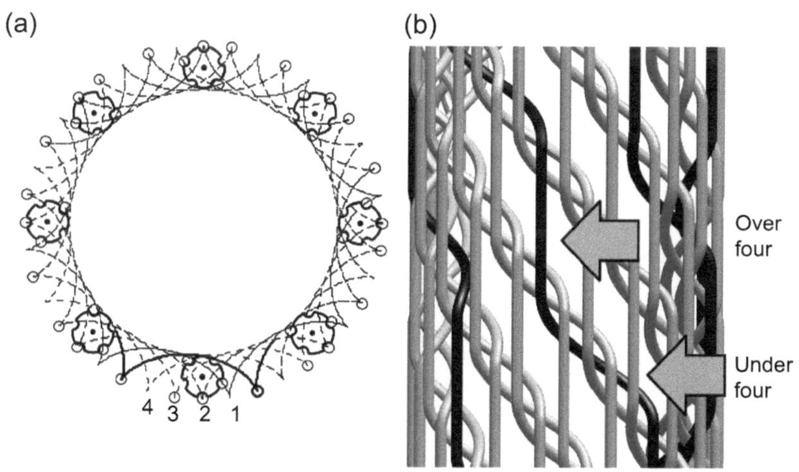

Figure 11.20 Spiral braiding machine with five-slot horn gears for producing of ropes with floating length of 4. (a) One carrier moves over four carriers between the two parking positions. (b) The black yarn floats under four and after then over four yarns (Kyosev, 2012).

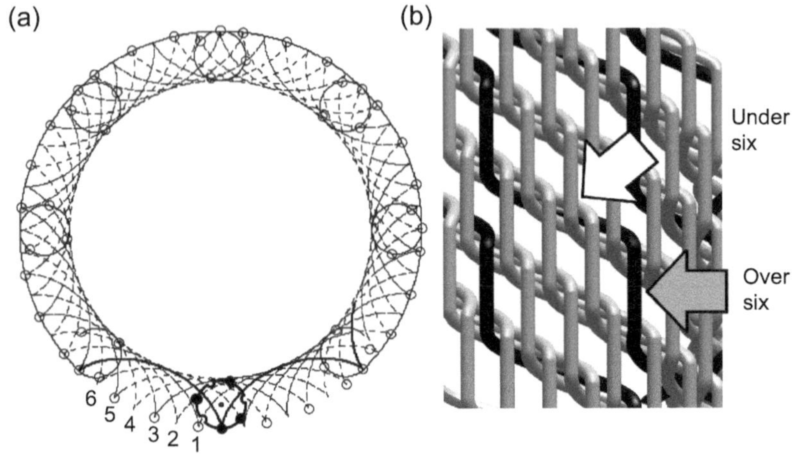

Figure 11.21 Spiral braiding machine with seven-slot horn gears for producing of ropes with floating length of 6. (a) One carrier moves over six carriers between the two parking positions. (b) The black yarn floats under six and then over six yarns (Kyosev 2012).

Assuming the most common case of carrier arrangement, namely 1 Full 1 Empty, then floating length is equal to the number of slots in each inner horn gear minus one:

$$\text{Floating\_length} = N_{\text{Slots\_horn\_gear}} - 1 \quad [11.14]$$

This rule derives from the motion of the carriers and the horn gears. Three-slot horn gears transfer each carrier from one parking position to another by moving them over two other carriers (Fig. 11.19a). During this motion the yarn of a moving carrier floats under two other yarns. While a carrier stays in a parking position, the horn gear moves two other carriers "over" it, in the course of which the yarn of the parked carrier floats over the yarns of the two moving carriers. In this way the yarn of each carrier floats under two and then over two other yarns. The floating length of this pattern is two. Because the horn gear has three slots, one of which takes the carrier being considered, only two other positions with carriers can be "operated," so the floating length is equal to the number of slots minus one. So, if horn gears with five slots are used, the floating length of the rope is 4 (Fig. 11.20), and for horn gears with seven slots, the floating length is 6 (Fig. 11.21).

11.7 Colour patterning of spiral braids

The number of the small gears determines the repeat of the pattern (Fig. 11.22). There are two kinds of ridges that can be recognized in the spiral braids – one

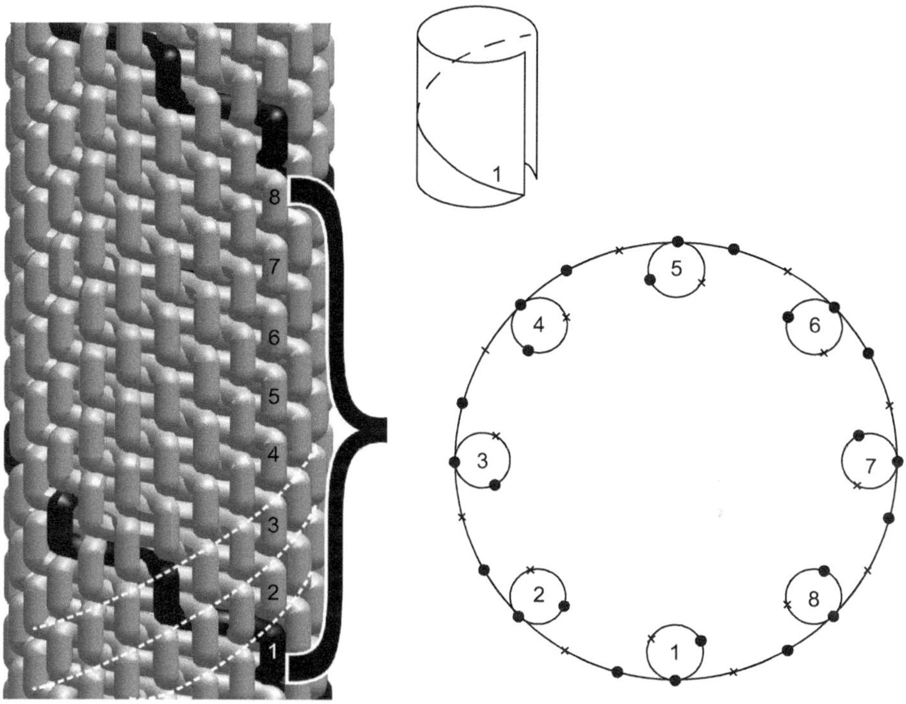

Figure 11.22 The number of the spiral ribs in one repeat corresponds to the number of the horn gears on the machine.

based on visible yarn parts which are arranged one after another following one axis and one based on visible yarn parts, which are parallel to each other (Fig. 11.23)

The rules for colour patterning of spiral braids are simpler to understand if the colour arrangement is described for each group of carriers rather than as a sequence (as is common for normal braiding). In Fig. 11.24a, the outer and the small horn gears are represented as circles, and the paths of the carriers are represented as lines. An idealized 3D view of the corresponding spiral rope is shown in Fig. 11.24b. Each longitudinal piece of yarn on the rope is taken up while its carrier stays in a parking position. In this example, the ratio between the radii of the large and small horn gears $k = 3$, and the curve traced by the motion of the carriers has three cusps. So each visible piece of yarn on the surface of the rope is taken up while a carrier is at a cusp of its path (in this case a deltoid).

Figure 11.25 shows the common variants for colour patterning of the same spiral braid. If only one carrier takes a yarn with a different colour (colour A is black, B is light grey), then this yarn will be visible in three places on the surface circumference of the rope (Fig. 11.25a). Adding a second yarn with the same colour on the same path leads to two pieces of yarn on one ridge (Fig. 11.25b). Putting different colours on

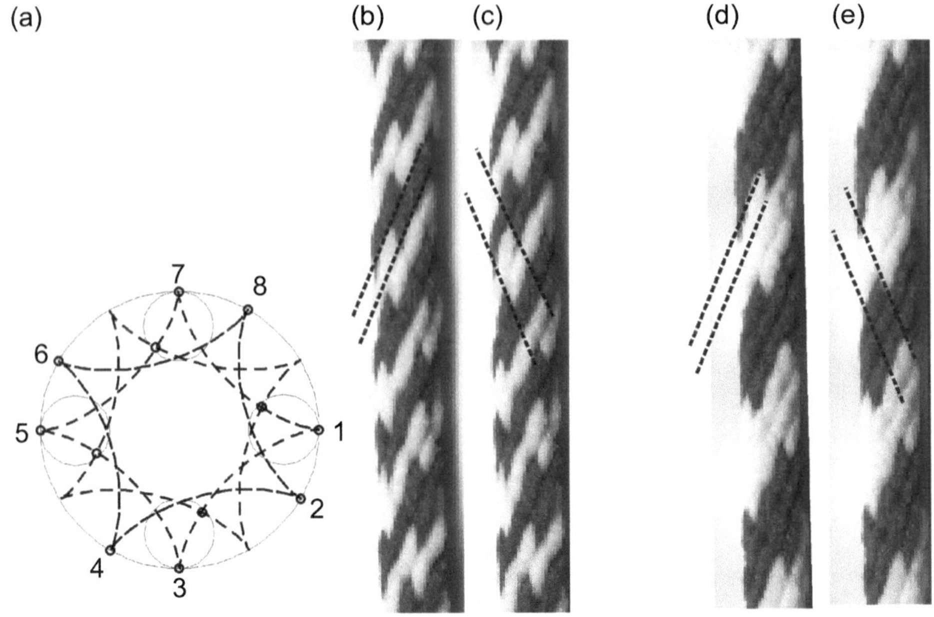

Figure 11.23 The number of the cliffy ribs in one repeat corresponds to the maximal number of the carriers which can stay on parking position (on the fixed horn gear) on the machine, according to Goseberg *et.al.* (1981). (a) Machine with four horn gears with their slots per horn gear have maximum 4*3 – 4 = 8 carriers in parking position; (b) and (d) cliffy ribs; (c) and (e) spiral ribs.

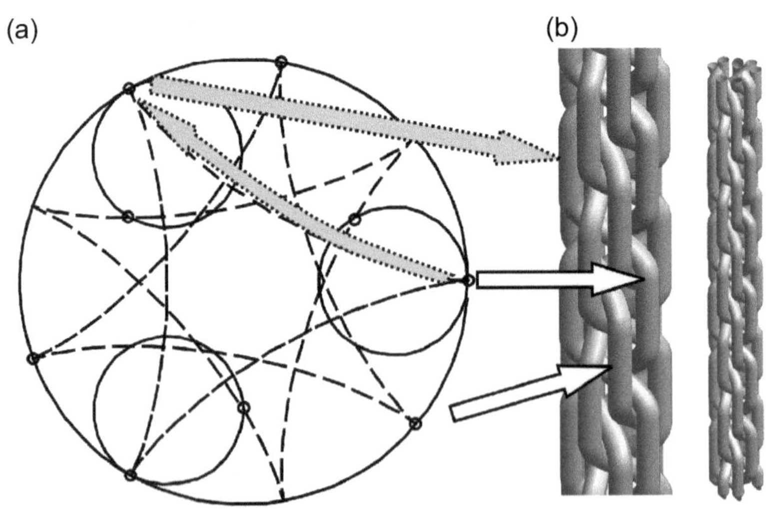

Figure 11.24 Relation between the path of the carriers (a) and the braided product (b). If one carrier stays in parking position, its yarn builds visible vertical piece.

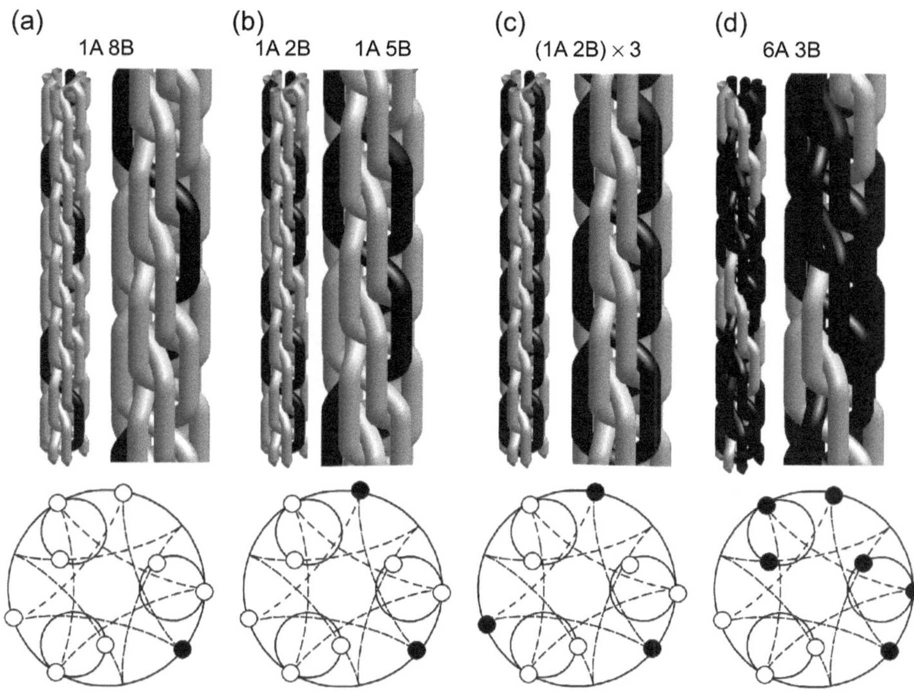

Figure 11.25 Colour pattern for spiral braids, produced on a machine with three horn gears with three slots per horn gear (Kyosev, 2012) The colour arrangement of the carriers is (a) 1 colour A and 8 colour B; (b) 1A 2B 1A 5B; (c) three times (1A 2B); (d) 6A 3B.

all three carriers on the same path (Fig. 11.25c) leads to three continuous lines with this colour appearing on the rope surface. Between these lines, the other two sets of carriers will place two white lines.

Spiral effects can also be produced on the rope surface. Figure 11.25d gives an example in which two groups of carriers have black yarns and one group has light grey. Each of these groups of yarns builds spiral stripes on the surface of the rope, and the black stripe is thicker in this case because it is formed of two groups of yarns.

Machines with four horn gears and three slots allow more combinations of colours, but the range of pattern elements is almost the same – single dashes, vertical lines or spiral effects. Figure 11.26 shows the basic pattern, and Fig. 11.27 shows variants with spiral effects. It should be mentioned that the 3A 3B arrangement – all the carriers around one horn gear having one colour and all those around the other horn gear having the other colour – also produces a spiral stripe (Fig. 11.27c), but this is not as clear as when alternating colours are used (Fig. 11.27b). For patterns with floating length of 4, horn gears with five slots are used (Fig. 11.28). The same principle can be used to derive patterns with floating length of 6, produced on machines with seven-slot horn gears.

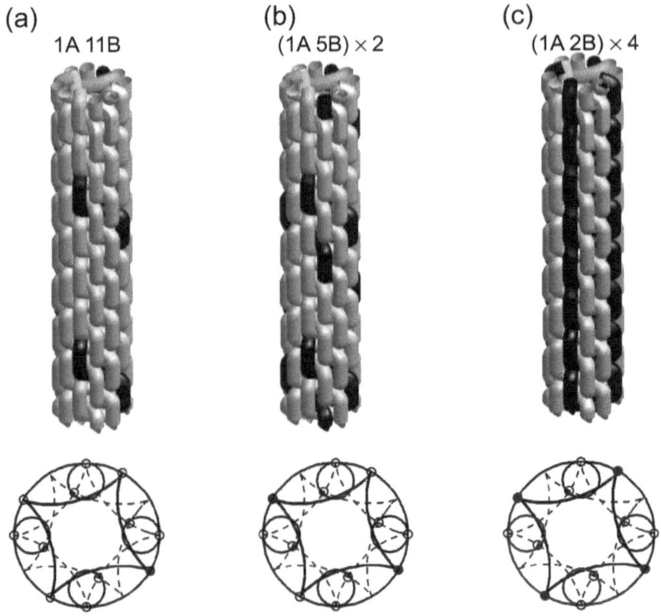

Figure 11.26 Colour patterns for spiral braids, produced on a machine with four horn gears with three slots per horn gear (Kyosev, 2012) Colour arrangement (a) 1A 11B, (b) (1A 5B) x 2; (c) (1A 2B) x4.

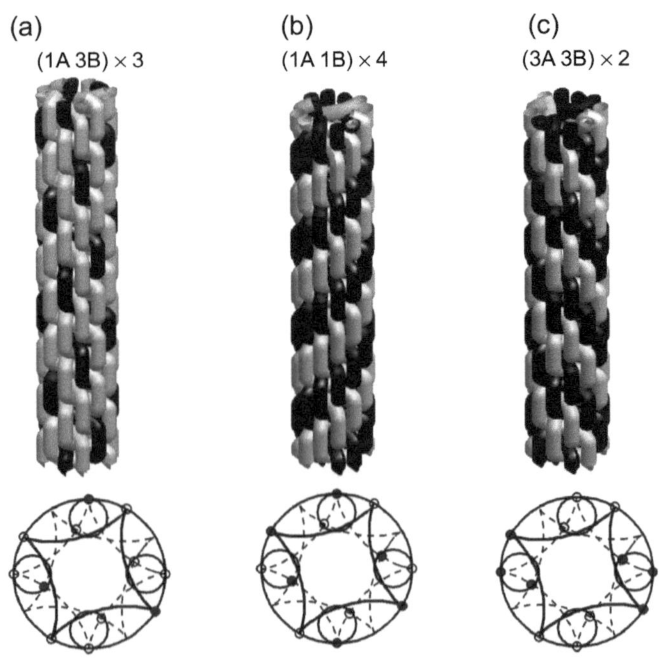

Figure 11.27 Colour patterns for spiral braids, produced on a machine with four horn gears with three slots per horn gear (continuation from Fig. 11.26) (Kyosev, 2012) Colour arrangement (a) (1A 3B) x 3; (b) (1A 1B) x 4 building Z stripes; (c) (3A 3B) x 2 building S stripes.

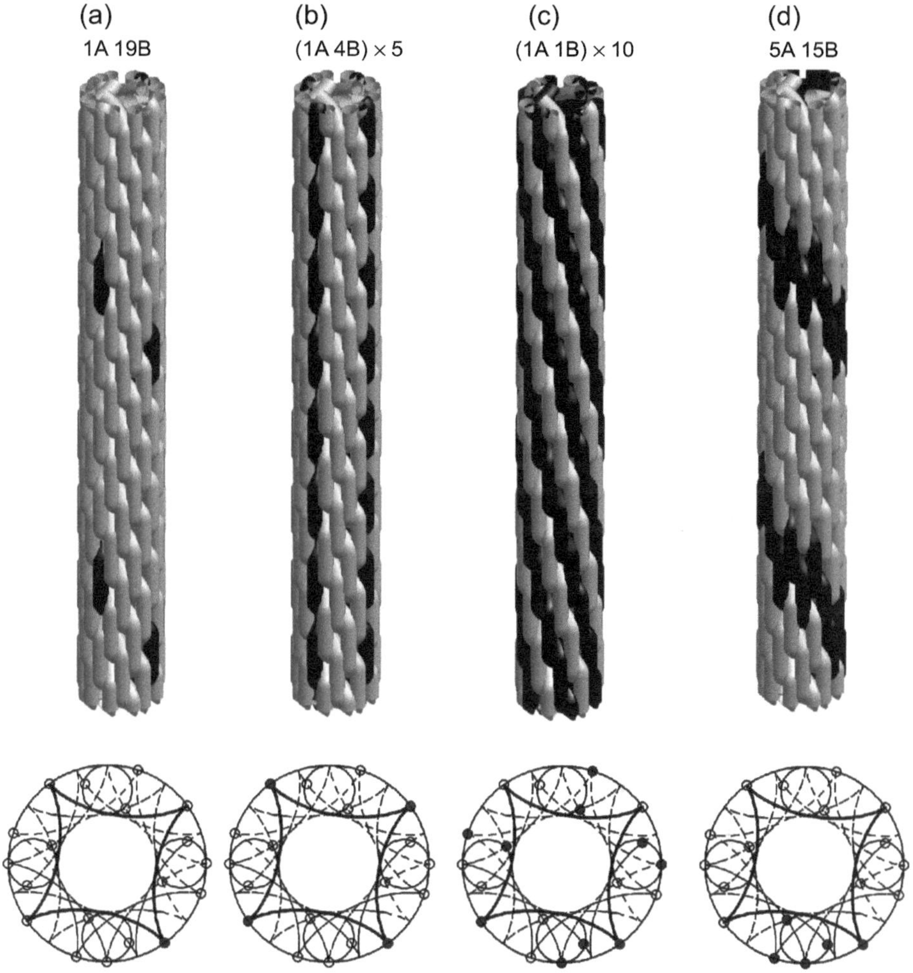

Figure 11.28 Colour patterns for spiral braids, produced on a machine with four horn gears with five slots per horn gear (Kyosev, 2012) Colour arangement (a) 1A 19B; (b) (1A 4B) x 5 building vertical stripes; (c) (1A 1B) x10 building spiral Z stripes; (d) 5A 15B building thicker Z stripes

11.8 Special properties and applications of spiral braids

Spiral braids have one very important advantage over normal braids, especially in their application as rope – yarn length is mostly orientated in the same direction as the loading force on the rope (Fig. 11.29a). The breaking force Q of the spiral braid is therefore proportional to the breaking force P of a single yarn, multiplied by the number of the yarns N:

$$Q_{\text{Spiral}} \approx N \cdot P \qquad [11.15]$$

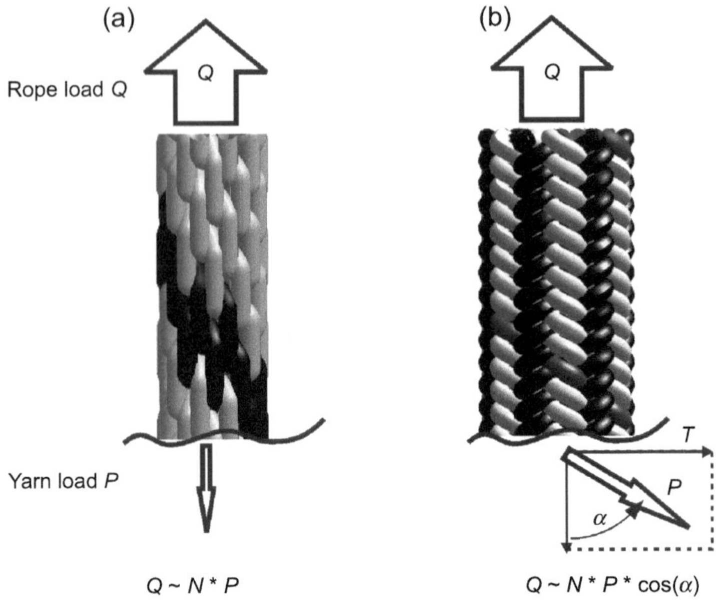

Figure 11.29 Large parts of yarn pieces in a spiral braid are orientated almost vertically, i.e. in the loading direction for the rope (a). Because of this, the spiral braids can take larger loads compared to the normal braids, where the yarns are under a certain angle to the loading direction (b).

By contrast, the breaking force of normal braids is significantly influenced by the braiding angle α, because the loading force on the rope is only a projection of the force on the yarns and is lower when the braiding angle is higher (Fig. 11.29b):

$$Q_{\text{NormalBraid}} \approx N \cdot P \cdot \cos(\alpha) \quad [11.16]$$

The braiding angle is normally in the range 20–40°. From Equation [11.16] it is easy to calculate the difference between the breaking forces of a normal tubular rope and a spiral rope, produced with the same number and type of yarns. For a braiding angle of 45° the difference is 30%! This means that spiral braiding ropes can be up to 30% thinner and lighter than normal braided ropes, with the same breaking strength. A climber therefore has 30% less weight of rope to carry, and several applications in industry benefit from more compact machinery and 30% smaller openings.

11.9 Machines

Spiral braiding machines are produced in similar configurations to those of normal braiding machines – they need to have their main plates, take-off devices, winding

Figure 11.30 View from a spiral braiding machine with three horn gears with three slots each. *Source*: Courtesy of Company Körting Nachfolger Wilhelm Steeger GmbH (Steeger, 2012).

units, etc. An example of such a standard machine is the two-head spiral braiding machine type SP 2/1/12–90 of the August Herzog company, Oldenburg, Germany, which has four horn gears each with three slots. The Steeger Körting Nachfolger Wilhelm Steeger company, Wuppertal, Germany, produces spiral braiding machines as well as single machines, as in the variant with replaceable braiding heads, which can be mounted on a table with a common drive for braiding heads of different types. Figure 11.30 shows such an arrangement with three horn gears, each with three slots.

Table 11.1 lists some possible configurations of spiral braiding machines, according to Goseberg *et al.* (1981). Only a small fraction of the configurations theoretically possible is realized in the standard machines normally built by manufacturers. Machines equipped with horn gears of five or seven (or more) slots can be used with partial occupation of the carriers to produce ropes that would normally be produced on smaller machines. The columns in Table 11.1 marked "full" list the parameters of the braids produced with all possible carrier positions filled (full occupation), and the columns marked "part" (partial occupation) list cases in which some of the carriers are removed.

11.10 Conclusions

Spiral braiding is a special technology for rope production, with advantages and disadvantages. The key advantage, which keeps the technology alive, is the structure of spiral braiding such that the yarns are orientated mainly in the direction of the loading force on the rope. This orientation is why spiral braided ropes can be up to

Table 11.1 **List of some possible machine configurations for spiral braiding machines, following Goseberg *et.al.* (1981)**

Slots per horn gear	Number of horn gears	Slots on the large horn gear	Machine/product number		Floating length		Number of cliffy ribs		Number of spiral ribs	Partical occupation arrangement
			Full	Part	Full	Part	Full	Part		(Carrier in the machine: Carrier out)
3	3	9	**9**		**2**		**6**		**3**	
3	3	11	**10**		**2**		**6**		**3**	
3	3	15	**12**		**2**		**6**		**3**	
3	4	12	**12**		**2**		**8**		**4**	
3	4	14	**13**		**2**		**8**		**4**	
3	4	16	**14**		**2**		**8**		**4**	
3	4	18	**15**		**2**		**8**		**4**	
3	4	20	**16**		**2**		**8**		**4**	
3	4	24	**18**		**2**		**8**		**4**	
3	5	15	**15**		**2**		**10**		**5**	
3	5	17	**16**		**2**		**10**		**5**	
3	5	19	**17**		**2**		**10**		**5**	
3	5	21	**18**		**2**		**10**		**5**	
3	5	25	**20**		**2**		**10**		**5**	
3	6	12	**15**		**2**		**12**		**6**	
3	6	14	**16**		**2**		**12**		**6**	
3	6	16	**17**		**2**		**12**		**6**	
3	6	18	**18**		**2**		**12**		**6**	
3	6	20	**19**		**2**		**12**		**6**	
3	6	22	**20**		**2**		**12**		**6**	
3	6	24	**21**		**2**		**12**		**6**	

3	7	21	**21**		**2**		**14**		**7**	
3	8	16	**20**		**2**		**16**		**8**	
3	8	24	**24**		**2**		**16**		**8**	
5	3	15	**15**	10	**4**	2	**9**	7	**3**	2:1
5	3	17	**16**	8	**4**	2	**10**	5	**3**	1:1
5	3	21	**18**	9	**4**	2	**12**	6	**3**	1:1
5	4	16	**18**	9	**4**	2	**10**	5	**4**	1:1
5	4	20	**20**	10	**4**	2	**12**	6	**4**	1:1
5	4	24	**22**	11	**4**	2	**14**	7	**4**	1:1
5	4	28	**24**	16	**4**	2	**16**	12	**4**	2:1
5	4	28	**24**	12	**4**	2	**16**	8	**4**	1:1
5	5	25	**25**		**4**		**15**		**5**	
5	6	24	**27**	18	**4**	2	**15**	12	**6**	2:1
7	3	21	**21**	14	**6**	4	**12**	8	**3**	2:1
7	3	21	**21**	7	**6**	2	**12**	4	**3**	1:2
7	3	27	**24**	16	**6**	4	**15**	10	**3**	2:1
7	3	27	**24**	12	**6**	2	**15**	6	**3**	1:1
7	3	27	**24**	8	**6**	2	**15**	4	**3**	1:2
7	3	33	**27**	18	**6**	4	**18**	12	**3**	2:1
7	3	33	**27**	9	**6**	2	**18**	6	**3**	1:2
7	3	39	**30**	20	**6**	4	**21**	14	**3**	2:1
7	3	39	**30**	15	**6**	2	**21**	11	**3**	1:1
7	3	39	**30**	10	**6**	2	**21**	7	**3**	1:2
7	4	24	**26**	13	**6**	2	**14**	8	**4**	1:1
7	4	28	**28**	14	**6**	2	**16**	10	**4**	1:1
7	4	32	**30**	20	**6**	4	**18**	12	**4**	2:1
7	4	32	**30**	15	**6**	2	**18**	10	**4**	1:1
7	4	32	**30**	10	**6**	2	**18**	6	**4**	1:2

The columns, marked "Part" (= partial occupation) are for the case where a part of the carriers is removed according the rule given in the last column.

30% thinner and lighter than normal braided ropes with the same breaking force. The disadvantage of the technology is the non-continuous motion of the carriers in the machine – having moved around the horn gears, the carriers pause in parking positions. The stop–start nature of this motion has detrimental effects on the dynamics of the machine and reduces production speed. With limited patterning possibilities and structural advantages, the future of spiral braiding will probably remain for the most part in the production of rope, where colour and patterning are of secondary concern for the user, and highest importance is assigned to breaking force and compactness.

11.11 Sources of further information and advice

The author of this book has not found any source in which spiral braiding is described, except for 12 pages (pp. 167–179) in Goseberg *et al.* (1981) and its slightly updated version in Engels (1994), both written in German.

References

The American Heritage(R) Dictionary of the English Language (2013). Available online at http://ahdictionary.com/word/search.html?q=spiral&submit.x=60&submit.y=14, checked on 01.2013.

Engels, H. (1994): Handbuch der Schmaltextilien. *Die Flechttechnologie. Teil I: Maschinen und Verfahren zur Erzeugung konventioneller Geflechte*. Mönchengladbach.

Goseberg, F., Brabender, K. and Moeller, P. (1981): Textiltechnik Maschinengeflechte. *Ausbildungsmittel Unterrichtshilfen*. Frankfurt am Mai: Arbeitgeberkreis Gesamttextil.

Kyosev, Y.K. (2012): TexMind Spiral Braider. Version 1.1: TexMind UG. Available online at www.texmind.com, checked on 2012.

Toponogov, V.A. (2006): *Differential Geometry of Curves and Surfaces. A Concise Guide*. Boston: Birkhäuser.

Wikipedia (2013): Hypocycloid. Available online at http://en.wikipedia.org/wiki/Hypocycloid, checked on 01.2013.

– August Herzog Maschinenfabrik GmbH & Co. KG, Am Alexanderhaus 160, D-26127 Oldenburg, http://www.herzog-online.com/_rubric/index.php?rubric=Products+Rope-braiding-machines+EN last access (02.2013).

– Körting Nachfolger Wilhelm Steeger GmbH & Co. KG, Wittensteinstrasse 208 – 212, 42283 Wuppertal, Germany, http://www.steeger-online.com/deutsch/programm_packungen_spiral.php last access (02.2013).

Square and other types of form braiding

12

12.1 Introduction

This chapter presents basic information on the production of braids with a solid square cross-section and with certain other, more complex forms using the maypole braiding technique, and on their application. Gaskets are a classic area of application for such braids, while recent ideas and research indicate an increasing interest in applying square braids as reinforcement in composite beam structures.

The following sections introduce the types of square braids used for gaskets, covering two-, three-, and four-track systems with different carrier arrangements. New developments in horn gears with variable tracks are then discussed and some of these tracks are described. Finally, there is an overview of some ideas, most of them patented, concerning the production of more complex maypole braiding structures.

Excluded from this discussion are tubular and flat cross-sectional forms, while the technique of 3D maypole braiding (known as rotary or two-step braiding), because of its complexity, will be the subject of several chapters in a future publication.

12.2 Terminology

12.2.1 *Form braiding, 2½D braiding, packing braiding, gasket braiding*

The machines and technique for producing special cross-sectional forms of maypole braiding are known by a variety of names in the literature (Engels, 1994a, b). Some researchers and machinery manufacturers use the term two-and-a-half-dimensional (2½D) braiding, because although such braids have a significant thickness to their cross-section, this thickness is less than that of three-dimensional (3D) braids. Other authors try to differentiate between solid braids and thin braids; gaskets are then solid braided structures. However, the value of the distinction is diminished because the same technique can also be used to produce thin structures, and even tubular and flat braids. A similar argument can be used against the usefulness of the term "packing" braiding. Further confusion arises from use of the terms 2D, 3D, and 4D in the gasket production industry, where D stands for "diagonal" rather than "dimension."

The intersection of several braiding tracks in such a way that the braided product adopts a particular, predetermined cross-sectional form is the common element in applications of this technique such as the production of gaskets and composite beams;

Braiding Technology for Textiles. http://dx.doi.org/10.1533/9780857099211.3.283

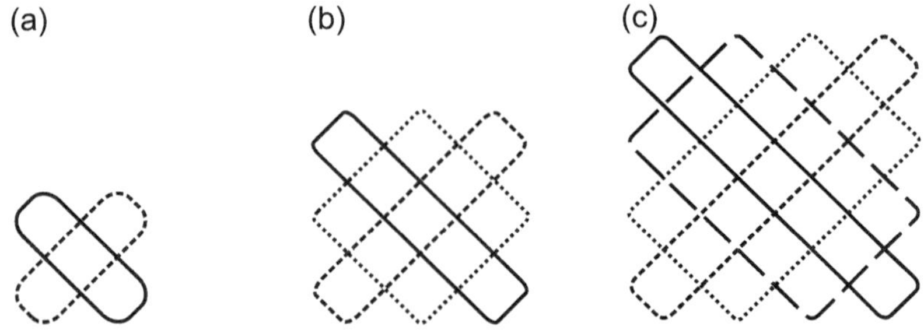

Figure 12.1 Idealized cross-section of squared braids for gaskets. (a) 2 Diagonal; (b) 3 diagonal; (c) 4 diagonal.

therefore, the most suitable term to cover the technique and the machines is *form braiding*. It should be noted that tubular and flat braids have, respectively, circular and flat cross-sections, and so are properly sub-groups of form braids. They can, furthermore, be produced on form-braiding machines, although this is neither the simplest nor the most efficient way to produce good quality tubular or flat braids.

12.2.2 Types of square braids for gaskets

There are special types of gaskets, of great length and with square cross-section, which the customer can buy in metres, cut according to the intended application. Depending on the required thickness, 2D (two-diagonal), 3D (three-diagonal) or 4D (four-diagonal) braids are used, where each diagonal is built by the yarns from a separate system of carriers, following one track. Figure 12.1a shows a simplified cross-sectional view of 2D braid – the braiding yarns are placed as the two diagonals of the square form. Figure 12.1b shows a 3D braid, which has an additional, middle system of yarns, and Fig. 12.1c shows a 4D braid, where two additional "diagonals" are used parallel to the two main diagonals.

Gasket users are not generally concerned with production technology, so they can keep the terms 2D, 3D, and 4D for their products, but for developers of braiding products it is more suitable to use the terminology of two-, three-, and four-track braids. Reference to the number of tracks connects directly to the carrier arrangement rules and assists in better understanding of the process.

Figure 12.2 shows an idealized cross-section of two-track braid and the horn gear configuration with the tracks for the carriers; it is immediately clear that one track corresponds to one diagonal. The square form of the gaskets and their mechanical properties are ensured by the use of inlay yarns, which are inserted through special pipes in the middle of the horn gears (see Chapter 6, Section 6.5.2). The inlay yarn can alternatively be inserted through the centre of the braid using a special pipe mounted on the braiding plate between the horn gears. For two-track braids, this inlay yarn coincides with the core of the tubular braid, though this is not the case for machines with more tracks. Figure 12.3 shows a simplified cross-sectional form of three-track (3D) square braid and the configuration of the braiding machine for its

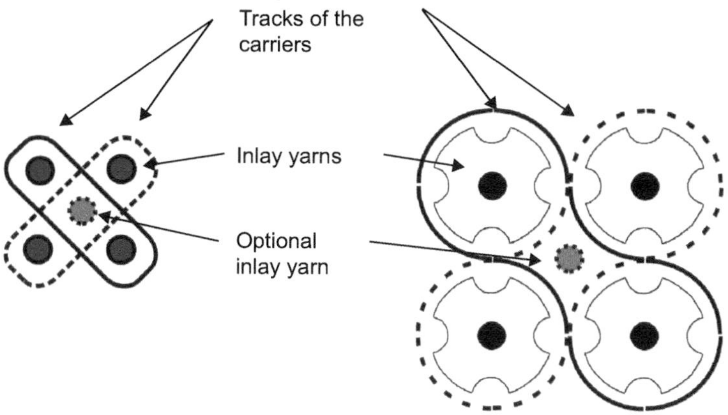

Figure 12.2 Two-track braiding of square braids.

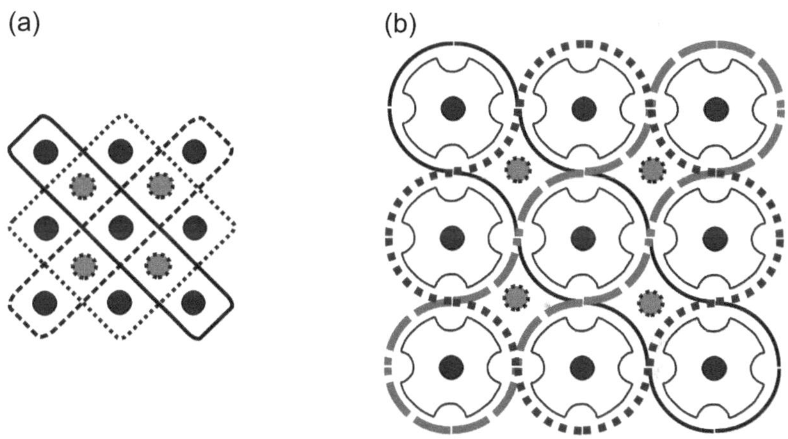

Figure 12.3 Three-track square braid: (a) idealized cross-section form; (b) machine configuration.

production. Figure 12.4 shows the corresponding configuration for four-track square braids.

Two-, three-, and four-track square braids are the main types used in the gasket industry, because they allow the production of as many different sizes of braided product as are needed. These configurations are not enough, however, for producing of composite beams, if beams with greater thickness are required. Theoretically, the system of two-, three-, and four-track square braiding can be extended to five, six, or more tracks (Fig. 12.5). Systems with an even number of tracks, N, will have an equal number of diagonals in both directions: $N/2$ parallel to one main diagonal, and $N/2$ parallel to the other main diagonal. Systems with an odd number of tracks will have $(N - 1)/2$ tracks parallel to each diagonal and one track which is symmetric to both and which connects the midpoints of the square sides. In practice, however,

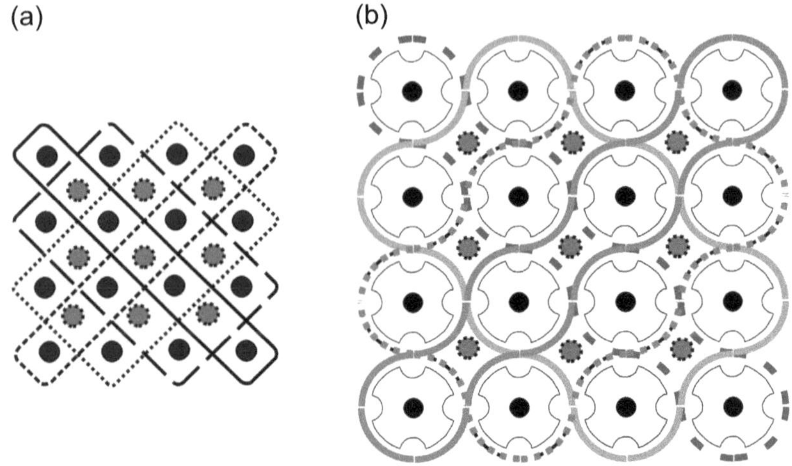

Figure 12.4 Four-track square braids: (a) idealized cross-section form; (b) machine configuration.

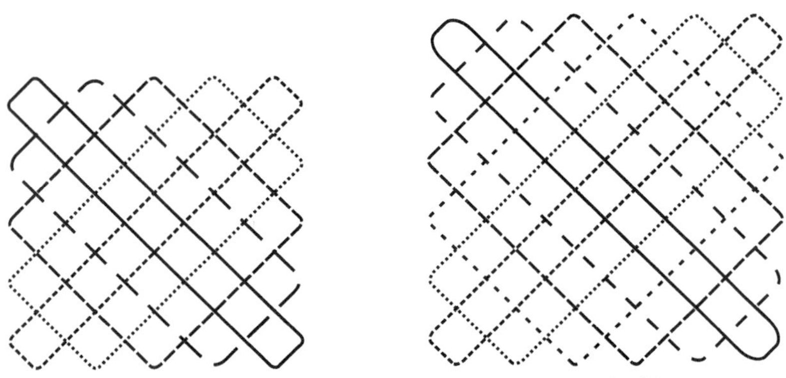

Figure 12.5 Possible track forms for five-track and six-track square braids.

such systems are not currently built, mainly because the machines would have to be very large. Machine dimensions would not scale linearly with the number of tracks, because more horn gears would be required for multiple-track square braids, and because the horn gears would need to have more slots to receive carriers for all these tracks. This issue is explained in more detail in the discussion below.

12.3 Applications of square-braided gaskets

More than 300 examples have been listed as applications of square-braided gaskets (W.L. Gore & Co. GmbH, 1986). Essentially, they serve to prevent the infiltration of liquid from a "wet room" into a "dry room" (Fig. 12.6). It is a common situation

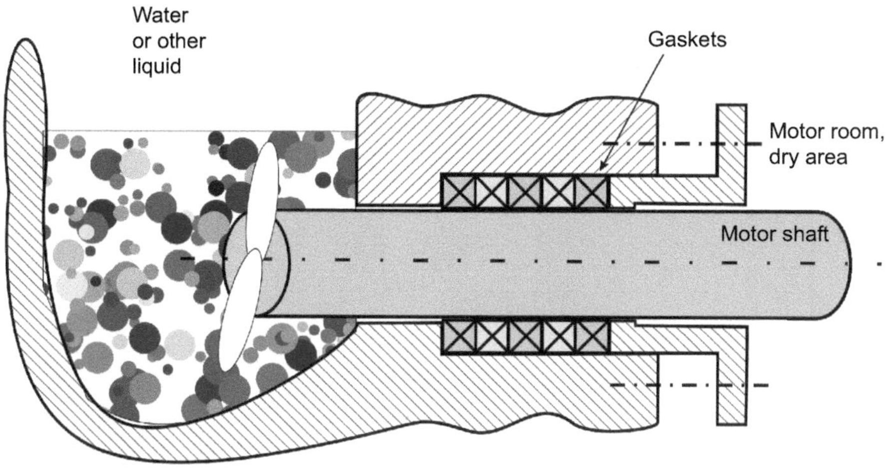

Figure 12.6 Example of application area of gaskets with square cross-section.

in the chemical industry, in textile colouration and finishing, in the food industry, and in ships, that there are machines that use a propeller or a similar device to move or homogenize some fluid. The propeller, or its equivalent, operates in a wet room, where the liquid could just as well be water or some dangerous alkali, acid, or other chemical, potentially also under high temperature and pressure. The shaft rotates the propeller, but the motor cannot be located in the same room – it is more effective and easier to put the motor drive in a dry room with the shaft going through an opening to the wet room. Gaskets are used to seal the opening.

Normally, the gasket is received as a long braid, which is then cut to a given length such that it forms a closed ring around the shaft (Fig. 12.7). Several gaskets are pressed at a specially prepared place around the shaft. The pressed gaskets will normally leave a small amount of space in the opening (less than 0.5 mm) so that mounting is easier. After pressing (by the disc on the right-hand side, which is then fixed by a screw onto the main body of the machine), the gasket deforms and fills the remaining space between the shaft and the machine body, but still allows the motion of the shaft. Several gaskets are mounted; these have flat surfaces, and because the contact between shaft and gasket is good, the liquid cannot infiltrate through the opening into the dry area. The gaskets must therefore have square cross-sections; tubular braids would have less contact area and there would be empty spaces between them.

The gaskets have to be changed regularly; depending upon the process, this might be every 3–5 weeks or longer, but still it is important for users that the change is simple and not time consuming. The surface of the braid has to be smooth, and special attention therefore needs to be given to the choice of braiding machine and to floating length and tension adjustments. To minimize friction between the gasket and the shaft, special materials such as PTFE or graphite, as well as special coatings, can be used. The company Gore supplies a wide range of different gasket types according to size and application, and they claim that the use of their Gore-GFO® yarn can extend

Table 12.1 Recommended number of braids density per (W.L. Gore & Co. GmbH, 1986)

Diameter (mm)	Density (number braids per 10 cm)
6	30
8	29
10	24
12	25
14	21
16	20
18	19
20	18

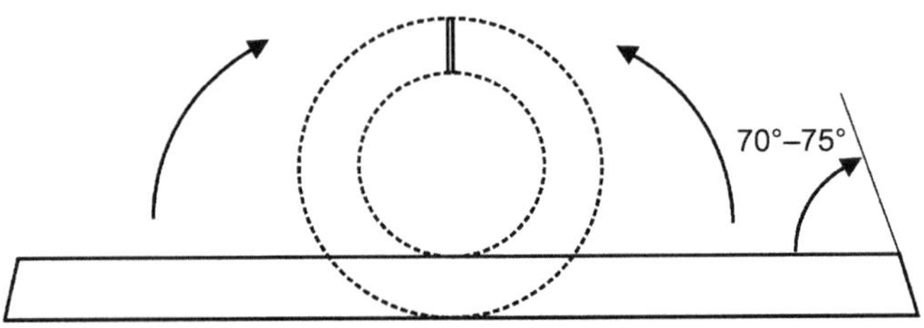

Figure 12.7 Cutting of gasket and folding up to the circular form (W.L. Gore & Co. GmbH, 1986).

Figure 12.8 Picture of one three-track gasket.

the life of a gasket several times (W.L. Gore & Co. GmbH, 1986). The Gore gaskets of up to 5 mm thickness use two-track braids (two-diagonal), those up to 10 mm use three-track braids, and from 12 mm thickness they use four-track braids. The inlays are only in the middle of the package and are used mainly to stabilize the sides of the braids; they do not have to fill more than 25% of the cross-sectional area. Table 12.1 gives an overview of the densities of the braids, as recommended for obtaining the best results, and Fig. 12.8 shows a photo of braided gasket.

12.4 Patterning for two-track braids

The patterning of two-track braids with square cross-section follows the same rules as the patterning of classical tubular braids, because two-track braids (two-diagonal braids) are also the smallest "tubular" form, even though their cross-section is more square than tubular. Figure 12.9a and 12.9b illustrate the carrier arrangements for producing braids with a floating length of 2 (2:2–1) using 1 Full 1 Empty carrier occupation of four-slot horn gears, and a floating length of 1 with a group of two yarns (1:1–2) if 2 Full 2 Empty carrier occupation is applied.

As explained in Chapter 2 (*Patterning of braided products*), the production of a pattern with a larger floating length requires horn gears with a greater number of slots. Recalling that the relation between carrier occupation and the number of slots is:

$$\text{Floating\_Length} = \frac{\text{Slots\_per\_horn\_gear}}{\text{Carrier\_Arrangement\_Repeat}} \quad [12.1]$$

and using the shortest possible repeat of the carrier arrangement (1 Full 1 Empty = 2), it can be seen that a floating length of 3 requires six-slot horn gears, and a floating length of 4 requires eight-slot horn gears (Fig. 12.10). Figure 12.11a and 12.11b demonstrate the arrangement of the carriers for producing braids with floating length of 3 (3:3–1) and 4 (4:4–1), and Fig. 12.12a and 12.12b demonstrate the carrier arrangements for producing braids with floating length of 1 but with three and four yarns in the group, respectively. Of course it is also possible to produce such braids on smaller machines with fewer slots and using ply yarns on the bobbins, but it is not then possible to guarantee, deterministically from the arrangement, the placement and exact orientation of all single yarns from the ply yarn. More details about the disadvantages

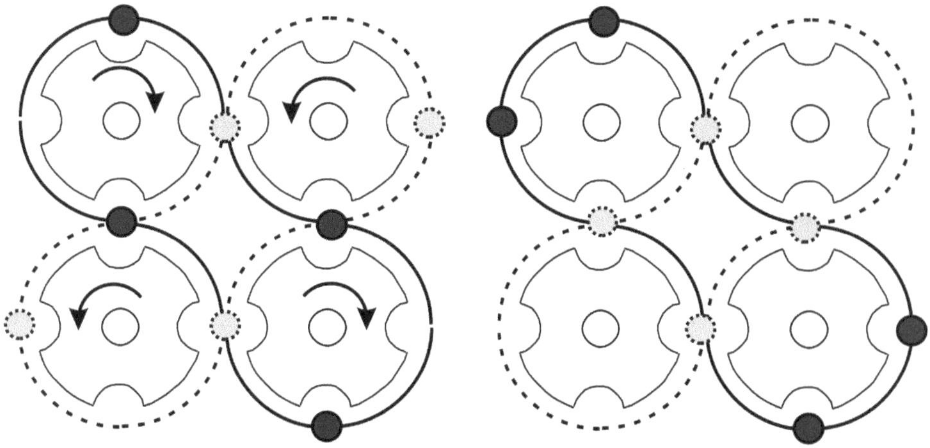

Figure 12.9 Carrier arrangements for producing braids with floating length of 2 (2:2–1) and one in a group of two yarns (1:1–2) on a four-slot horn gear machine.

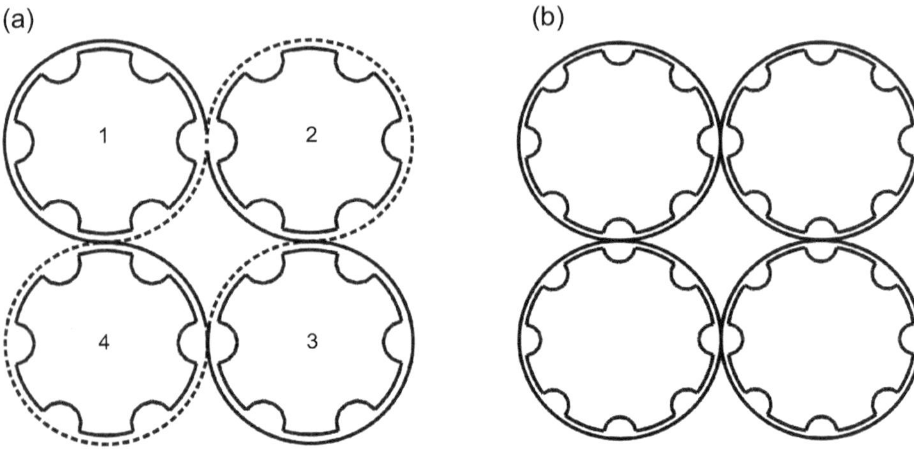

Figure 12.10 Configurations with six-slot horn gears (a) and eight-slot horn gear (b).

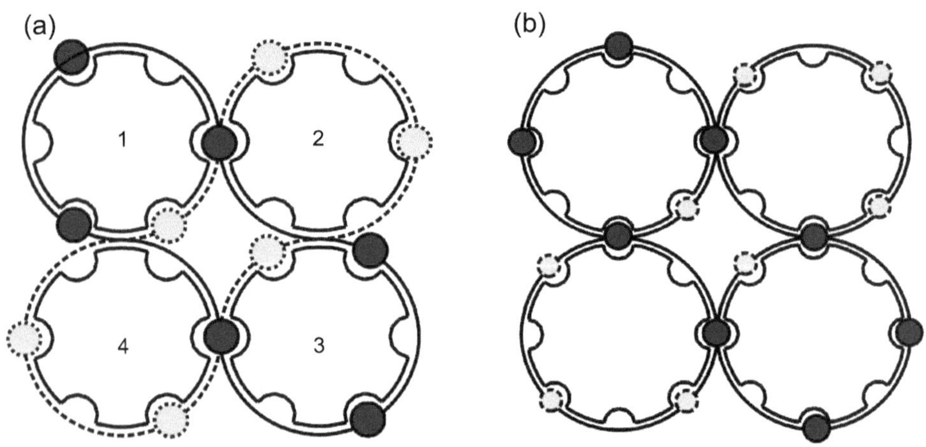

Figure 12.11 Carrier arrangements 1 Full 1 Empty for realization of braids with (a) floating length of 3 (3:3–1), and (b) floating length of 4 (4:4–1) on the corresponding horn gears.

and advantages of using multiple carriers or a single carrier with multiple yarns are discussed in Chapter 2.

12.5 Patterning for three-track braids

Producing braids with three "diagonals," or three-yarn systems, requires three tracks, and production of a square cross-section in this case requires nine horn gears. Figure 12.13 shows a configuration with the two tracks extended on a 3 × 3 set of horn gears; this configuration can, theoretically and practically, be used for "braiding," or

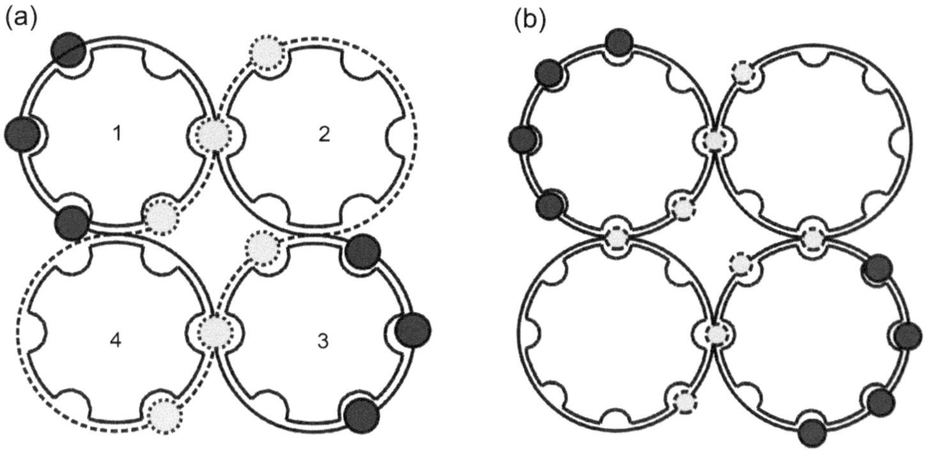

Figure 12.12 Carrier arrangement for getting braids with multiple yarns in the group (a) floating length of 1 group with three yarns in the group (1:1–3) and (b) floating length of one group with four yarns in the group (1:1–4).

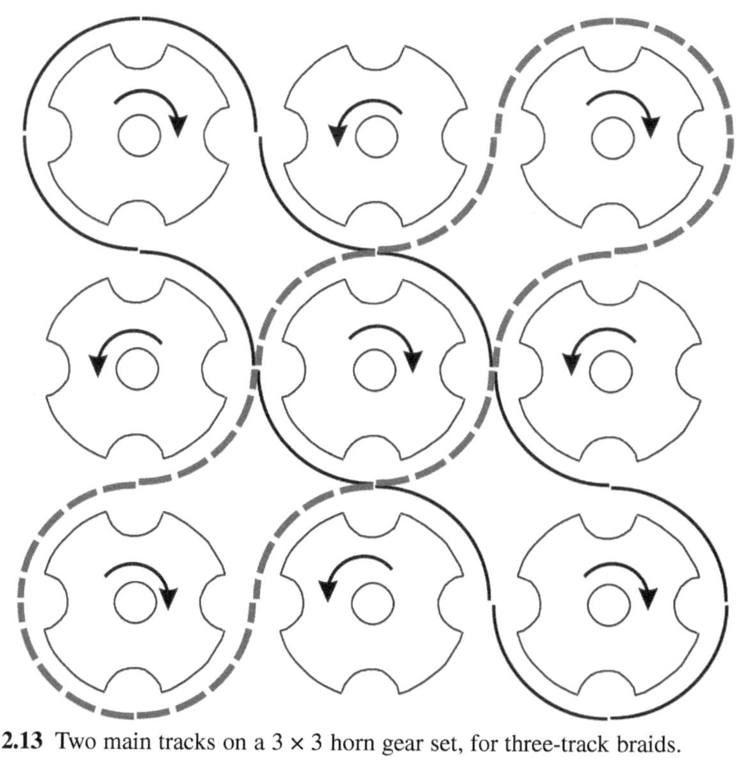

Figure 12.13 Two main tracks on a 3 × 3 horn gear set, for three-track braids.

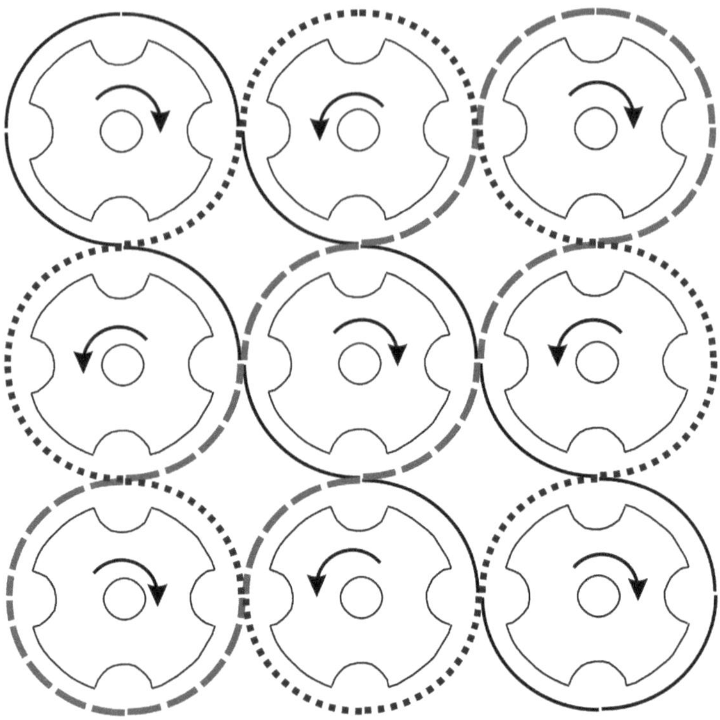

Figure 12.14 Tracks for three-track square braids.

more accurately the twisting or covering of a core. The reader will recognize that the carriers on both tracks in this configuration move in the same direction, so the two available systems of yarns will not interlace in the typical way for the braids. They will be just be twisted or cover an inlay yarn or yarns inserted through the openings of the horn gears. If an additional track is added (Fig. 12.14), its carriers move in the opposite direction to that of the first two tracks, and then a normal three-track braid with a square cross-section is produced. The carrier arrangement 1 Full 3 Empty for producing braids with floating length of 1 (1:1–1) on such a machine with four-slot horn gears is shown in Fig. 12.15.

On three-track machines where the horn gears have four slots, only the 1 Full 3 Empty arrangement is possible. Figure 12.16 shows why the 1 Full 1 Empty arrangement is not possible in this case. The carriers on two of the tracks can be arranged without problem, but the carriers on the third track will collide with some of the carriers on the first two tracks. This also means that structures with a floating length of 2 cannot be produced on this machine. If a floating length greater than 1 is required, horn gears with more slots have to be used. The problem is also illustrated in Fig. 12.17: when square braids are produced on three-track machines, each track crosses two other tracks. Crossing another track means that at any given time there is an oncoming carrier on the first track and then a carrier on the track is intersected.

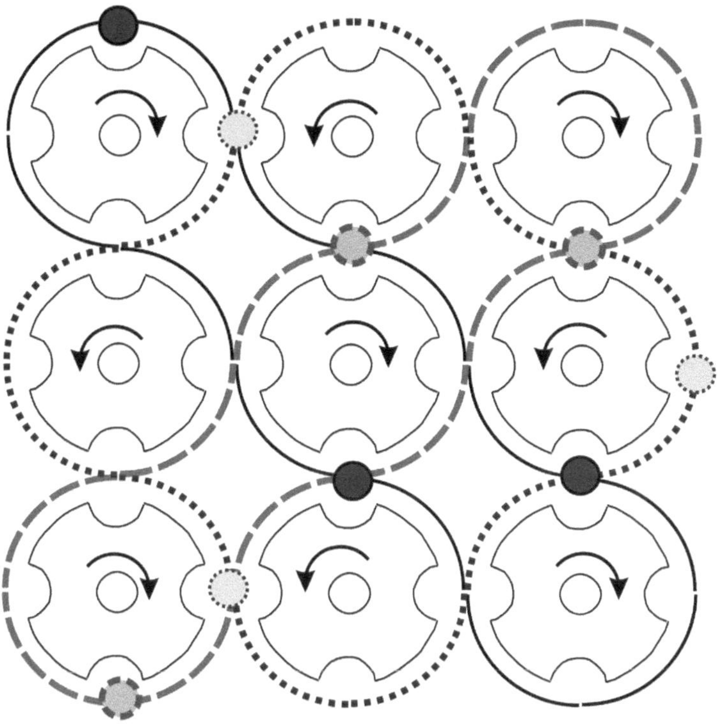

Figure 12.15 Carrier arrangement 1 Full 3 Empty on a three-track braids with square cross-section form on a machine with four-slot horn gears.

When the carrier on the intersected track approaches a slot, this slot has to be free of other carriers.

If one track intersects two other tracks, there must be at least one position for the carrier on that track and at least two empty positions for the carriers on the intersected tracks. Additional positions can be increased with empty positions until the full repeat, equal to the number of slots in one horn gear; but these cannot be filled with additional carriers. For four-slot horn gears, the repeat of the carrier arrangement is four, so only the 1 Full 3 Empty arrangement works, because the three empty positions are more than the two required for the two tracks that are crossed. This configuration can also be achieved, in theory and in practice, on three-slot horn gears, again with the arrangement 1 Full 2 Empty, which is illustrated in Fig. 12.18. This configuration can be "doubled" with six-slot used in place of three-slot horn gears, in which case the arrangement becomes 2 Full 4 Empty (Fig. 12.19). The floating length is then 1, with two yarns in the group (1:1–2), because:

6 slots: (2 Full + 4 Empty) = 6: 6 = 1.

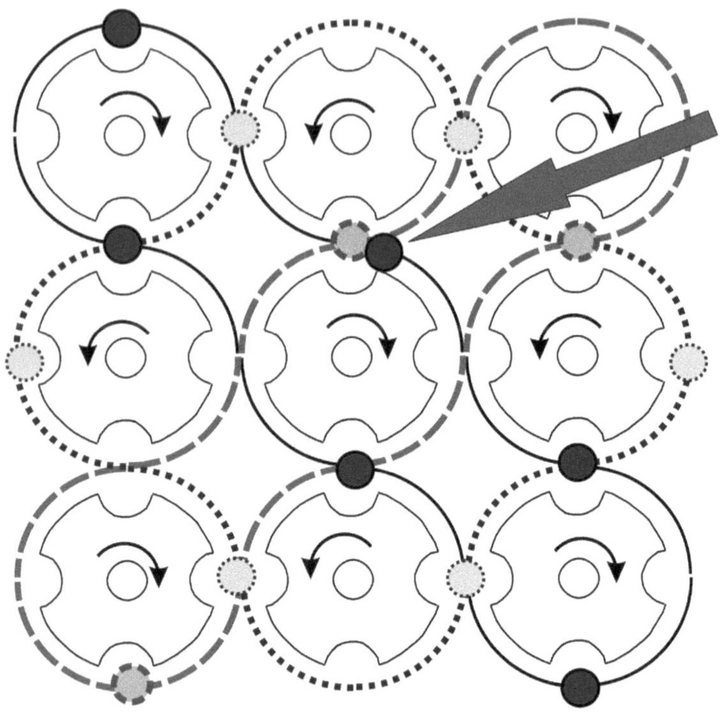

Figure 12.16 Trial to arrange three-track machine with four slots horn gears with carrier arrangement "1 Full 1 Empty." The carriers of the third track collide with some of the carriers of the first or second track.

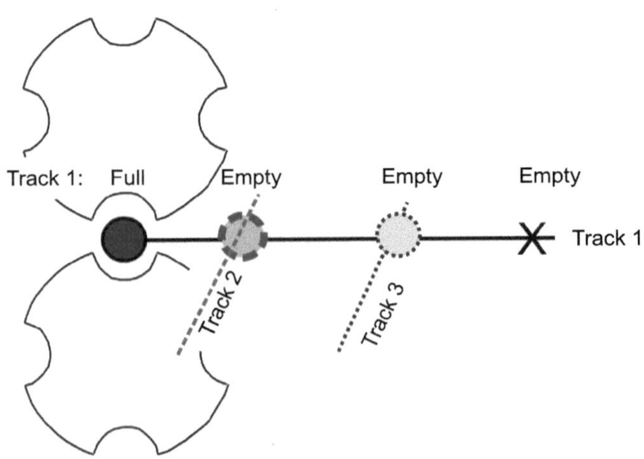

Figure 12.17 On a machine where one track intersects two another tracks, at least two empty positions are required in the carrier arrangement for the transfer to be successful.

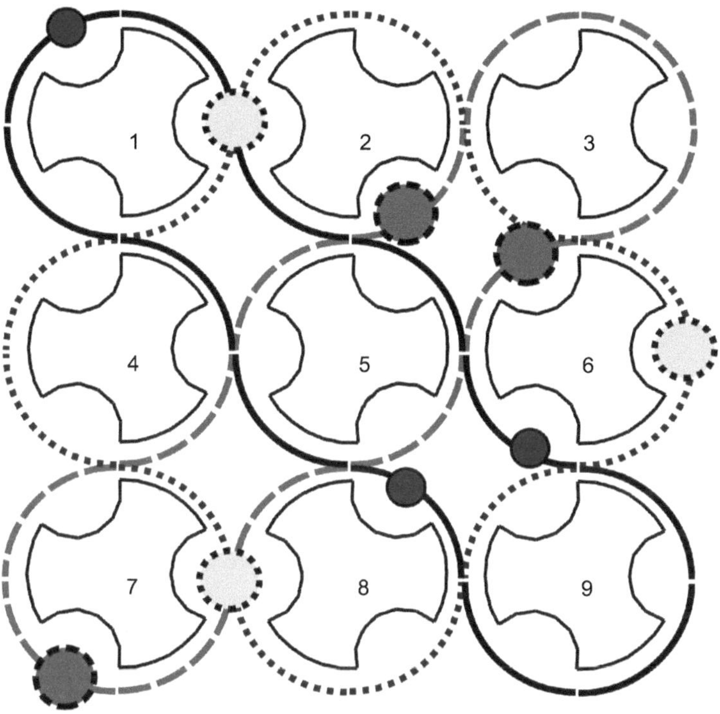

Figure 12.18 Three-track configuration for producing braids with square cross-section on machine with horn gears with three slots.

Using the Basic Rules for Structural Patterning, presented in Chapter 2, other types of structure can be generated using different combinations of horn gears and carrier arrangements.

The track configuration controls the position of the yarns in the braid. The carriers moving along the main diagonals place yarns that are visible only on the edges of the braid. The carriers following the middle track place yarns that are visible only along the middle of the braid (Fig. 12.20). Depending on the direction of rotation, the visible yarn pieces are orientated on each of the ridges. This is important not so much for colour design, because square braids are generally not used in clothing or other accessories, but more for selecting appropriate materials for the different working areas of the gaskets. As Figs 12.8 and 12.21b and c demonstrate, the edges of the braid can be produced from a material with high abrasion resistance, and the middle of the braid can be produced from a material that improves the sliding contact between the gasket and the rotating shaft. Carriers for the first material follow the diagonal tracks, and carriers for the second material follow the middle track. If two materials with alternating colours are used on all the tracks, then it is possible to create horizontal stripes (Fig. 12.21c).

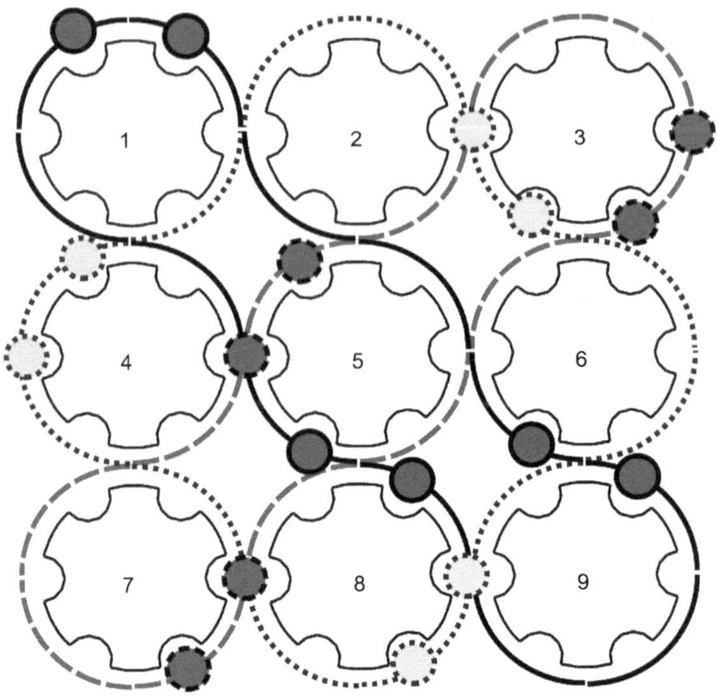

Figure 12.19 Configuration for three-track braids with square cross-section on machine with six-slot horn gears. The structure will be 1:1–2, the carrier arrangement 2 Full 4 Empty.

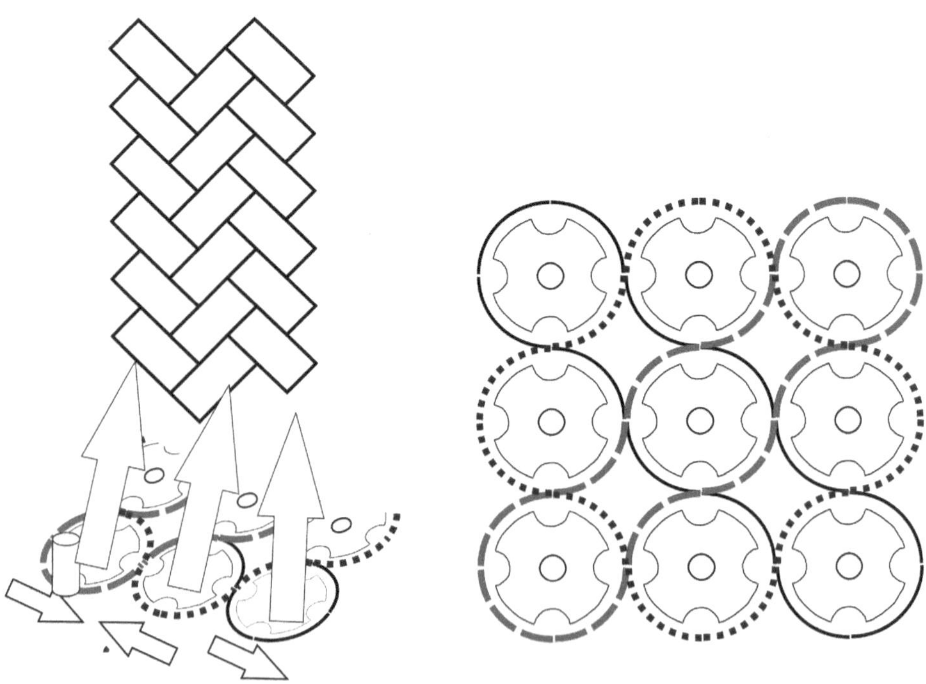

Figure 12.20 Track configuration determines the place of the yarn in the braid.

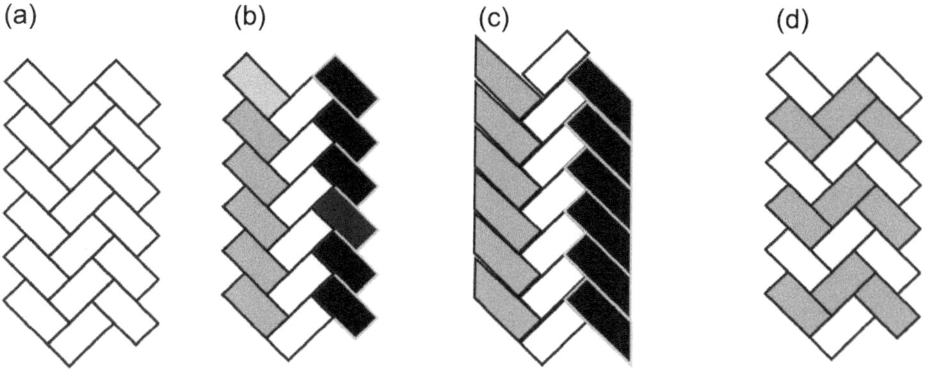

Figure 12.21 Colour or material selection (a) basic grid for the visible ridges on one side of the three-track gasket with square cross-section (b) if the middle track is using white material, and the two diagonals – grey and black–then vertical stripes will be created; (c) presentation of figure (b) but including the yarn pieces at the borders extending the visible pieces with triangle; (d) horizontal stripes.

The correct representation of the ridges would consist of a rectangle for the completely visible yarn pieces on one side and a triangle for the yarn piece which is on the border and which changes its position from one side to the next (Fig. 12.21c). However, colour design is generally not important for such products, so the creation of colour pattern pictures for these braids is not common and is therefore not discussed further here.

12.6 Patterning for four-track braids

For thicker braids the three diagonals are not sufficient. The use of thicker yarns with three diagonals would lead to products with a rough surface, neither dense nor smooth enough for gaskets. The solution is to increase the number of tracks, because thinner yarns will be more regularly interlaced, with smaller variations in thickness arising from yarn geometry. Figure 12.22 shows the four-track configuration for square braids, which requires a 4 × 4 set of horn gears. As explained in the previous section, with one track between the full positions, the number of empty positions needs to be at least equal to the number of (other) tracks which this track intersects. In the case of four-track braids, each of the tracks intersects with three other tracks, so the only possible carrier arrangement with four-slot horn gears is 1 Full 3 Empty, as shown in Fig. 12.23. The braid will then be produced with a floating length of 1 (1:1–1).

If larger floating lengths are required, the horn gears will need a higher number of slots. Figure 12.24 shows a configuration with eight-slot horn gears, with which,

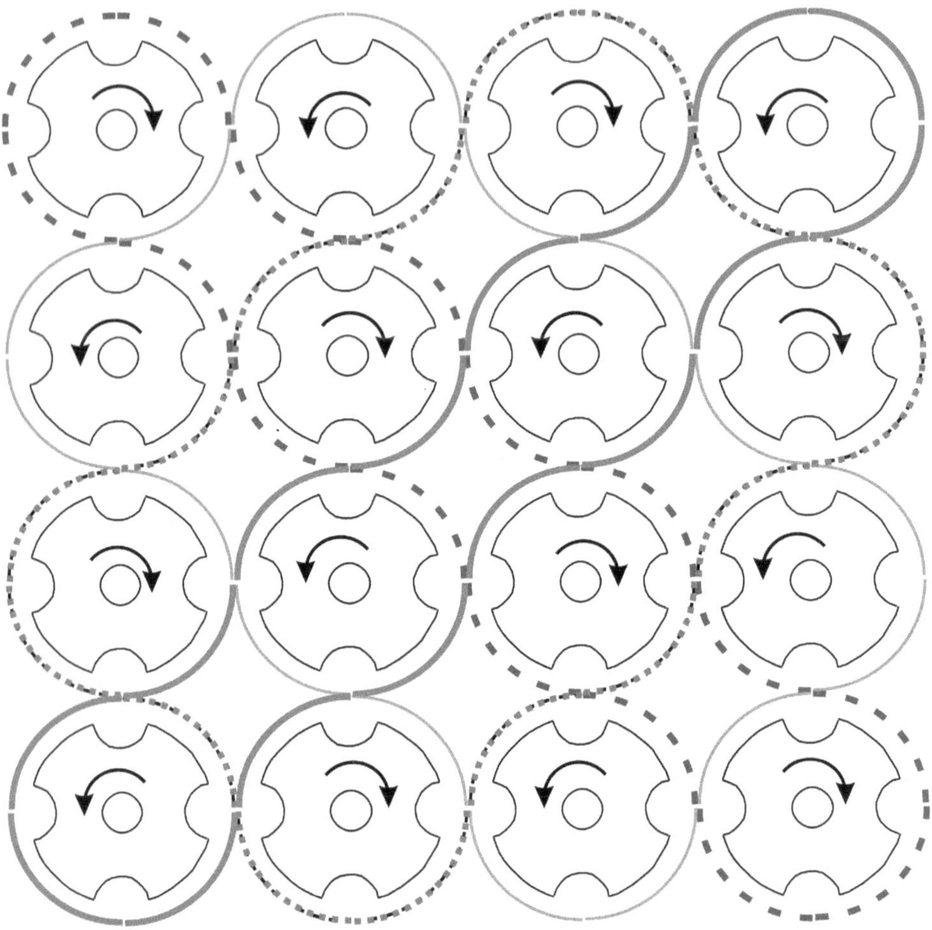

Figure 12.22 Four-track configuration for producing gaskets with square cross-section.

for example, a floating length of 2 (2:2–1) is possible using the carrier arrangement 1 Full 3 Empty:

8 slots/(1 Full + 3 Empty) = 8 /4 = 2

A floating length of 1 (1:1–2) can still be achieved, using a 2 Full 6 Empty arrangement, which is left as an exercise for the reader.

Patterning of four-track structures with different colours or materials can be achieved by a method analogous to that for three-track structures. If all carriers moving along a track have one colour, this track will produce visible stripes with movement around some of the outer horn gears. An example of such vertical stripes for a structure with a floating length of 1 is shown in Fig. 12.25.

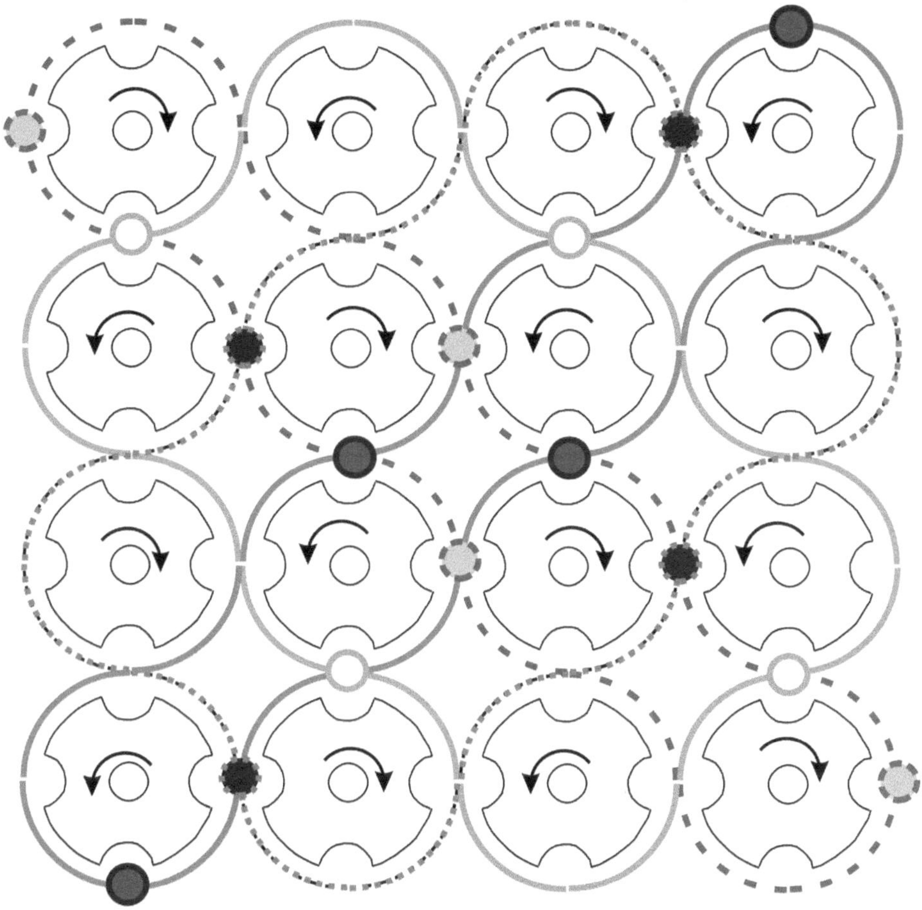

Figure 12.23 Carrier arrangement of the four-track configuration with 1 Full 3 Empty for producing braids with floating length of 1 (1:1–1).

12.7 A braiding machine with variable tracks

The VF (VarioFlechter, variable braider) machine has been developed by the company Herzog (see Chapter 6) from the 3D Braiding machine (Büsgen, 1993; Schneider, 2000; August Herzog, 2013), using one module with switchable cams between the horn gears. This machine has 16 horn gears, mounted on a 4 × 4 grid, the tracks between all the pairs being determined by the position of the switchable cams. Figure 12.26 shows the horn gears and slots in initial position. Different tracks can be defined and different products can be produced, including all of the gaskets described in the previous section, by rotating the switches. The advantage is that more than ten different cross-sectional forms are achievable on a single machine after brief

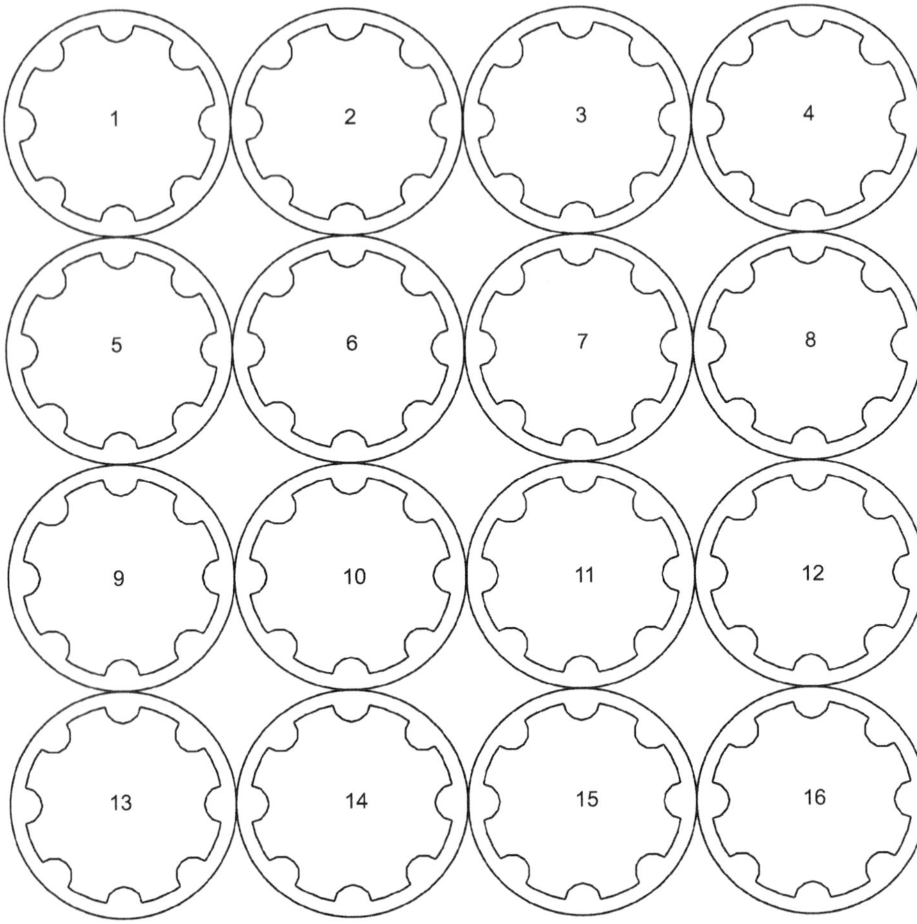

Figure 12.24 Configuration for four-track product with square cross-section and eight-slot horn gears.

rearrangements, which saves costs and space, compared to the alternative of having ten separate machines.

Additionally, such a machine can, for example, produce a two-layer braid, if the middle four horn gears are arranged to create a two-track square braid (actually a tubular braid) and the outer horn gears are arranged to produce the tubular structure (Fig. 12.27). The switches are not shown in the figure, because the reader will recognize their positions by the direction of the tracks: where a track changes from one horn gear to another, the switch there is in transfer position; where a track continues following the same horn gear, the switch is in the "keep" position.

The machine can produce tubular braids using 4, 6, 10 or 16 horn gears (Figs 12.28 and 12.29), and can also produce multiple tubular braids simultaneously, as shown by the track configuration in Fig. 12.28d for simultaneous production of four small

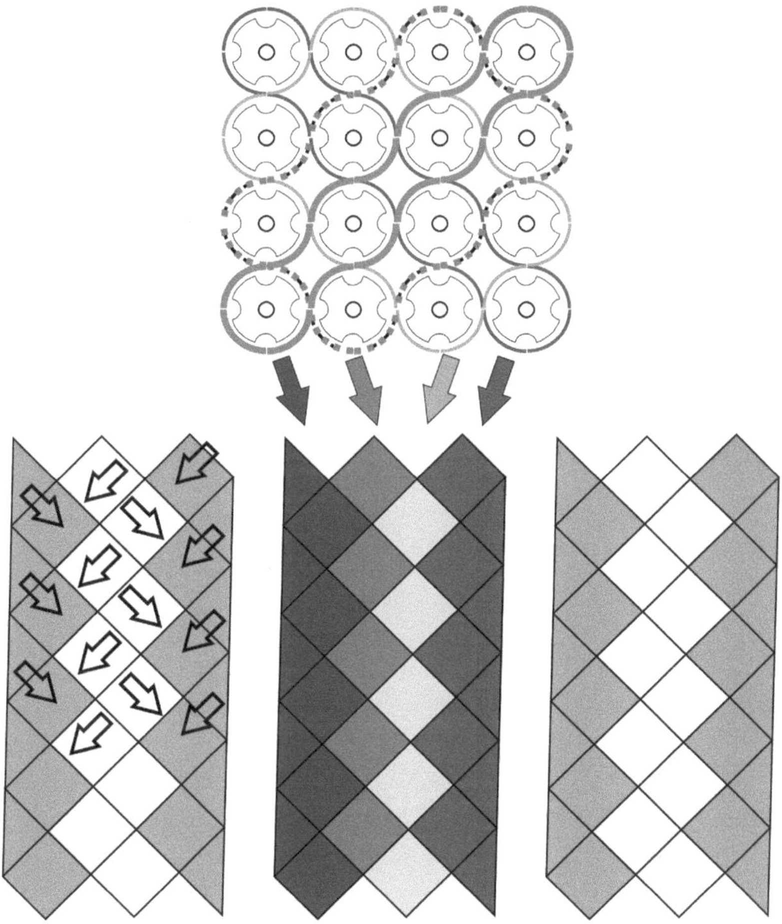

Figure 12.25 Colour or material patterning using different materials or colour on the different tracks.

braids. The next-generation form of this machine has several of the switches under computer control, which enables the machine to switch from producing one tubular braid to producing two tubular braids (Müller *et al.*).

Further adjustments are possible, leading to a variety of products. Tubular braid can be produced with greater floating length on one side, for instance to influence the bending direction of a rope or candle. The modified configuration for this is shown in Fig. 12.30. One of the tracks can be extended from the other side to the upper right corner and there intersect with an additional track; a Y-shaped profile can then be produced on three tracks (Fig. 12.31). A configuration for producing braids with a U-shaped cross-section on four tracks is shown in Fig. 12.32.

Tubular braid using all 32 carriers (32 carriers) and the 1 Full 1 Empty arrangement can be produced on the same machine, but the track geometry is not trivial.

Figure 12.26 Initial configuration of vario braider of the August Herzog company (August Herzog, 2013).

For clarity, only the first track is shown in Fig. 12.33a; the second track is added in Fig. 12.33b, and 12.34 presents the complete machine configuration with both tracks and the carrier arrangement. The range of products that can be produced on such machines increases with the number of horn gears. The possibilities for creating braided profiles with different cross-sectional forms are explored in the work of Li (2013).

12.8 Form braids with more complex forms

Further configurations of maypole braiding machines for producing profiles with different forms can be found in the scientific literature and in patent documentation.

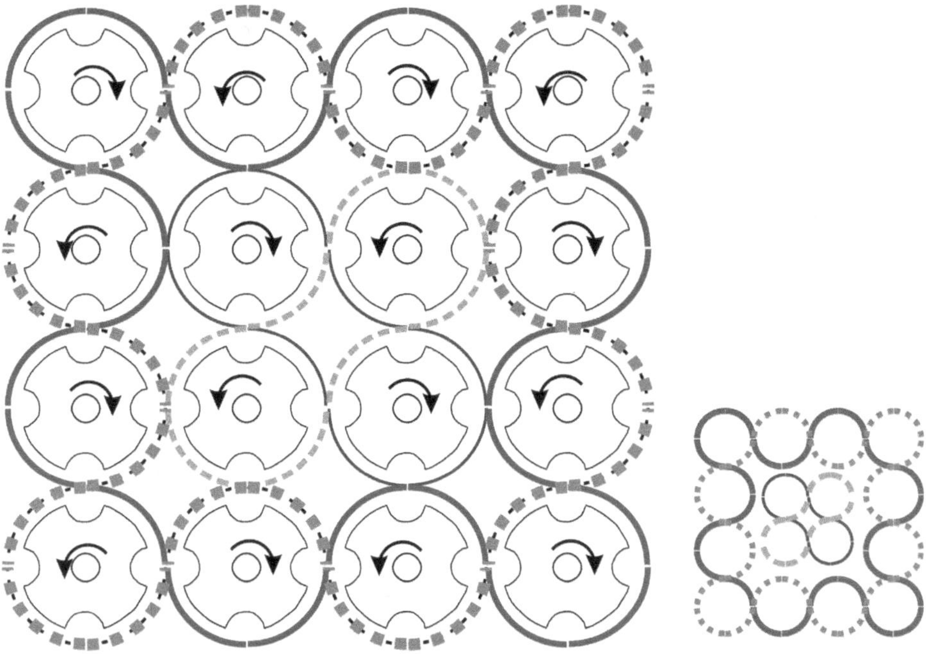

Figure 12.27 Configuration for two-layer braid.

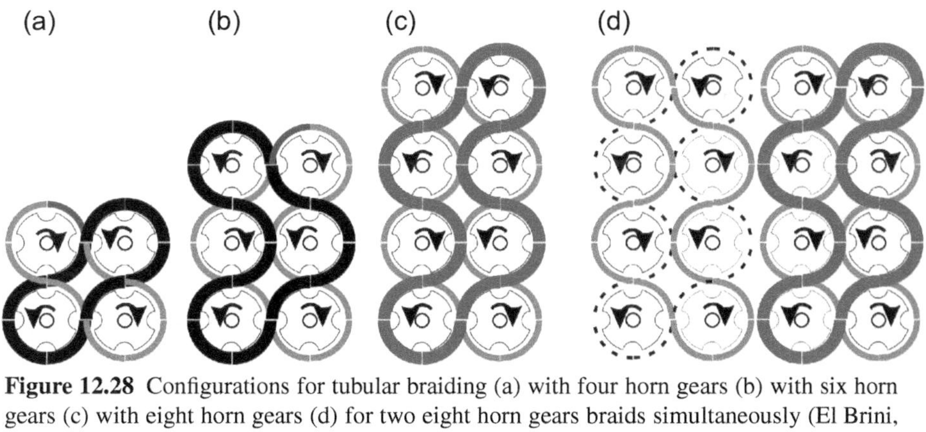

Figure 12.28 Configurations for tubular braiding (a) with four horn gears (b) with six horn gears (c) with eight horn gears (d) for two eight horn gears braids simultaneously (El Brini, 2010).

Normally, such machines are created and built for a specific form and allow fewer variety of products. An example is a mechanism for the production of double-T (I-beam) forms described in the patent Akiyama *et al.* (1995), shown in Fig. 12.35. There, two linear sections (as in flat braiding machines) are connected with a third. It is not clear how the machinery would be designed to switch carriers between tracks at the connection points where one track is connected to two, but this might possibly

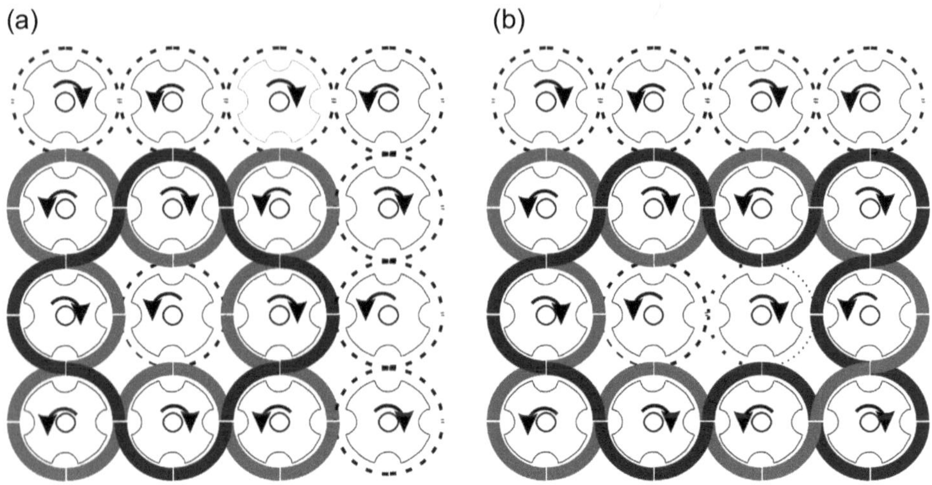

Figure 12.29 Configurations for tubular braids with (a) eight and (b) ten horn gears.

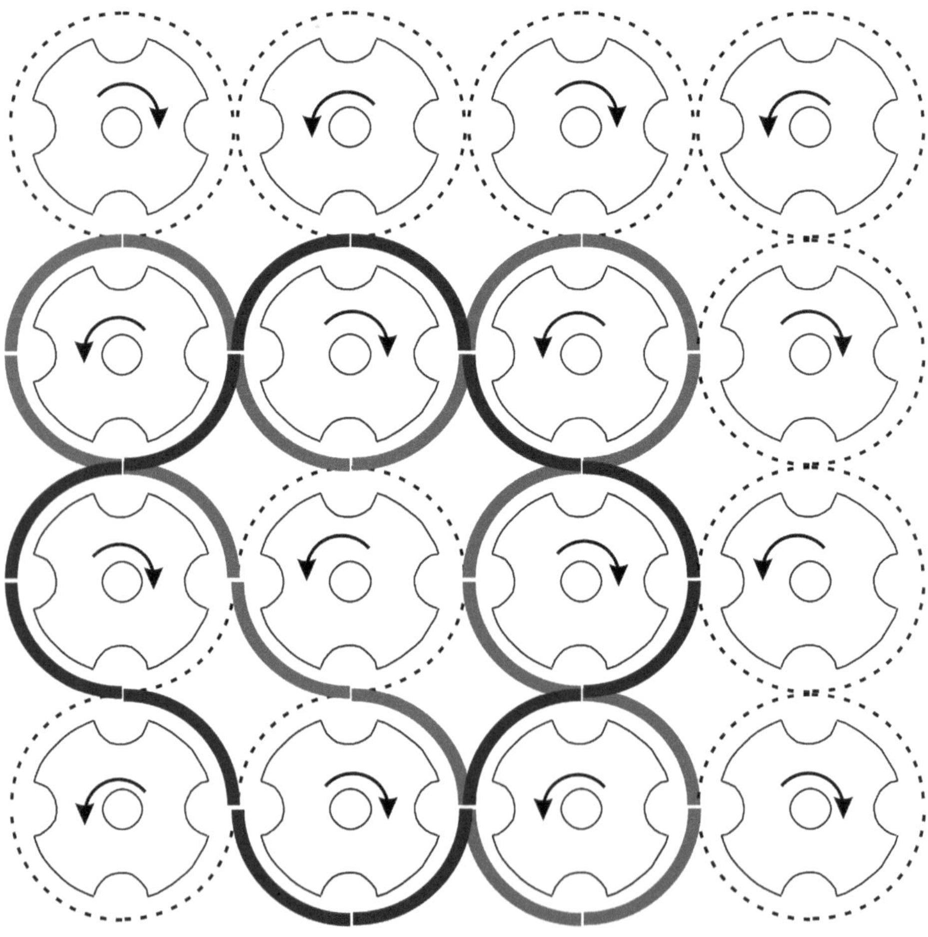

Figure 12.30 Tubular braid with one side of different floating length.

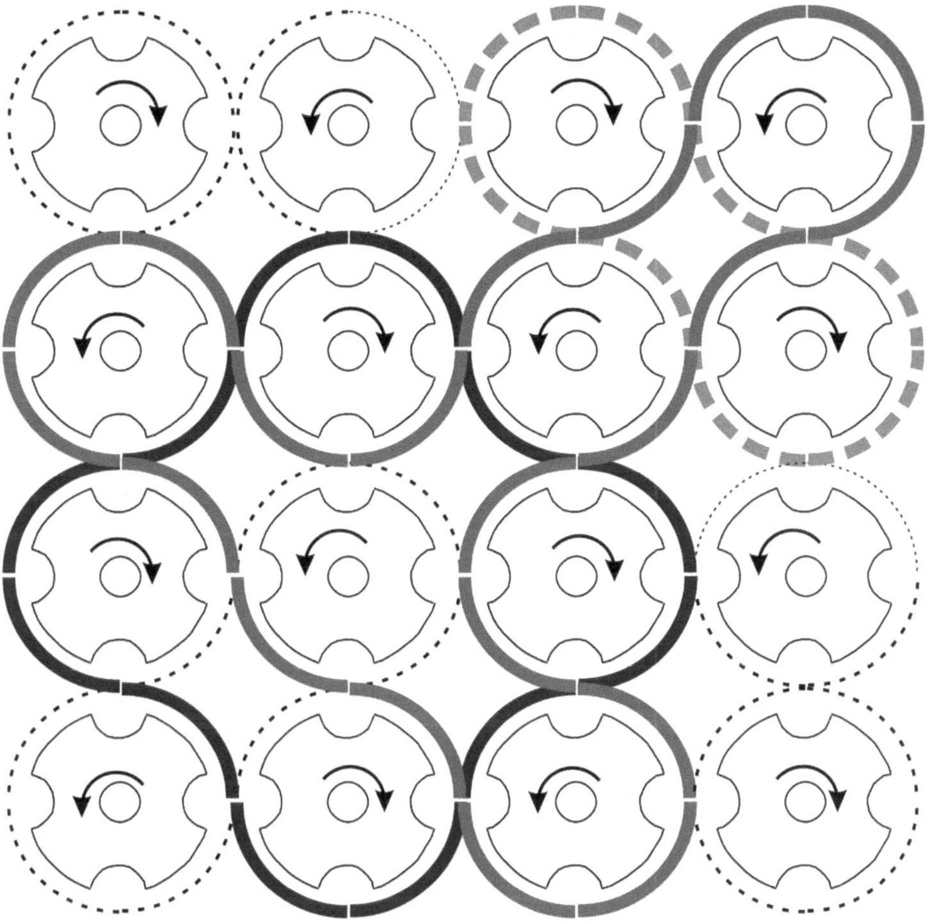

Figure 12.31 Four-track Y-like profile.

be achieved using switch cams. Figure 12.36 shows an idea for interlacing yarns on several tracks to braid a solid structure for composites (Brookstein, 1991; Bilisik, 2013). This method could be applied for linear and for radial structures, as shown in the pictures in the patent Temple (1984) (Fig. 12.37).

Figure 12.38 shows an interlacing three-layer structure, where the interconnection of the single layer is less dense than the interlocked structure (Uozomi *et al.*, 2001). Brookstein *et al.* present an interesting idea for combining a set of plates with tracks to create the basic elements of braided structures (Figs 12.39 and 12.40), which can be used to make larger machines for braiding thick profiles for composites.

Machines have been built on the basis of these and other patents, but they are exceptional cases, either single experimental pieces or unique machines for very specialized production. They are not common because of their very large dimensions and

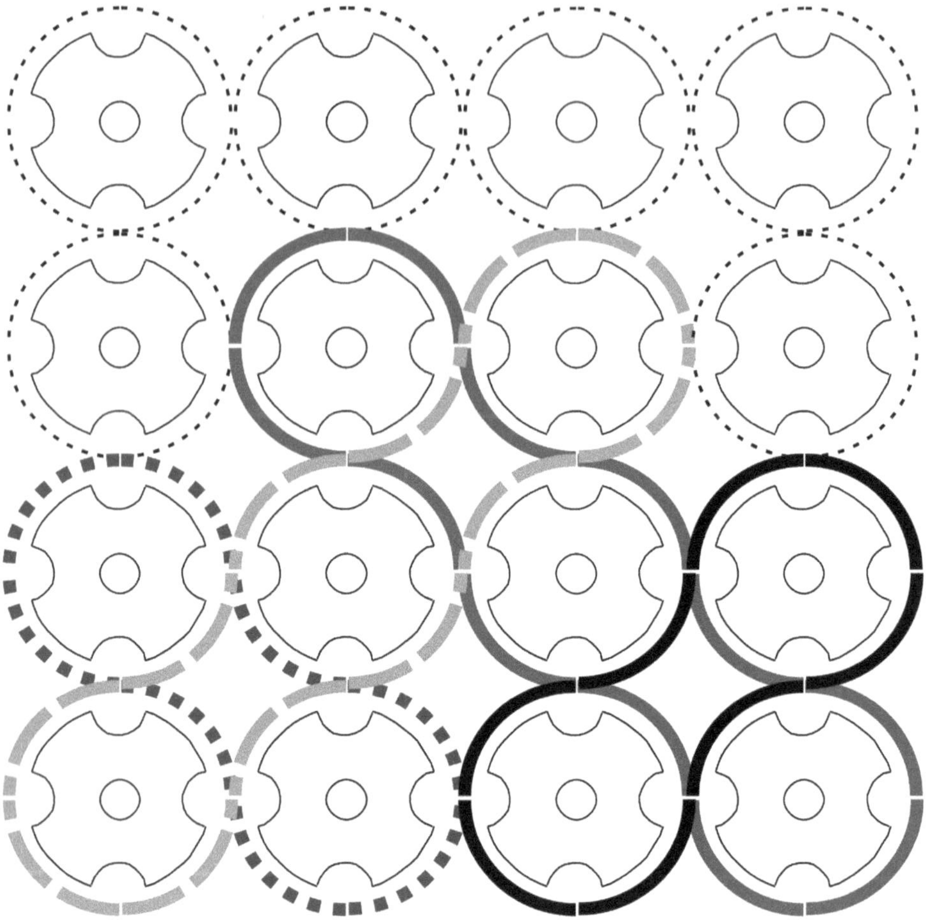

Figure 12.32 Configuration for U-shaped cross-section (Becher, 2010).

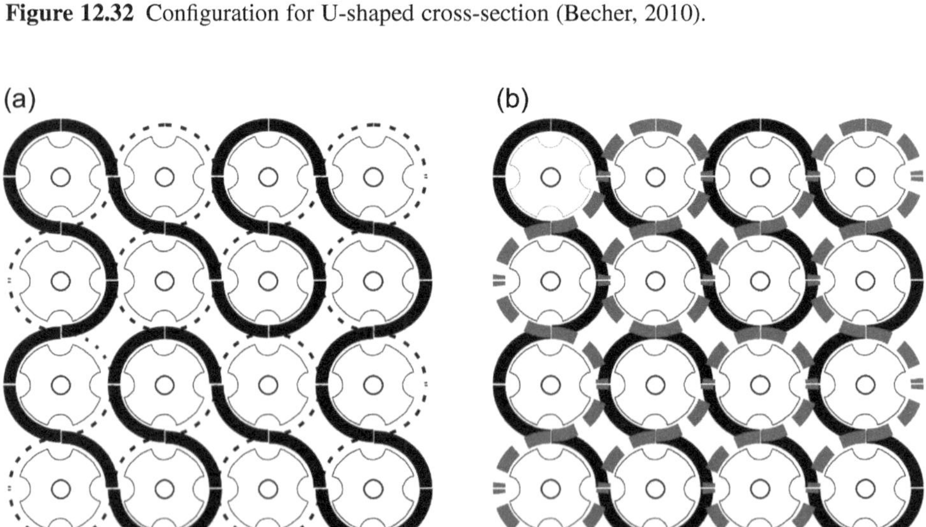

Figure 12.33 Configuration for tubular braids with 32 carriers (a) first track (b) both tracks (Becher, 2010).

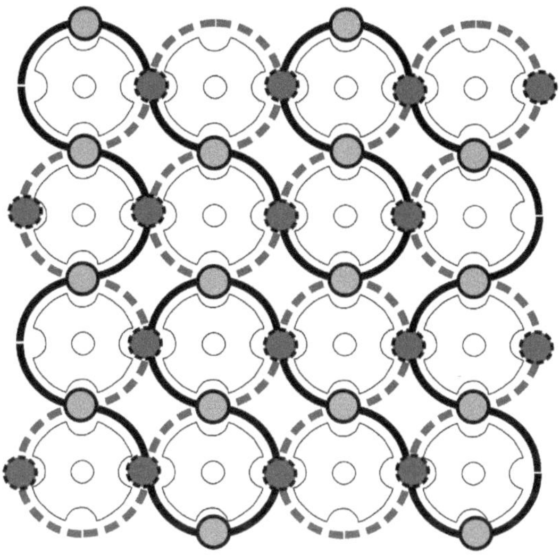

Figure 12.34 Configuration for tubular braids with carrier arrangement 1 Full 1 Empty (Becher, 2010).

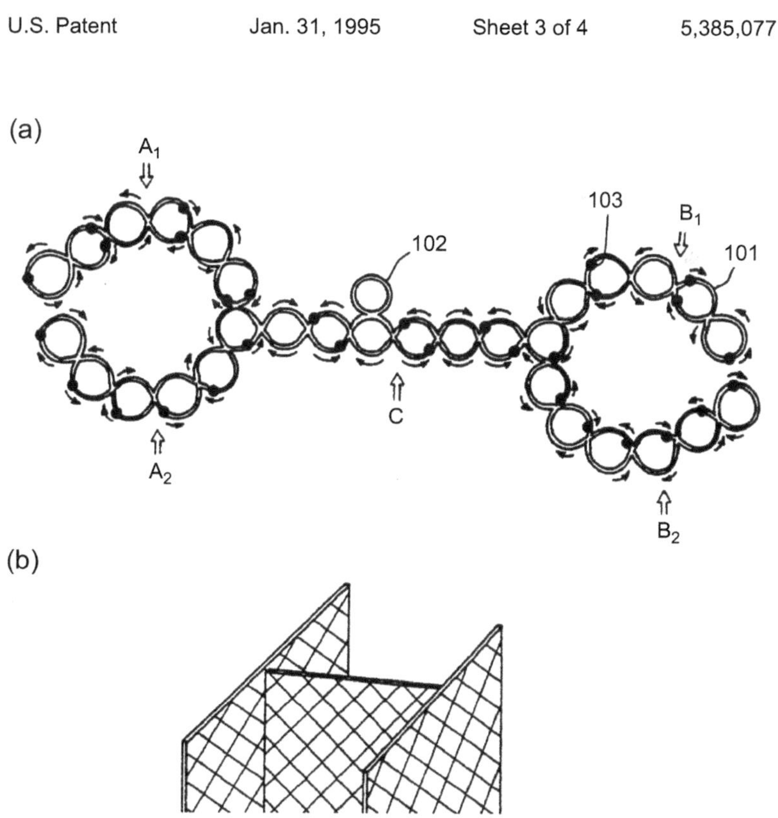

Figure 12.35 Configuration for producing I-profile (double-T) (Akiyama *et al.*, 1995).

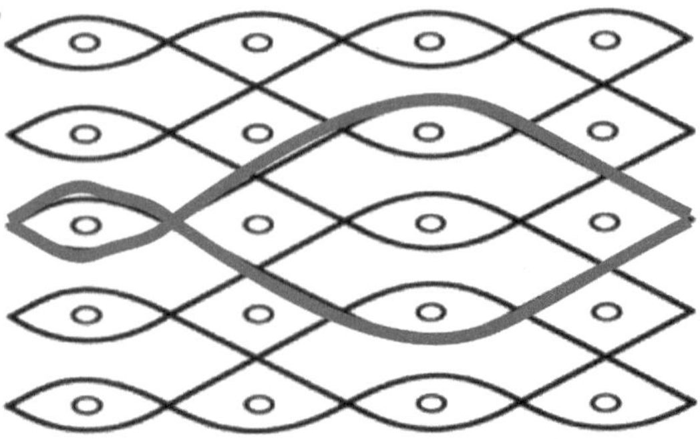

Figure 12.36 Unit cell of interlock braided structure (Brookstein, 1991; Bilisik, 2013).

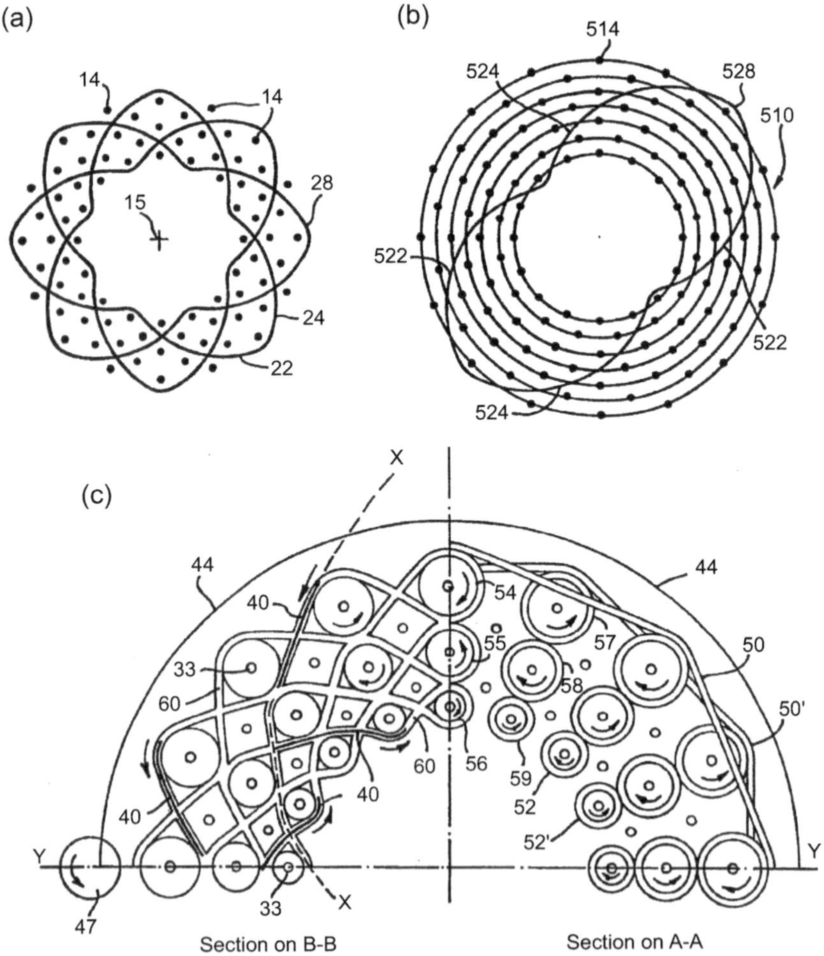

Figure 12.37 Tubular braiding machine for interlock braided structure (Temple, 1984) (a) and (b) show different track configurations on such machine; (c) simplified machine configuration.

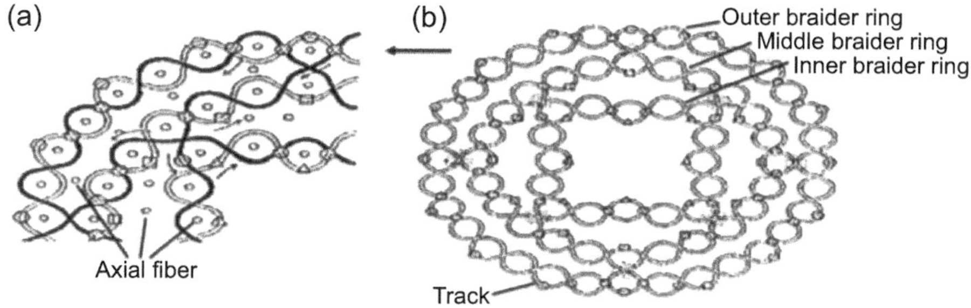

Figure 12.38 Three-layer braiding machine (Uozomi *et al.*, 2001) (a) part of the tracks making the conection between the layers better visible (b) view of the complete machine.

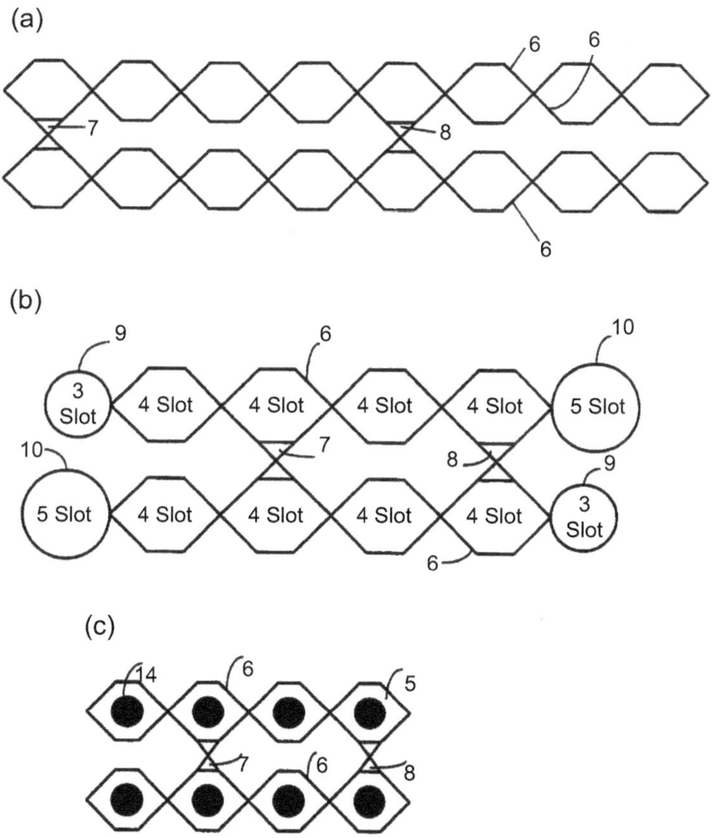

Figure 12.39 Track of one unit (Brookstein *et al.*, 1994). (a) Possible idealized track; (b) view of possible horn gear arrangement; (c) track with inlay yarns.

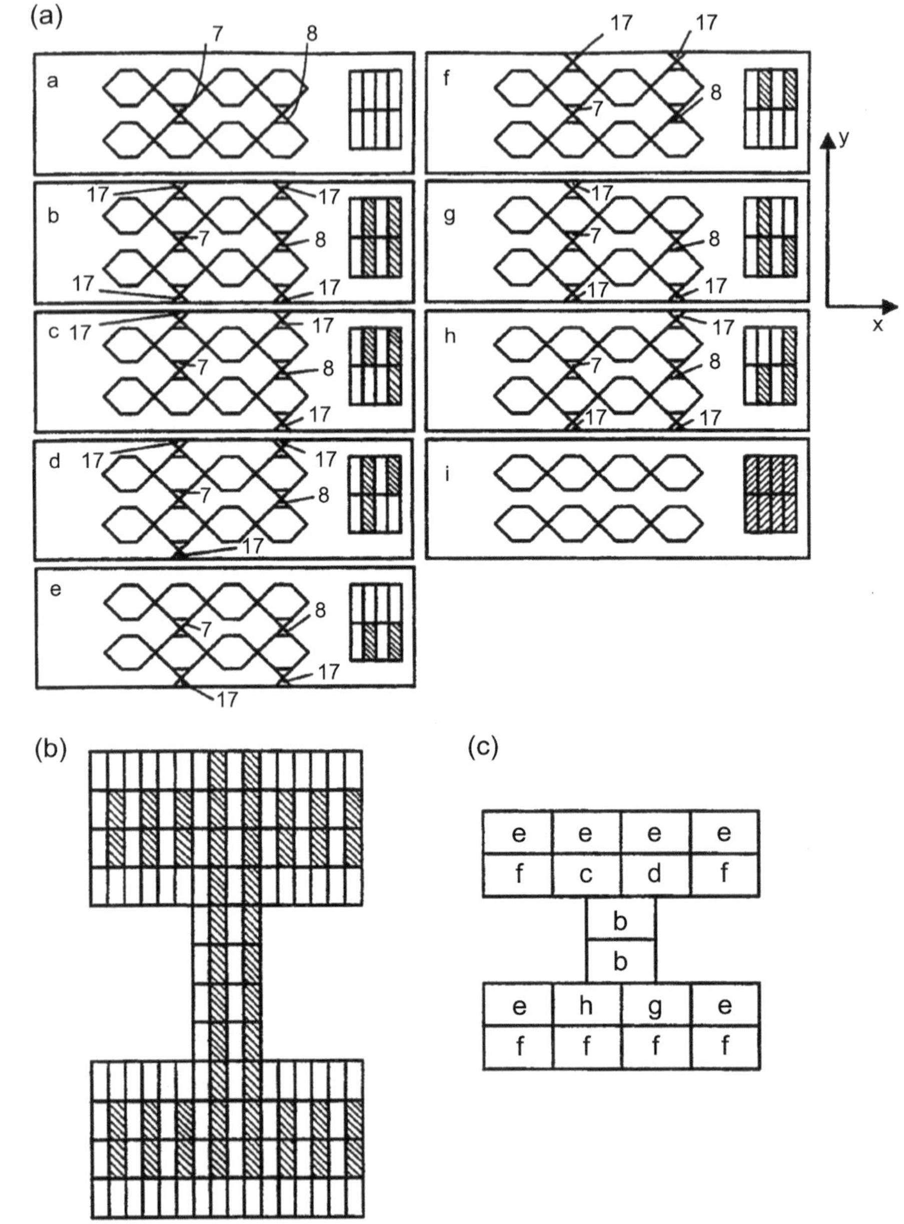

Figure 12.40 Combination of blocks with different tracks for I-profile (Brookstein *et al.*, 1994). (a) Possible configurations as separate blocks; (b) double-T profile produced by a combination of such blocks; (c) arrangement of the block types for such a profile.

very limited variety of products. Most such machines can be used only for one profile or type of profile, and even small modifications can be costly, for example replacing four-slot horn gears with five-slot to keep the tracks consistent.

12.9 Conclusions

The braiding of gaskets or more generally profiles with square form is a classic technology, and the design of such machines and products is well developed. The rules for floating length that apply in the cases of tubular and flat braids remain valid in this case as well. To produce braids with greater thickness, using more tracks, horn gears with more slots are required, which increases the size of the machine. Braiding machines for thicker profiles and special cross-sectional forms are therefore seldom built, and exist mainly in patents and papers. However, the overview of some of the profiles possible with the variable tracks on the VF braiding machine, from August Herzog, demonstrates that this particular technique significantly expands the range of the products that can be produced with a single braiding machine and will probably find a place in braiding production.

References

Akiyama, Y., Hamada, H., Maekawa, Z., Uratani, Y. and Yokoyama, A. (1995). *Braid and Braiding Method*: Google Patents. Available from: http://www.google.es/patents/US5385077.

August Herzog (2013). *Product Catalogue* [online], August Herzog Textilmaschinenfabric GmbH & Ko KG, Oldenburg. Available from: www.herzog-online.com (Accessed 01/2013).

Bilisik, K. (2013). Three-dimensional braiding for composites: A review. *Textile Research Journal*, **83** (13), 1414–1436.

Brookstein, D.S. (1991). A comparison of multilayer interlocked braided composites with other 3D braided composites. *In: 36th international SAMPE symposium*. Anaheim, USA.

Büsgen, A. (1993). *Neue Verfahren zur Herstellung von dreidimensionalen Textilien für den Einsatz in Faserverbundwerkstoffen: Dissertation.*

El Brini, O. (2010) Analyse der Zusammenhänge zwischen Maschineneinstellungen und Geflechtseigenschaften, Hochschule Niederrhein, Mönchengladbach.

Engels, H. (1994a). *Handbuch der Schmaltextilien. Die Flechttechnologie. Teil I: Maschinen und Verfahren zur Erzeugung konventioneller Geflechte*. Mönchengladbach.

Engels, H. (1994b). *Handbuch der Schmaltextilien. Die Flechttechnologie. Teil II: Maschinen und Verfahren zur Erzeugung von Flechtprodukten mit speziellen physikalischen und chemischen Anforderungen*. Mönchengladbach.

Li, M. (2013). *Braiding for Composites – Investigation About the Braiding of Products with Non-Tubular Cross Section.* Hochschule Niederrhein.

Müller, L., Milwich M., Gruhl, A., Böhm, H. Gude, M. and Haushahn,T. Biomimetisch optimierte verzweigte Faserverbundstrukturen mit hoher Tragfähigkeit. *Melliand Textilberichte* Nr. 02 Part: Band- und Flechtindustrie, 07.06.2013 Page 088.

Schneider, M. (2000). *Konstruktion von dreidimensional geflochtenen Verstärkungstextilien für Faserverbundwerkstoffe*. Aachen: Shaker Verlag.

Temple, S. (1984). *Woven Tubular Structure*, European Patent EP0113196 A1.

Uozomi T., Iwahori Y., Iwasaw S. *et al.* (2001). Braiding technologies for airplane applications using RTM process. *In: Proceedings of the seventh Japan international SAMPE symposium*.

W.L. Gore & Co. GmbH (1986). *Gore-GFO(R) Packungshandbuch*. Putzbrunn b. München: W.L.Gore & Co. GmbH.

Part Four

Computer assisted design (CAD), other software and productivity calculations for braiding

Computer assisted design (CAD) software for the design of braided structures

13

13.1 Introduction

Braiding is, technically speaking, relatively undifferentiated compared to other textile production methods, such as weaving and knitting; machines of similar construction, using the same principles, are used in braiding for completely different applications to satisfy different requirements.

For software producers, this means that there is no large market for software, and thus a limited commercial base to cover the development costs for large software packages; this is why only very few products are available "ready to use." Most notable among these is the Computer Aided Braiding (CAB) Design application (August Herzog Textilmaschinenfabric GmbH & Ko KG, Oldenburg, 2013), developed and maintained by the braiding machinery producer August Herzog to simplify and speed up the work of their customers. The TexMind Braiding Suite (Kyosev, 2012), developed by the author of this book to meet the needs of teaching and consulting, was used for crreating most of the 2D and 3D pictures of braids in this book. There are also some small freeware tools, designed to support hand braiding work; evaluation of these has shown that they are in the very early stages of development, and are not sufficiently stable for use under industrial conditions; therefore of this, these tools are not presented here.

There are several papers on modelling and simulation tools for braiding, developed by individuals, labs or companies, but none seem to be available on the market for normal users. So, these will be presented in another book.

The chapter is divided into sections covering colour patterning, production calculations, 3D geometry and braiding machine configurations. It covers the state of the art of software in 2012–2013. While not being an exhaustive review of these programs, their main features are outlined. A careful reader should be able to perform all colour design operations and the calculations – explained in detail in separate chapters – without any software, but using such software can speed up the development process significantly. This chapter does not set out to compare these programs, even though some features bear comparison in producing the same results while being differently designed. Those readers looking for software comparisons are recommended to contact the producers and compare their requirements to the actual software products.

Braiding Technology for Textiles. http://dx.doi.org/10.1533/9780857099211.4.315

13.2 Colour design of braided structures

13.2.1 CAB Design Herzog

On starting the colour design part of CAB Design, a braid property selection menu should appear. The user should then select between tubular and flat braiding, before selecting the type of pattern. The numbers in brackets show the occupation of the carriers as "Full – Empty," while the text preceding them describes the structure. Translated into the language of the book, the first menu item "one over two (1–1)" denotes a floating length of 2, corresponding to a "regular braid" (2:2–1); "one over one (tandem) 2–2" is a floating length of 1 as a group of two (1:1–2), and "one over one (half occupation)[1–3]" is a floating length of 1 as a group of one (1:1–1).

Next, the number of the carriers has to be selected, and the braiding angle of the structure (Fig. 13.1).

Once the initial settings are confirmed, the main window for the colour design appears. During braiding of tubular structures, the carriers are separated into two tracks; the user should select the track to which the yarn colours are assigned (Position 1, Fig. 13.2) before selecting the colour for the next carriers (Position 2, Fig. 13.2). Clicking the radio buttons with the carrier numbers immediately sets the yarn of this carrier with the selected colour (Position 3, Fig. 13.2). The colours can

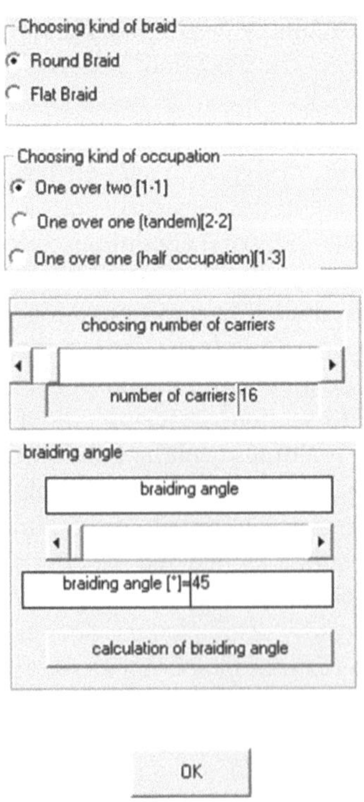

Figure 13.1 Starting form of CAB Design.

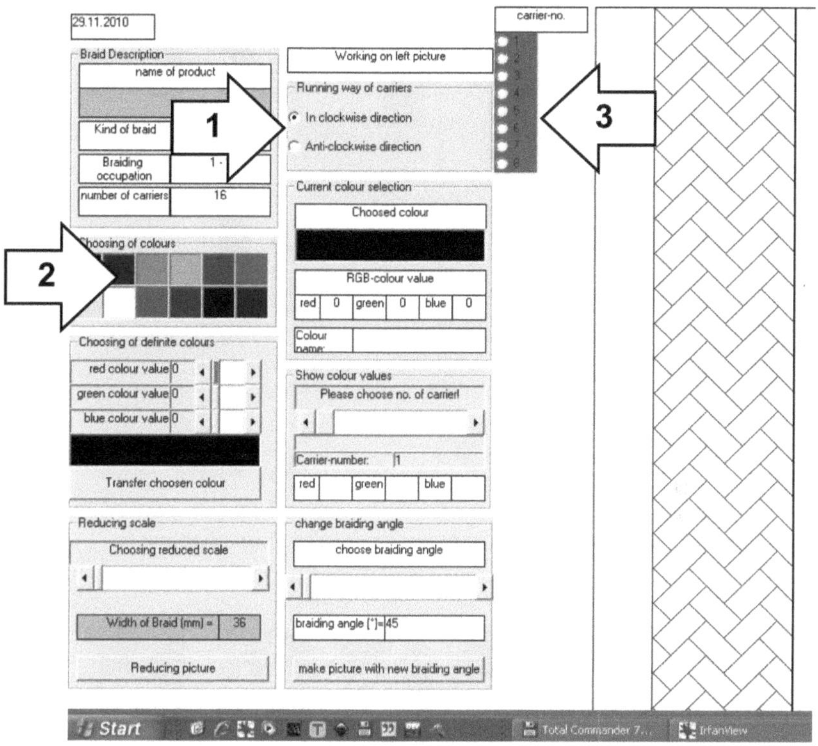

Figure 13.2 Window for design of tubular braid.

also be selected as RGB (red, green, blue) values from the standard Windows colour selection dialog.

Figure 13.3 shows the colour black being assigned to the first four carriers in one track and the resulting changes to the appearance of the braid. In Fig. 13.4, the process is continued as the colour grey is assigned to all eight carriers of the second track. Figure 13.5 is of a window showing the tubular structure with two yarns in a group (1:1–2) for a machine with 48 carriers, while Fig. 13.6 shows some developed samples.

The braids with a floating length of 1 (1:1–1) require a simple square mesh – shown in Fig. 13.7 – where all other options are identical. Figure 13.8 presents some steps in developing such samples.

The flat braided structures have only one track, and this is the main difference in the user interface – all carriers are accessible at once (Fig. 13.9). Figure 10.10 presents the first three steps in developing a flat braided sample and some possible end solutions. The braiding angle for both tubular and flat structures can be changed while the program is running – as shown in Fig. 13.11 for one tubular braided fabric – so that the appearance of the fabric is adjusted to be closer to that of the real product. One convenient option is that the program can show two samples at the same time (assuming identical machines and arrangements), and the user can choose one – left

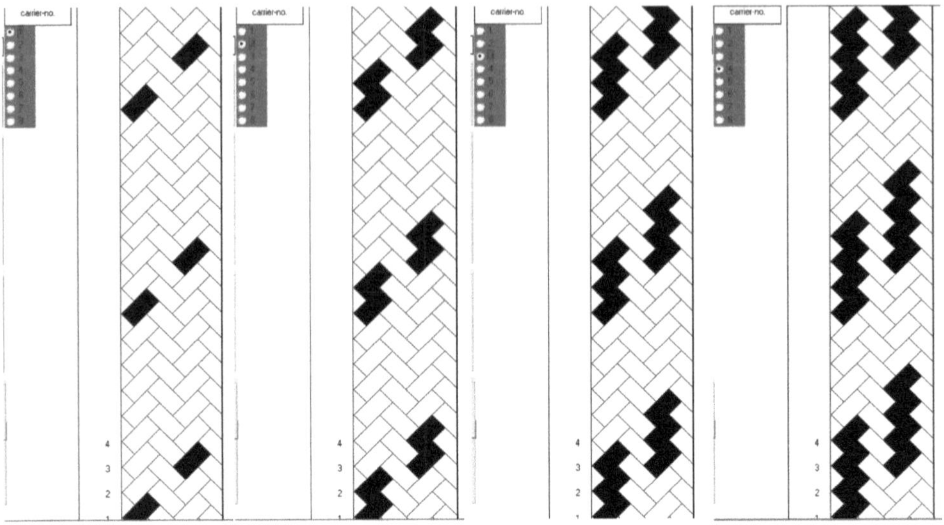

Figure 13.3 Setting the first four carriers in one track with black as colour.

Figure 13.4 Sequence for setting grey as colour for all carriers in the second track.

or right – to put into production. This allows easy comparison between the variants (Fig. 13.12).

The sample pictures can be copied into the clipboard as bitmap images, and used in other applications or printed on standard printers; the colour arrangement can also be printed. The omission of both print preview and some information about the current colour of each carrier during the work (from the version used) is a little confusing. The user interface has other problems – large folder names are not visible, the windows are not really resizable, etc., all requiring some degree of user adaptation to the software. For standard structures, and after some adaptation (for instance, setting

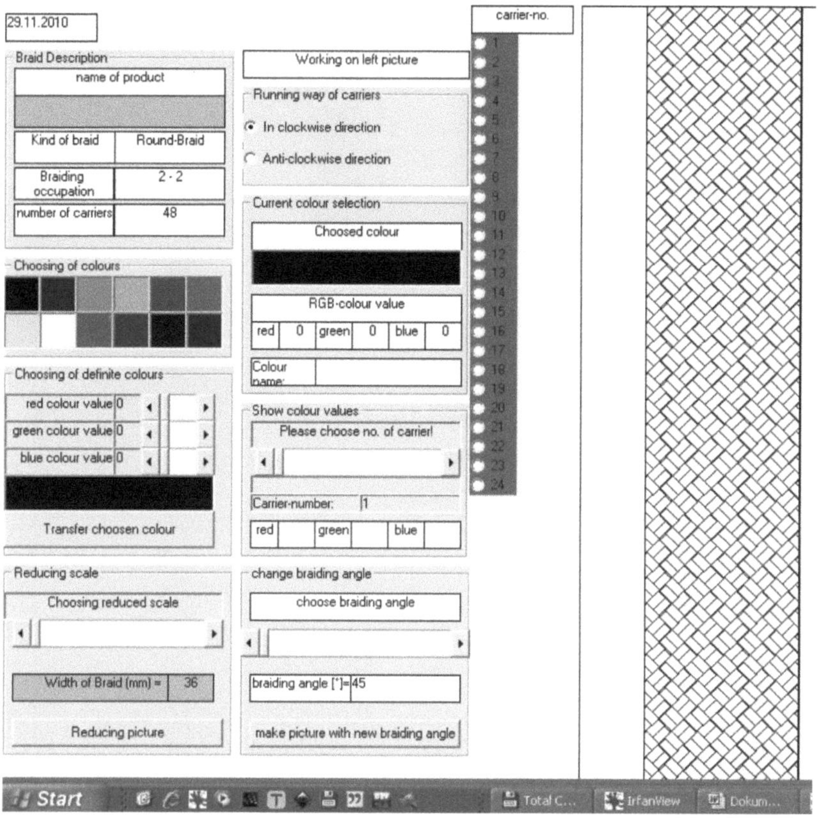

Figure 13.5 Main windows for editing of tubular braid with two yarns in a group and floating length of one group, (1:1–2).

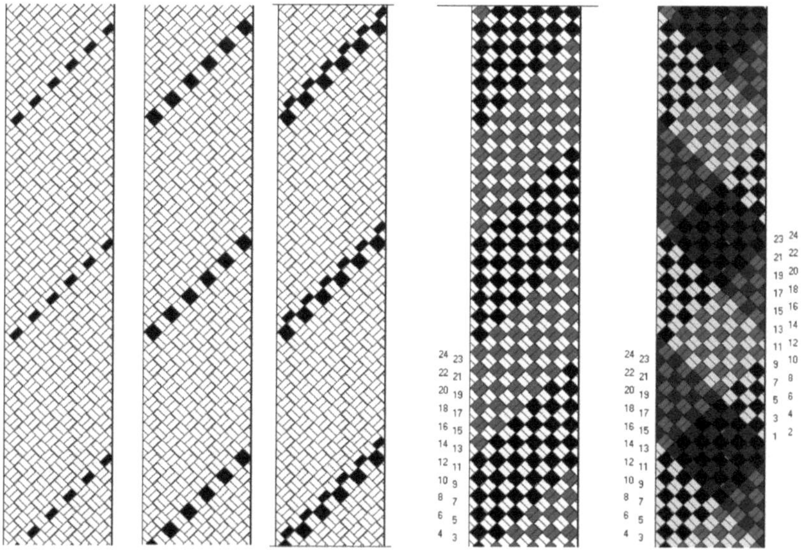

Figure 13.6 First three steps of the colour selection and two final variants for 1:1–2 braid.

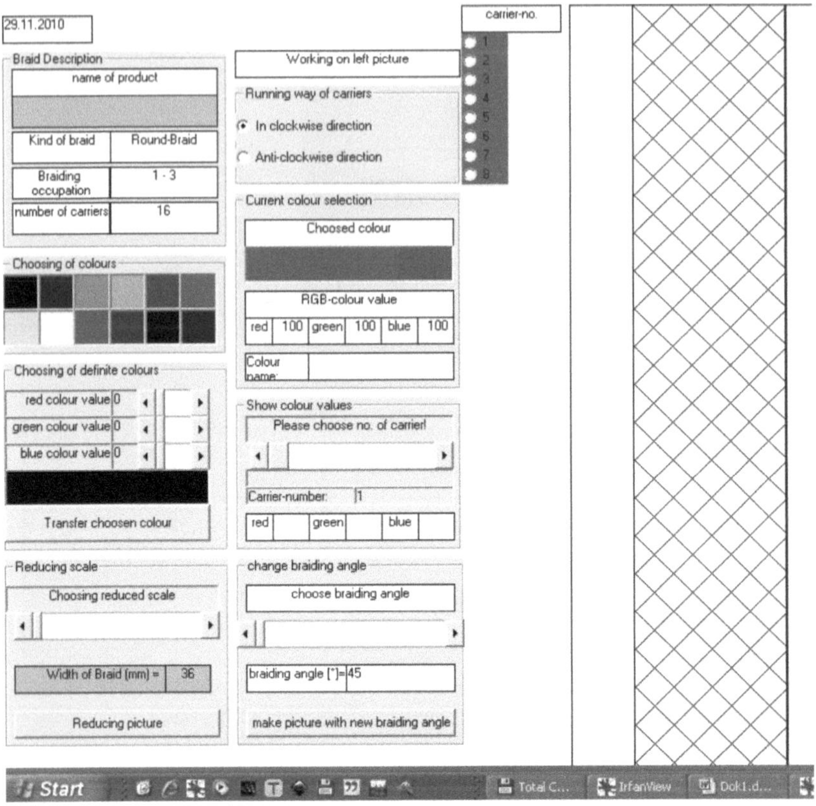

Figure 13.7 Main windows for colour selection of tubular braid with floating length of 1 (1:1–1).

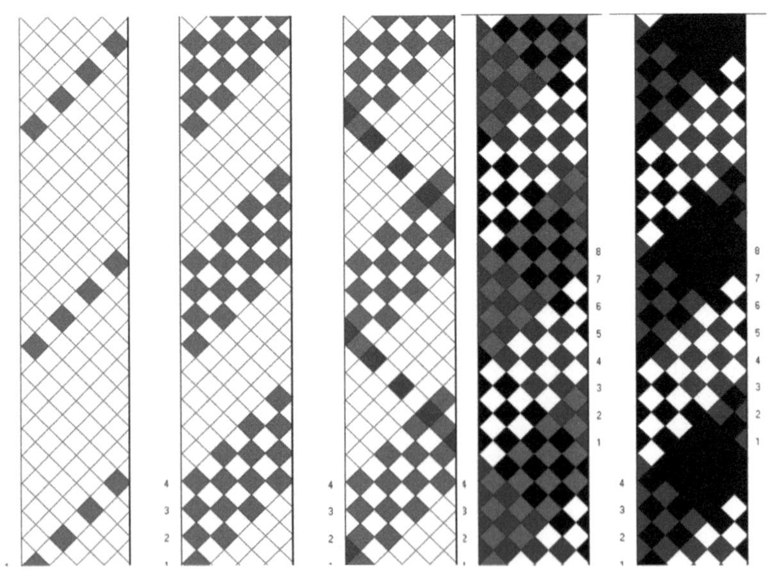

Figure 13.8 Some steps during the colour design of 1:1–1 braid.

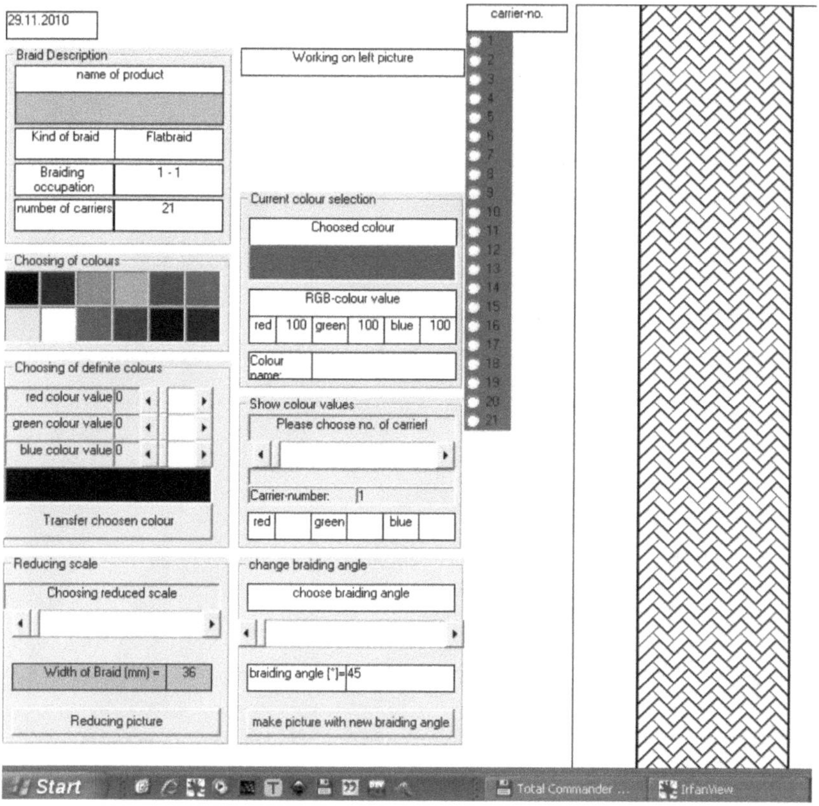

Figure 13.9 Main window prepared for flat braid with floating length of 2 (2:2–1).

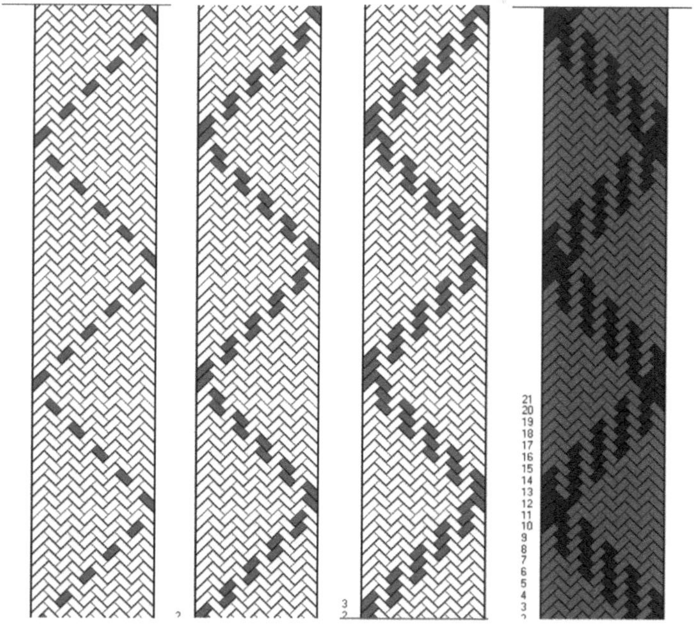

Figure 13.10 First three steps and some final variant of colour design of flat braid with structure 2:2–1.

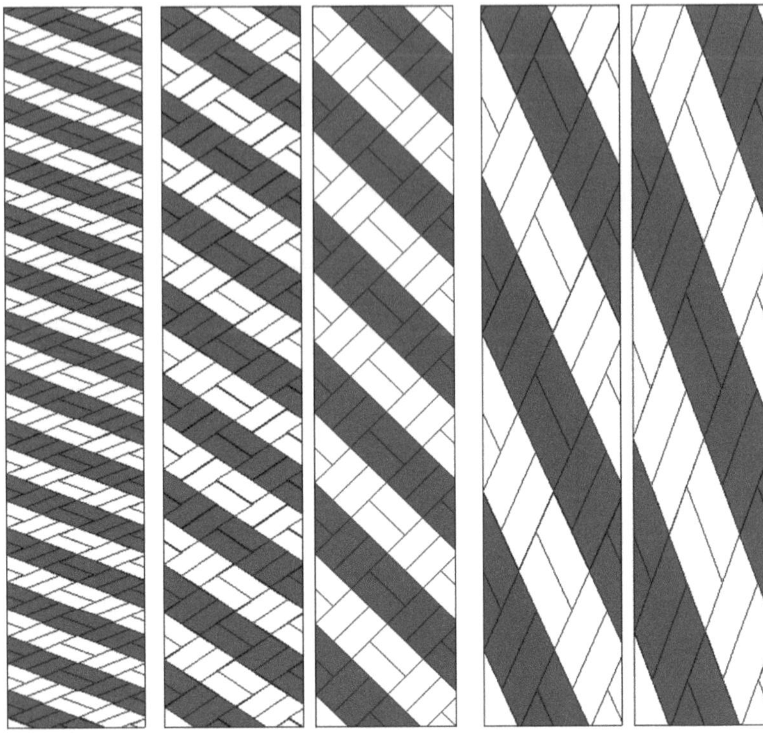

Figure 13.11 Tubular braid in 1:1–2 with different drawn at braiding angles.

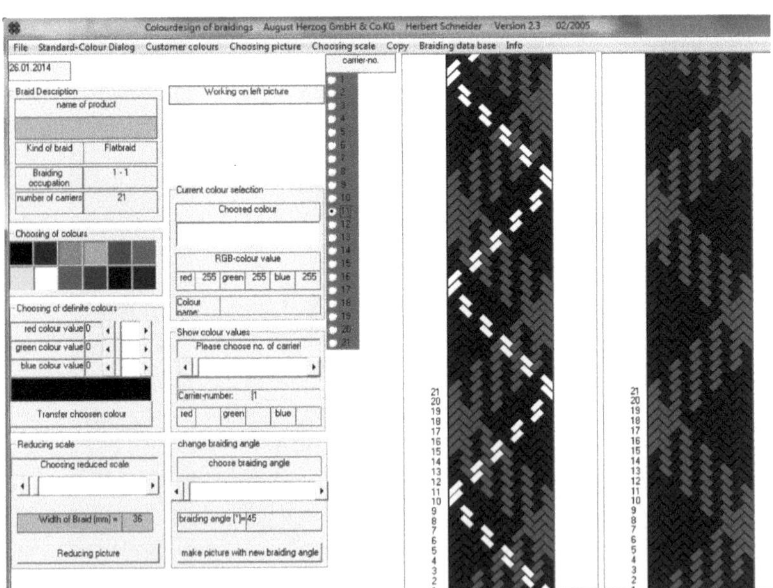

Figure 13.12 The sample can be copied in a second panel for comparison.

a PDF printer as standard printer), the program can give reliable support during the design process of braided structures.

13.2.2 TexMind 2D braid designer

The program "TexMind Colour Braids 2D" was created in 2012 to address some issues in CAB Design; these improvements included the ability to export the pictures in high quality vector format, create structures with various floating lengths of more than two, and a more intuitive user interface (that was especially necessary in teaching braiding pattern design).

After selection of the type of the braid (tubular or flat), two types of work are possible – using standard or customized machine specifications. A set of standard machines is available for direct loading and use for colour patterning. Machine customization in the main window requires the main sample properties (Fig. 13.13) – such as the number of carriers on the machine, number of horn gears, floating length and number of yarns in a group – to be specified. After the machine specification is complete, the user can select colours for each carrier by clicking over the specific cells with the right mouse button. Here, the selection is comparable to CAB – the user has to first specify the colours that will be used in the sample (Fig. 13.14) in a colour

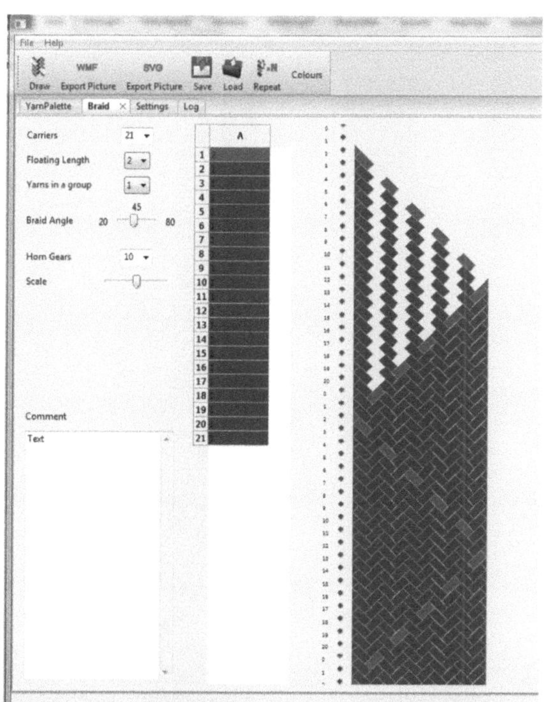

Figure 13.13 Colour sample of flat braid with 21 carriers and one carrier with different colour.

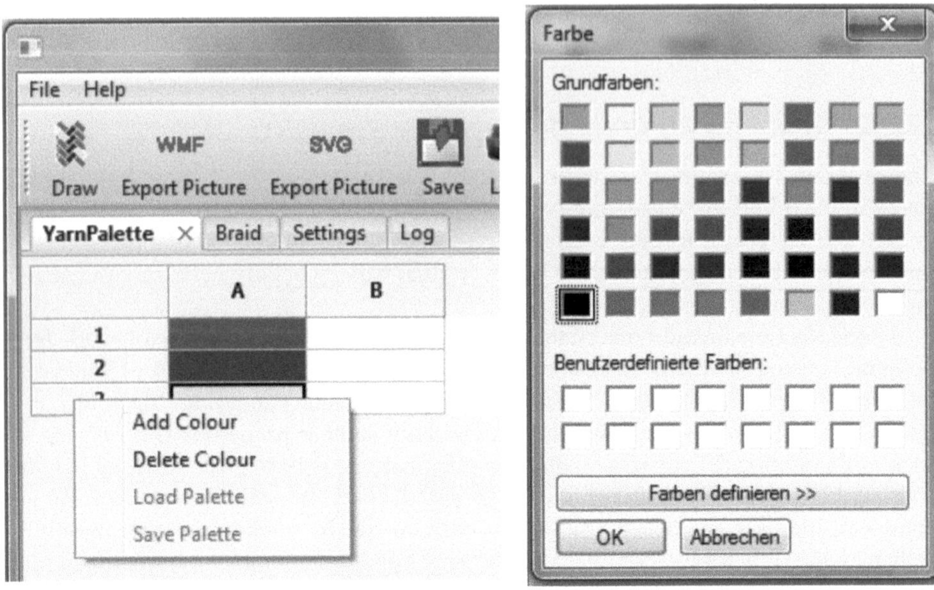

Figure 13.14 The yarn palette collects the colours for the current pattern. The colours can be selected using standard colour dialog.

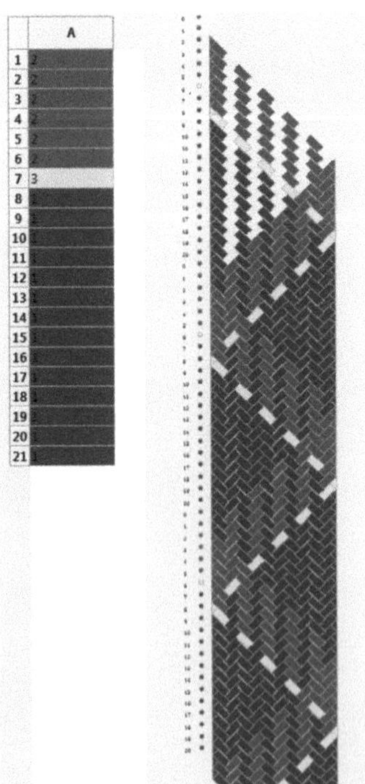

Figure 13.15 Modified sample from Fig. 11.13 with three colours.

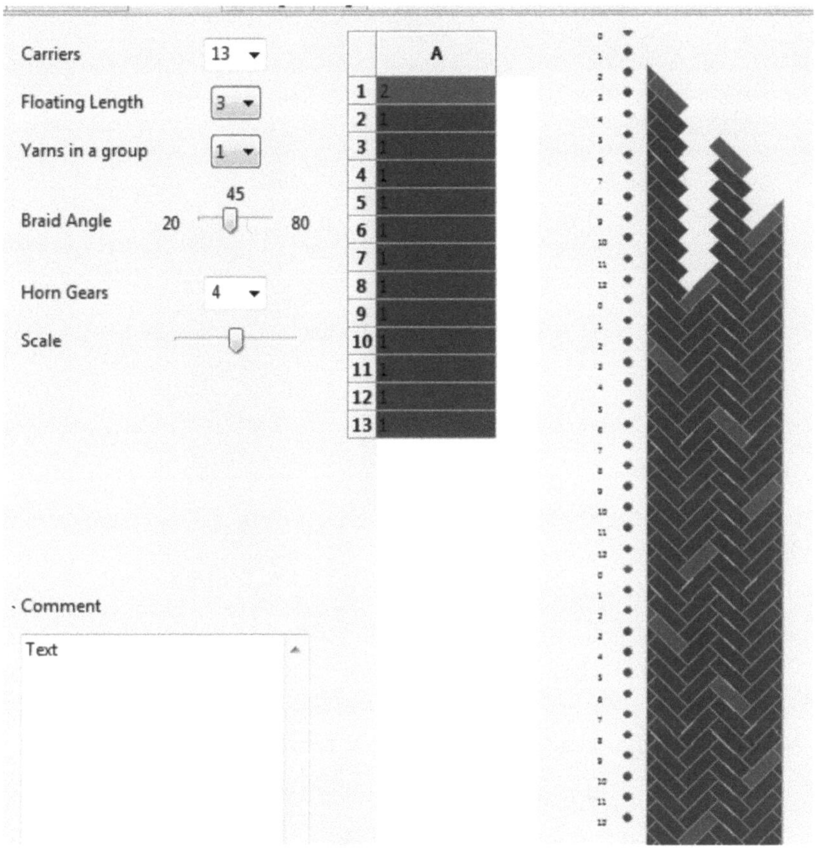

Figure 13.16 Flat braid with floating length of 3 (structure 3:3–1).

palette. After that, every subsequent mouse click over one cell, corresponding to one carrier, changes the colour to the next one from the colour palette.

Figure 13.15 shows the colour patterning of a flat braided sample with three colours and floating length of 2 (2:2–1), while the sample in Fig. 13.16 has a floating length of 3 (3:3–1). However, this – and the following figures – shows colour patterning for machines that are seldom built, as they requires larger number of slots per horn gear.

Figures 13.17 and 13.18 show samples with a larger number of yarns in one group, and Fig. 13.19 shows a sample with the smallest braid with three yarns. Changes to the braiding angle of the structure can be applied in every case, and some variants with different braiding angles are shown in on Fig. 13.20.

For all samples, the first repeat is presented without additional yarns; this seems to facilitate the understanding of the yarn path in the braids, which is important from a training perspective. Cropping of this first part for presentation is not problematic, and has to be configured in the settings.

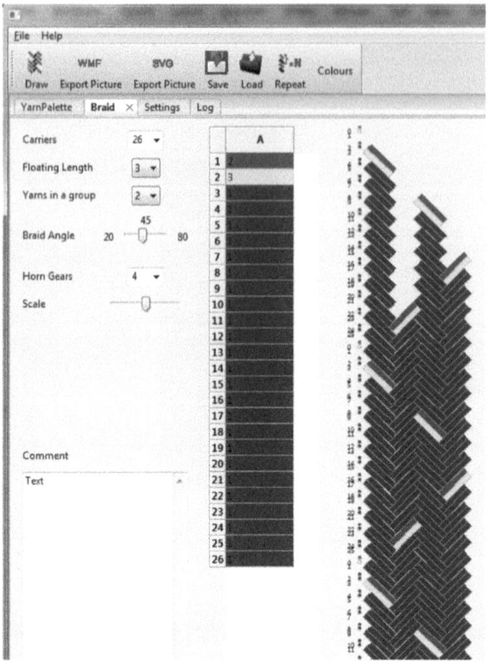

Figure 13.17 Flat braid with two yarns in the group and floating length of 3 (3:3–2).

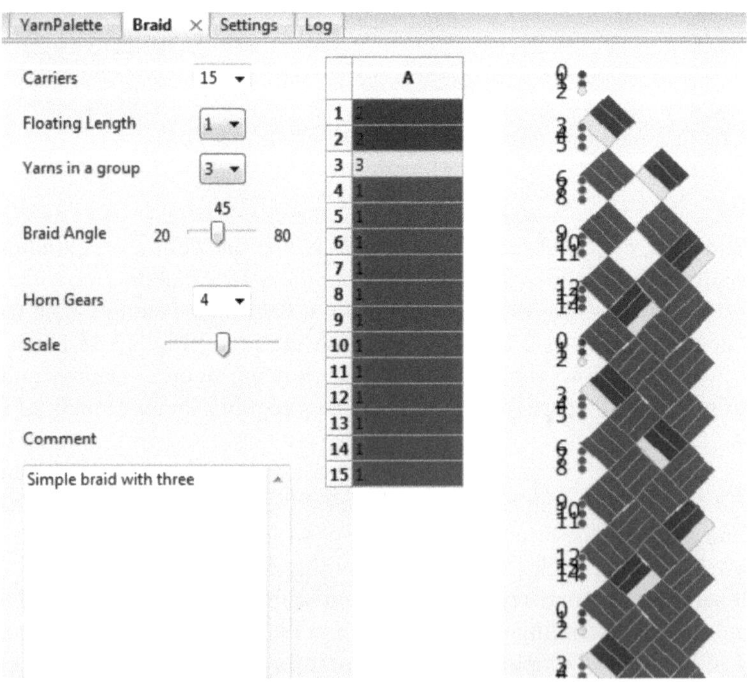

Figure 13.18 Flat braid with floating length of one and three yarns in a group (1:1–3).

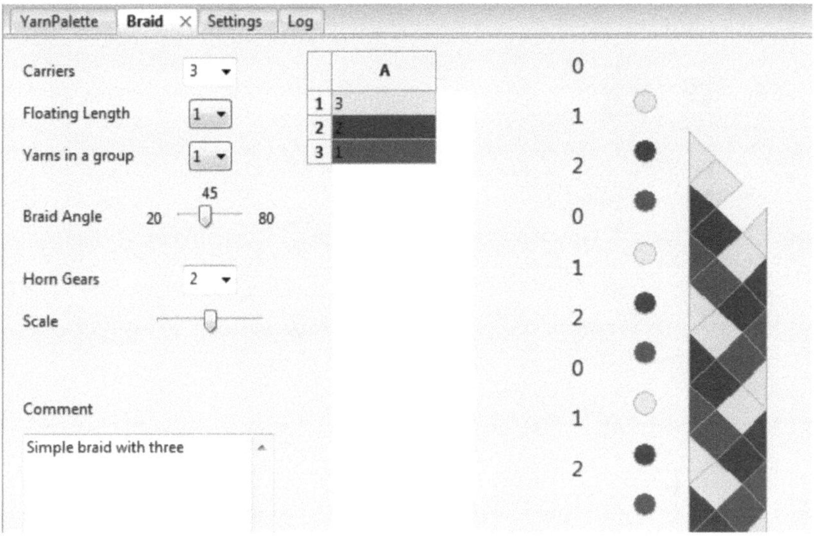

Figure 13.19 Pigtail sample with three carriers.

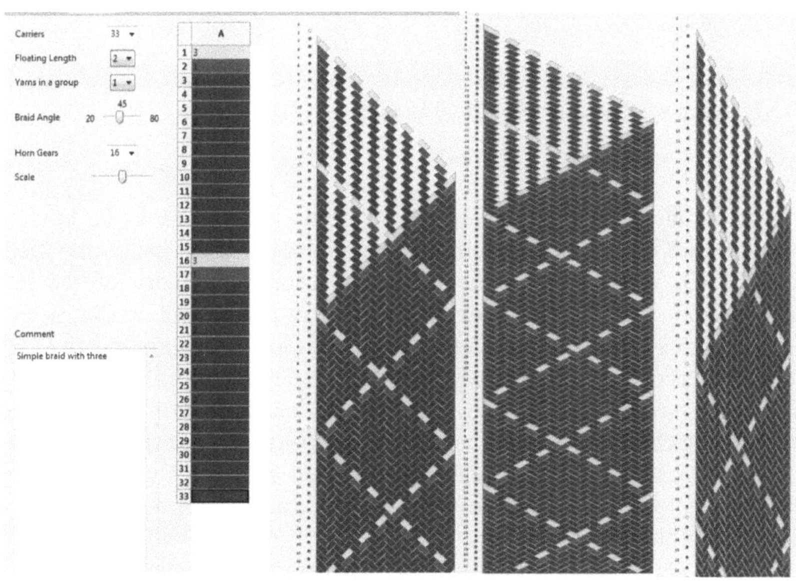

Figure 13.20 Changes in the braiding angle are immediately visible.

13.3 3D geometrical models

A survey of the literature reveals several tools used for generating the three-dimensional geometry of braided structures, but most of these seem either to be performing

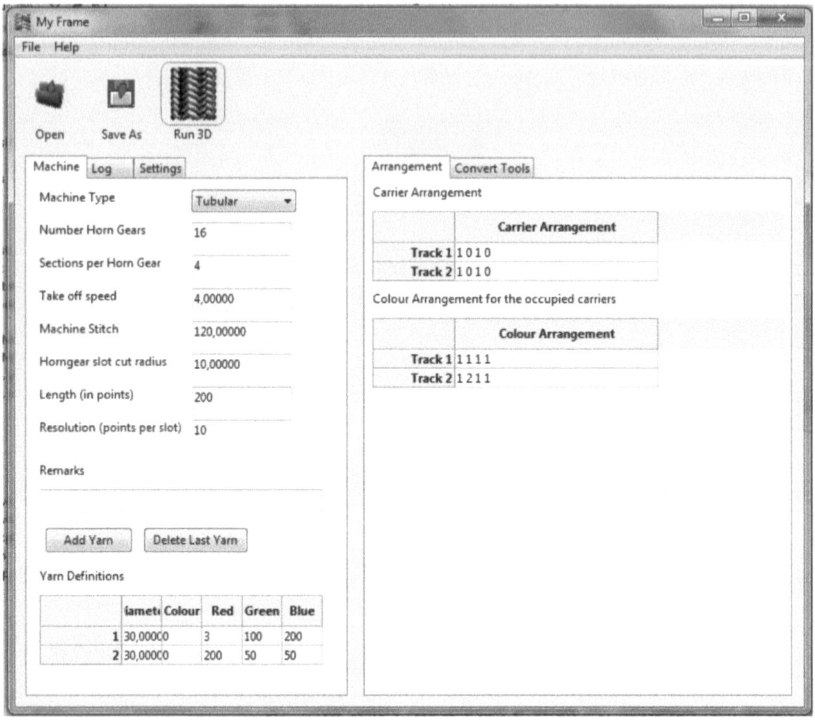

Figure 13.21 Main window of the TexMind Braider Geo 3D.

limited tasks in university environments, or to be bespoke solutions for companies. The only software currently available for general, commercial use is the TexMind Braider Geo 3D, which creates the three-dimensional geometry of the braided structures based on geometrical models and emulates carrier motion (Kyosev, 2012, 2013a). Figure 13.21 presents the starting screen of the program for classical tubular and flat braided structures. For these structures, the definition of the horn gear position and carrier arrangement can be implemented analytically in software, so that only the main machine parameters have to be specified. The carrier arrangement is given per track, where the empty places are marked with zeros and the full places with ones. The colour arrangement of the occupied tracks is defined separately, with manual entry of colour numbers.

Figure 13.22 shows three different tubular structures generated by the software, while Figs 13.23 and 13.24 show flat braided structures and pigtail-braided structures, respectively. As visible for experienced readers, the created geometry is based on pure geometrical calculations using emulation of the carrier path. Because of this, the geometry is idealized and cannot be used directly for mechanical calculations. The program also creates a figure with horn gear configurations and carrier arrangement (Fig. 13.25); this figure allows the user to check if the carrier arrangement is suitable, or whether the starting slot of one of the carriers has to be shifted in order to avoid

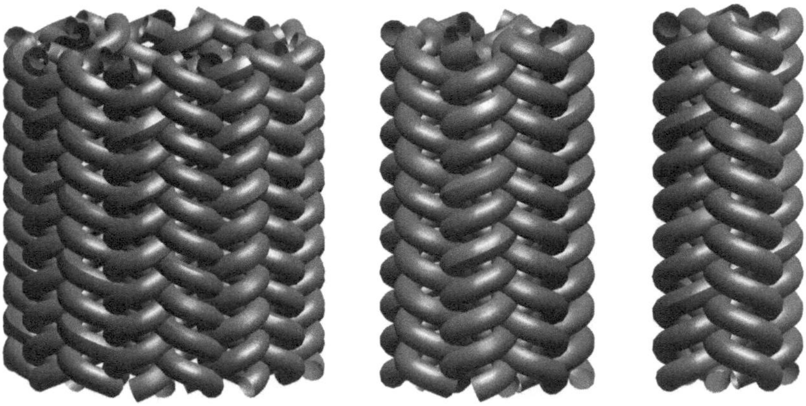

Figure 13.22 Tubular braided structures simulated with TexMind Braider Geo 3D.

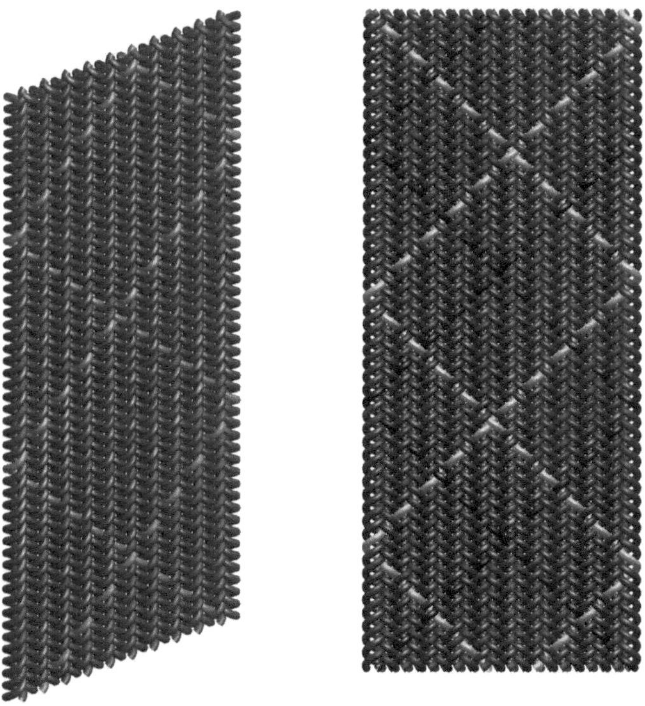

Figure 13.23 Flat braids, generated with TexMind Braider Geo 3D.

collision of the carriers. There are currently no automated checks for collision in this software.

The program uses open XML format for the structure definition (Fig. 13.26), allowing easy exchange, not only with the other tools of the TexMind Braiding Suite, but with any external software in which XML parsing may be implemented.

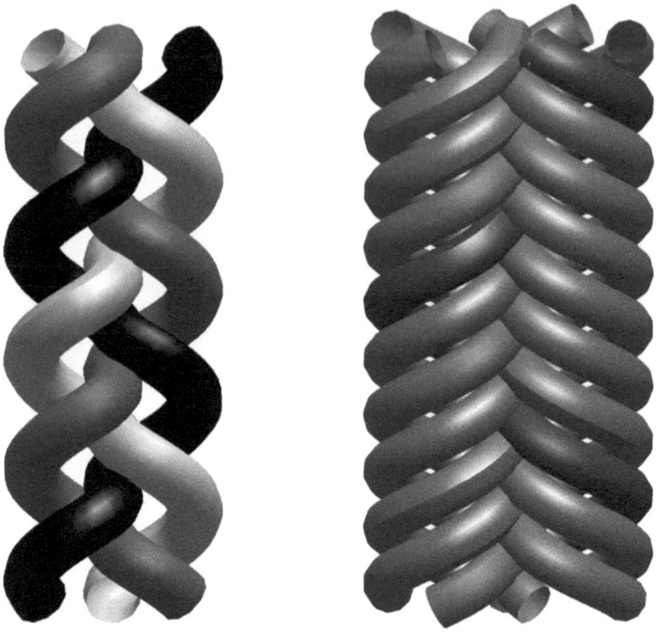

Figure 13.24 Pigtail braids, generated with the TexMind Braider Geo 3D.

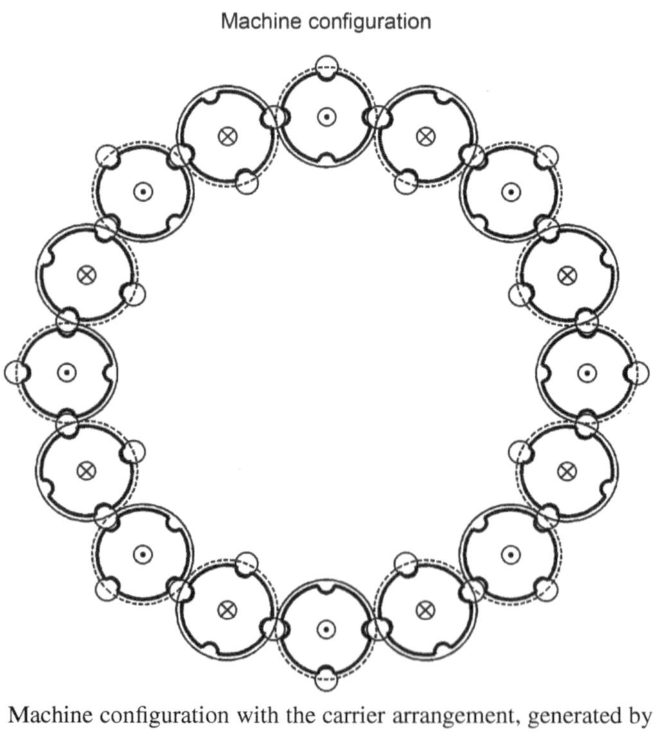

Figure 13.25 Machine configuration with the carrier arrangement, generated by TexMind Braider Geo 3D.

```
<?xml version="1.0"?>
<Format Company="TexMind www.texmind.com" Type="Braid2D" Version="1" />
<Braid FloatingLength="2" Yarns_in_group="1" Nr_carriers="21"
Nr_horngears="10" Nr_tracks="1">
    <data2D B="20" BraidAngle="45" />
    <ColourPalette Nr_Colours="3">
        <Colour-0 Number="0" Red="128" Green="0" Blue="255" />
        <Colour-1 Number="1" Red="255" Green="0" Blue="128" />
        <Colour-2 Number="2" Red="255" Green="255" Blue="0" />
    </ColourPalette>
    <Yarns>
        <Yarn-0 Nr="0" ColourNr="2" />
        <Yarn-1 Nr="1" ColourNr="2" />
        <Yarn-2 Nr="2" ColourNr="2" />
        <Yarn-3 Nr="3" ColourNr="2" />
        <Yarn-4 Nr="4" ColourNr="2" />
        <Yarn-5 Nr="5" ColourNr="2" />
        <Yarn-6 Nr="6" ColourNr="3" />
        ...
    </Yarns>
</Braid>
```

Figure 13.26 Example of the XML structure for flat braided sample calculation.

13.4 Custom machine configurator

For non-standard machines, a configurator is available, where the position and size of each horn gear can be specified individually (Fig. 13.27) (Li, 2013). It is planned (Kyosev, 2013b) to use these configurations to simulate the braiding structure in Braider Geo 3D in the near future; currently, the tracks have to be entered manually, making the process slow and inconvenient.

13.5 Calculations for braiding

13.5.1 CAB design

CAB Design consists of a large set of single calculation tools, where different aspects of the braiding product and process are considered. Figure 13.28 shows one of these; using the length of the yarn on the bobbin, the take-off speed (output) and the braiding angle, machine running time and rope length are estimated. Figure 13.29 shows the calculations for double-braided rope, where the bobbin sizes for the machines for cover and core have to be selected; based on the angles and data for the both products (core and cover) the main properties and data for production are calculated.

13.5.2 TexMind braiding calculator

The Braiding Calculator of TexMind can directly read data for 3D Simulation settings with the Braider Geo 3D (Fig. 13.30), and allows the printing, saving and exporting

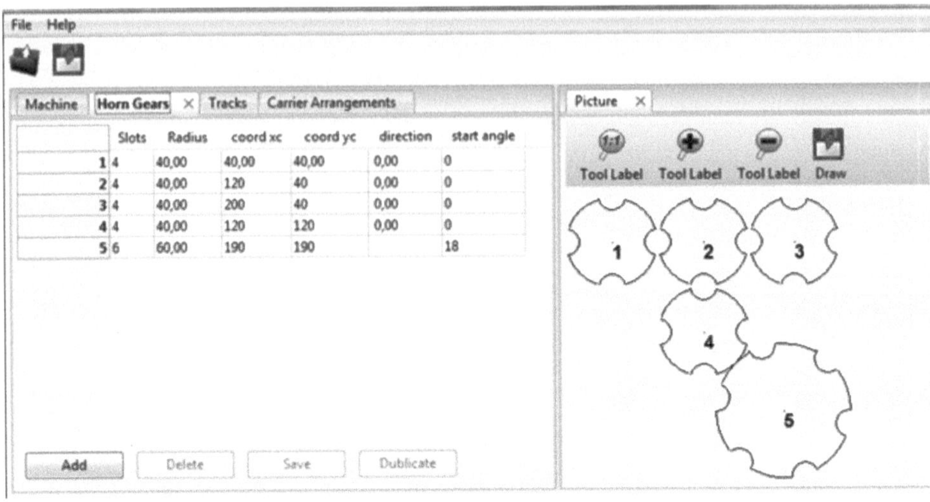

Figure 13.27 TexMind braiding machine configurator.

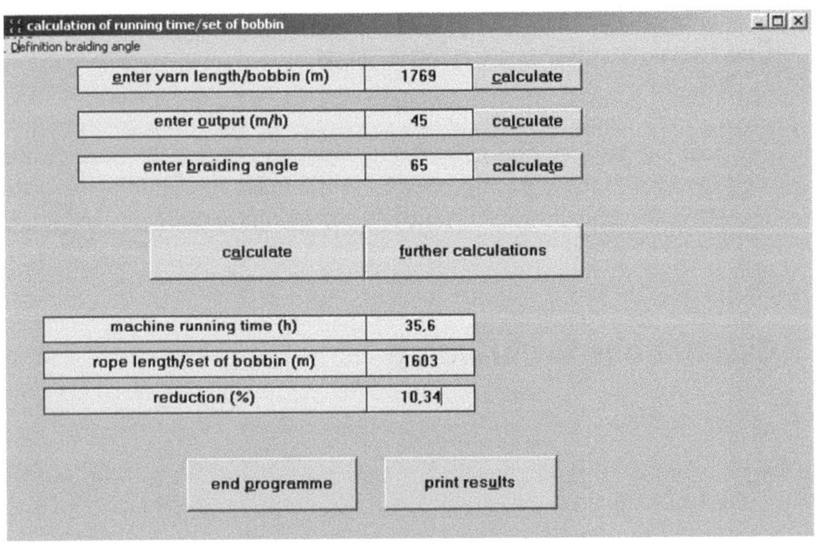

Figure 13.28 Calculation of the running time and the rope length within CAB Design.

of data. In CAB Design, each calculation is prepared in a separate window, while in TexMind most of the calculations are shown in a window resembling a grid in Microsoft Excel. In this grid, the user can immediately see the result of any change of one parameter. In its advanced version the calculator recognizes several dimension units for the parameters directly, converting between the dimensions internally.

The calculations follow the equations given in Chapter 14, "Productivity calculations in braiding." The number of carriers, braiding type, the speed of the horn gears and the width of the braid (for flat fabrics, or the width of the pressed tubular

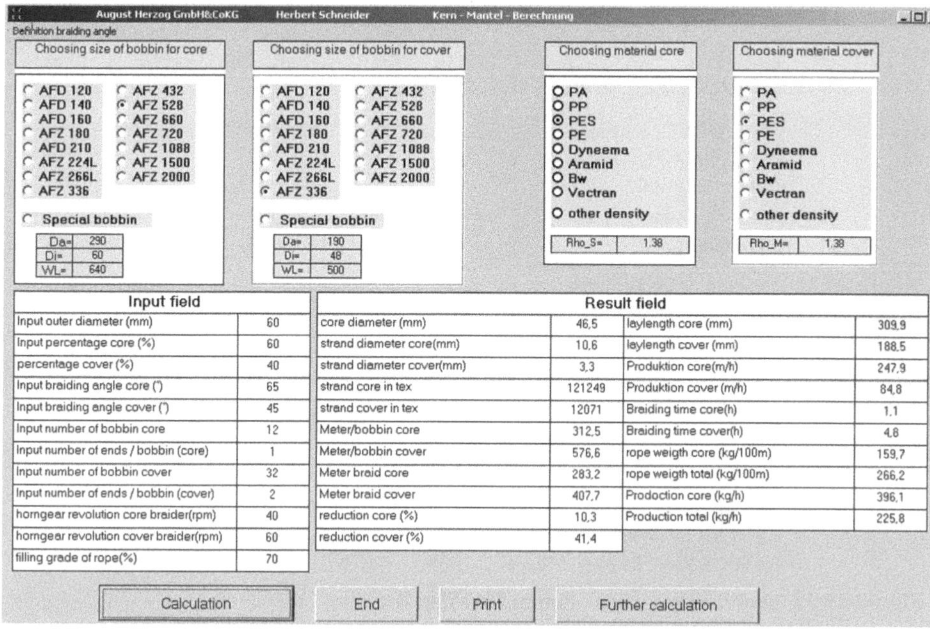

Figure 13.29 Calculations for core-cover (double layered) braided products within CAB Design.

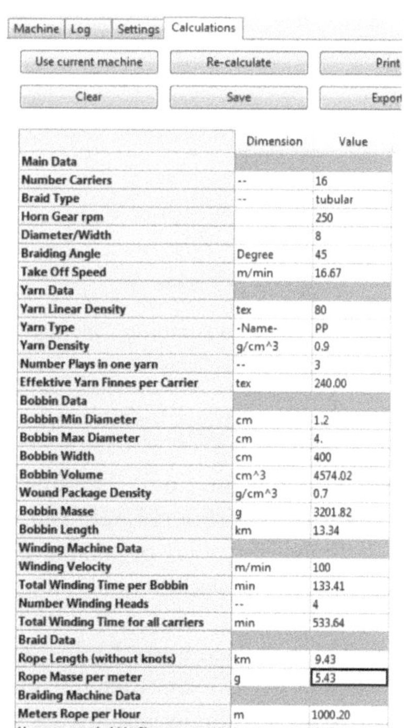

Figure 13.30 Main window of the TexMind Braiding Calculator.

	Dimension	Value
Main Data		
Number Carriers	--	16
Braid Type	--	tubular
Horn Gear rpm		250
Diameter/Width		8
Braiding Angle	Degree	45
Take Off Speed	m/min	16.67

Figure 13.31 Take-off speed calculation for given machine and width of the braid in the TexMind Braiding Calculator.

Take Off Speed	m/min	[illegible]
Yarn Data		
Yarn Linear Density	tex	80
Yarn Type	-Name-	PP
Yarn Density	g/cm^3	0.9
Number Plays in one yarn	--	3
Effektive Yarn Finnes per Carrier	tex	240.00

Figure 13.32 Yarn fineness calculation in TexMind Braiding Calculator.

Effektive Yarn Finnes per Carrier	tex	240.00	
Bobbin Data			
Bobbin Min Diameter	cm	1.2	
Bobbin Max Diameter	cm	4	
Bobbin Width	cm	400	
Bobbin Volume	cm^3	4574.02	1
Wound Package Density	g/cm^3	0.7	
Bobbin Masse	g	3201.82	2
Bobbin Length	km	13.34	3

Figure 13.33 Main windows of the TexMind Braiding Calculator.

braid, see Chapter 14) and the braiding angle together allow the take-off speed of the machine to be calculated (Fig. 13.31).

The yarn effective fineness based on the single yarn fineness and the number of ply yarns is calculated in a separate section (Fig. 13.32). The yarn density given there is used for calculating the yarn diameter, but is not presented in the figures.

The fineness of yarn is used together with the bobbin geometric data to determine the length of the yarn in one bobbin (Fig. 13.33). For proper calculation, the wound package density is required, which is used to determine the yarn mass in the bobbin (Position 2 on Fig. 13.33); this density is lower than the material density and the yarn density, once the air gaps between the single yarn pieces on the wound package are taken into account.

Bobbin Length	km	13.34
Winding Machine Data		
Winding Velocity	m/min	100
Total Winding Time per Bobbin	min	133.41
Number Winding Heads	--	4
Total Winding Time for all carriers	min	533.64

Figure 13.34 Main windows of the TexMind Braiding Calculator.

Braid Data		
Rope Length (without knots)	km	9.43
Rope Masse per meter	g	5.43
Braiding Machine Data		
Meters Rope per Hour	m	1000.20
Hours to next bobbin Change	hour	9.43

Figure 13.35 Summary of the calculations from TexMind Braiding Calculator.

Having the length of the yarn in one bobbin, the time for winding of one bobbin (or all bobbins, in the case of a machine with multiple winding heads) is calculated (Fig. 13.34).

The summary for the product is given in Fig. 13.35, where the rope length is calculated, considering the yarn length per bobbin and the braiding angle and the machine productivity. For braids with inlay yarns and cores, the sheet consists of an additional two columns for these separated systems.

13.6 Summary

Both CAB Design and TexMind Braiding Suite cover the topics of colour patterning and main productivity calculations, and can speed up the development process significantly. CAB Design is closely based on the machines produced by August Herzog, where some of the settings can be easily selected; however, it also allows specifying custom data types and may be applied to other manufacturers' models of similar construction. TexMind Tools are designed to be machinery-independent, and allow, for instance, colour patterning for structures with higher floating length. The tools have different designs, but with both of these the typical braider will be able to select the colour arrangement for any given design, and to make appropriate productivity calculations.

In 3D braid visualization, only the BraiderGeo3D is currently available "ready to use," where the geometry is created on the basis of the emulation of carrier motion without considering the forces between the yarns in the braiding area. It can be expected that, in the near future, geometrical simulation will be available for any custom machine, defined with the configurator tool; such models can currently be created only with the use of general CAD and FEM Software, not specialized for braiding.

References

August Herzog Textilmaschinenfabric GmbH & Ko KG, Oldenburg (2013). *CAB Design: Computer Aided Braided Design* [online]. Available from: http://www.herzog-online.com/_rubric/index.php?rubric=Products+Software+EN.

Kyosev, Y. (2013a). Computer aided colour and structural design of braided structures. *In:* X. Chen and J. Hearle, eds. *CD-ROM Proceedings of the First International Conference on Digital Technologies for the Textile Industries*. Manchester: TexEng Software LTD.

Kyosev, Y.K. (2012). *TexMind Braider, www.texmind.com* [online]. Mönchengladbach. Available from: www.texmind.com.

Kyosev, Y.K. (2013b). Some computational and modelling aspects about the multifilament modeling of braided preforms. *In:* S. Lomov, ed. *Proceedings, Composites Week@Leuven and TexComp-11 conference*. Leuven.

Li, M. (2013). *Braiding for Composites – Investigation About the Braiding of Products with non-Tubular Cross Section*. Hochschule Niederrhein.

Productivity calculations in braiding

14

14.1 Introduction

The productivity of a braiding machine depends on three groups of parameters, describing the machine, the yarn material and the structure. The machine parameters are the speed, number and size of the horn gears; higher speed indicates faster production, while a larger number of horn gears will produce wider (but not longer) braids. Finer yarns or wires take longer to produce each linear metre of braid compared to braiding with coarser, thicker yarns. Finally, the production of braids with higher braiding angles takes longer compared to that of similar braids (same material, produced on the same machine) with lower braiding angles.

An understanding of these relations is fundamental. This chapter has the goal of illustrating these relations numerically, so that the time required for the production of a given amount of braid, and the amount of yarn material required for this braid, may be readily calculated.

The following section presents some geometric relations within the braided products such as repeat length, braiding angle and yarn length per repeat. The crimp of the yarns is threaded as simply as possible to make the equations calculable even with a pocket calculator. In the second part of this chapter, the calculation of the weight per metre of the braid and the estimation of the time for the braiding of one repeat or given length are presented. The chapter ends with some calculated examples.

14.2 Required yarn length

14.2.1 *Yarn length in braids*

The length of one yarn, required for one repeat in the braids, depends directly on the braiding angle. L_{repeat} is the length of one repeat of the structure (the distance between the points A and B in Fig. 14.1, where one yarn starts at the same place). Half of the path of this yarn is orientated to the product axis at angle α, with the another half at angle $-\alpha$; if this second half is drawn as a prolongation of the first, then the total yarn piece in one repeat will be of length AC, where AB is the length of one repeat of the braid.

Braiding Technology for Textiles. http://dx.doi.org/10.1533/9780857099211.4.337

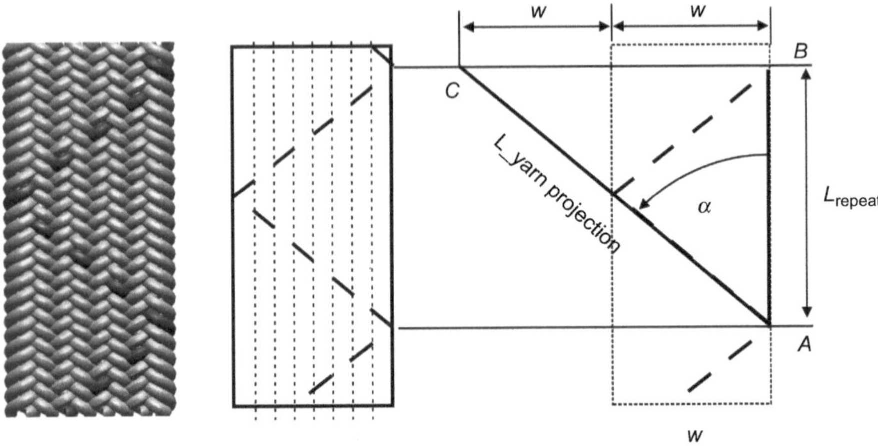

Figure 14.1 Geometric relations in one repeat of flat braid.

The relationship between the yarn length (without taking the yarn crimp into account at this stage) and the repeat length from the triangle ABC is then:

$$\cos \alpha = \frac{L_{\text{repeat}}}{L_{\text{yarn}}} \qquad [14.1]$$

For a 1 m braid, then, a yarn length of:

$$L_{yarn} = \frac{1}{\cos \alpha} \qquad [14.2]$$

is required; the width of the braid here does not influence the result.

Figure 14.2 shows the relation of Equation [14.2] graphically, where computed pictures of braids parallel to the curve with corresponding braiding angles are presented. It can be seen that, at higher braiding angle, the repeat length decreases, and thus more repeats are required per metre braided. Braids have, ordinarily, braiding angles between 30° and –70°, and for each metre braided, between 1.3 up to 3 m of yarn is required. For tubular braids the relations from Equations [14.1] and [14.2] remain valid, as shown in Fig. 14.3.

14.2.2 Repeat length and repeats per unit length

Accurate measurement of the braiding angle usually requires a magnified image of the braid. This is more involved and time-consuming compared to length measurements alone, and so is unsuitable for fast calculations under industrial conditions. Instead of this, the width *w* of the flat braids can be measured, or the diameter of the flat pressed tubular braid can be measured. In addition to width, repeat length has to

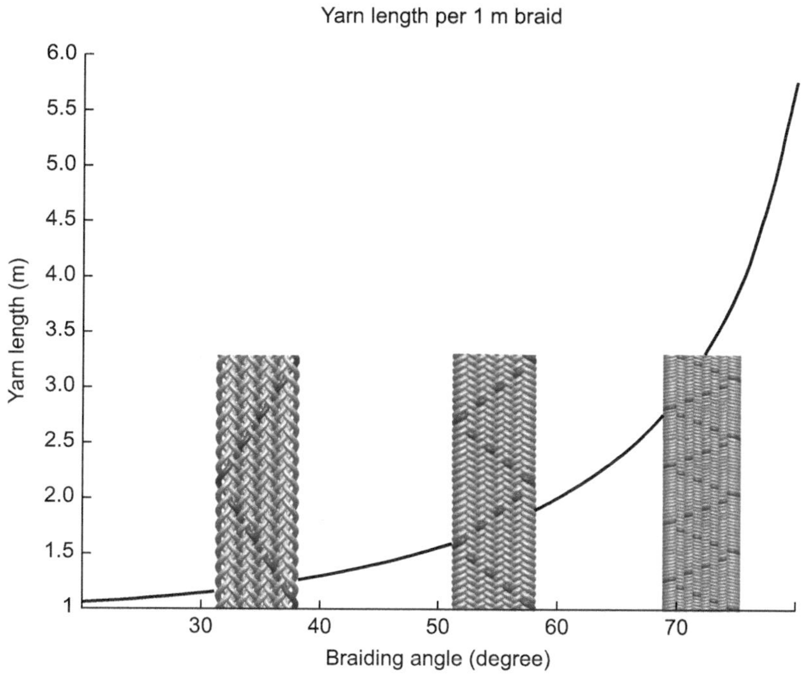

Figure 14.2 Geometric relations in one repeat of flat braid. 3D Braids courtesy of TexMind.

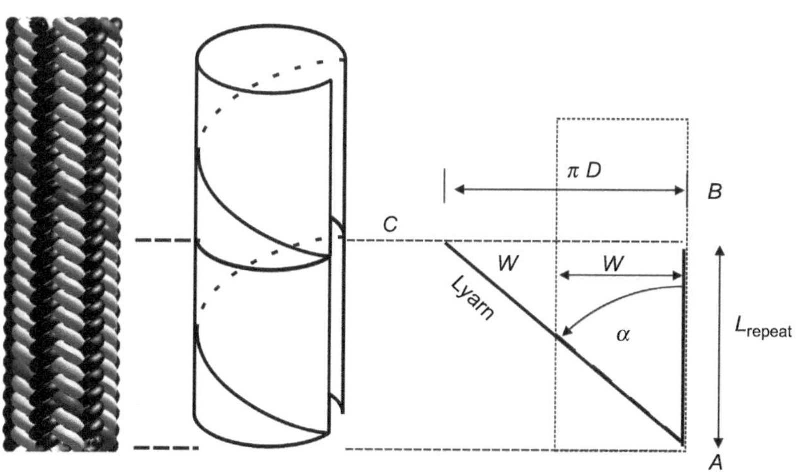

Figure 14.3 Geometric relations in one repeat of flat braid.

be known in order to calculate the braiding angle. From Fig. 14.1, triangle ABC can be then found for flat braids:

$$\tan\alpha = \frac{2 \cdot w}{L_{\text{repeat}}} \tag{14.3}$$

For tubular braids, the same equation also applies, although the part of the equation that doubles the braid width has to be replaced by the braid circumference, following Fig. 14.2:

$$\tan\alpha = \frac{\pi \cdot D}{L_{\text{repeat}}} = \frac{2 \cdot w}{L_{\text{repeat}}} \qquad [14.4]$$

Inside one unit cell, Pythagoras' theorem can be used to find one of the unknown dimensions, for instance the yarn length per repeat:

$$L_{\text{yarn\_per\_repeat}} = \sqrt{L_{\text{repeat}}^2 + 4 \cdot w^2} \qquad [14.5]$$

It has to be emphasized that the yarn length depends on two parameters that are **not** independent. Changes to braiding angles automatically lead to changes in the repeat length, so that both parameters on the right-hand side of [14.1] have to be re-determined (Fig. 14.2a). To simplify, it is assumed that the width of the braid does not change significantly (related to the change of the repeat length); the length of the yarn can then be calculated from the braiding angle as follows:

$$L_{\text{yarn}} = \frac{2 \cdot w}{\sin\alpha} \qquad [14.6]$$

For this idealized case, the relationship between yarn length per repeat and the braiding angle (Equation [14.6], and that between the repeat length and the braiding angle (Equation [14.3]), for a flat braid of constant width of 5 mm, are shown

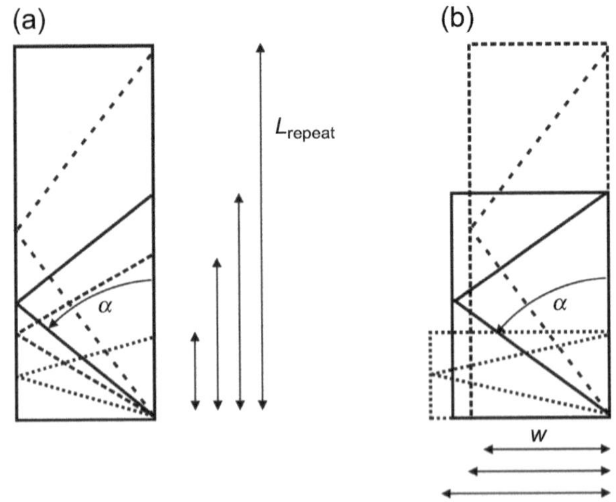

Figure 14.4 Changes of the repeat length under assumption of constant width (a), and possible real situation (b).

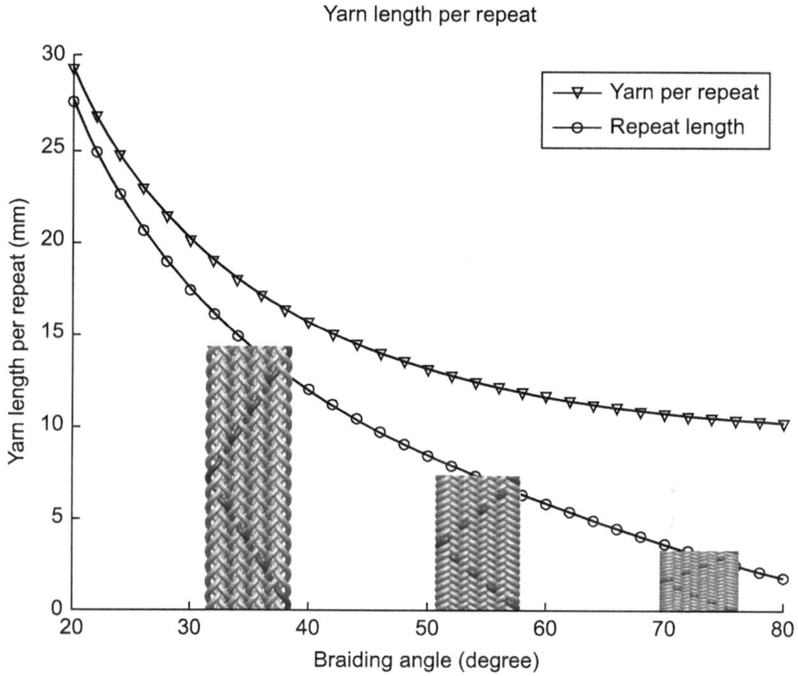

Figure 14.5 The yarn length per repeat decreases at higher braiding angle.

in Fig. 14.4. The Matlab code for the computation is given in the Appendix. As the diagrams (and equations) show, both repeat length and the quantity of yarn per repeat decrease with incremented braiding angles. Low braiding angles mean that the yarns are positioned closer to the braid axis, making the repeat longer (Fig. 14.5).

The length of the yarn in one repeat can be used as the basis for several calculations. In 1 m of braid there are $N_{\text{repeats\_1\_1m}}$ repeats:

$$N_{\text{repeats\_in\_1m}} = \frac{1}{L_{\text{repeat}}} \qquad [4.7]$$

14.2.3 Yarn crimp factor

Yarn crimp depends on a large number of parameters, and is generally a complex process; it is difficult to account for it effectively in production calculations, as no adequate algorithm yet exists. How two interlacing yarns deform their axes in the contact area depends on a combination of production parameters (e.g. braiding angle and yarn tension during the braiding), yarn properties (cross-section, lateral compression properties, e.g. twisted/untwisted, monofilament yarn, etc, as well the yarns' bending rigidities), and structure type (the pattern 1:1, 2:2, 1:1–2 etc, also influences the crimp, as the longer floating in the pattern reduces the yarn crimp).

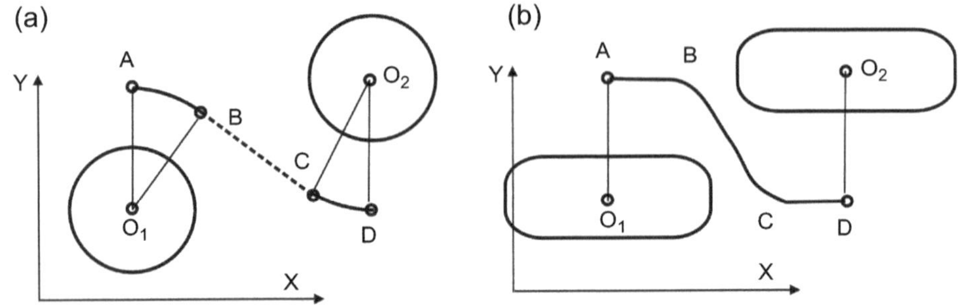

Figure 14.6 Yarn axis in a half repeat of braided structure (a) for yarns with circular cross-section (b) in the case of flat yarns.

Including all these parameters in some model is possible, but is an advanced computational topic, which can be read in almost any good paper about woven structure or braided structures. One good approach, based on yarn mechanics and taking into account the mechanical properties of the yarns, is implemented in the Wisetex software for both woven and braided structures (Lomov *et al.*, 2002, 2012). Some more titles with information about these topics are given at the end of this chapter.

In this chapter, a simplified treatment of the calculations is preferred, so that some of the important calculations may be performed without any in-depth examination of modelling and simulation. While it may be objected that the simplification used is too coarse, it must be remembered that the exact calculation of yarn length requires the value of elliptic integrals, which cannot be integrated analytically. Therefore, any more exact method for calculating yarn length will either be connected with specific assumptions about yarn geometry – and will be less general – or will require iterative calculation procedures that are not recommended for fast production calculations.

As presented in Fig. 14.6, the geometry of the interlacement can differ very significantly, depending on the yarn type. With twisted yarns, monofilaments and wire, an assumption about the circular cross-section of the yarns can be made (Fig. 14.6a). In case of untwisted multifilament yarns, tape, or even of twisted but more deformable yarns, flatter cross-sections must be used (Fig. 14.6b). The braiding of the yarns in both directions is commonly identical, so the crimp of both systems is normally the same. In the case of circular cross-sections, the yarn length between two interlacing points (line ABCD) can be presented as the sum of two arcs and a connecting line BC. For flat cross-sections, there are two lines and two arcs.

As a good enough approximation, flat yarns are considered with distance t between their axes and thickness $2h$, where h is the distance between the lowest and highest points of the yarn axes in one unit cell (Fig. 14.7). In such a case, the AD yarn length, approximated with a straight line between two interlacing points, will be:

$$AD = \sqrt{t^2 + h^2} \qquad [14.8]$$

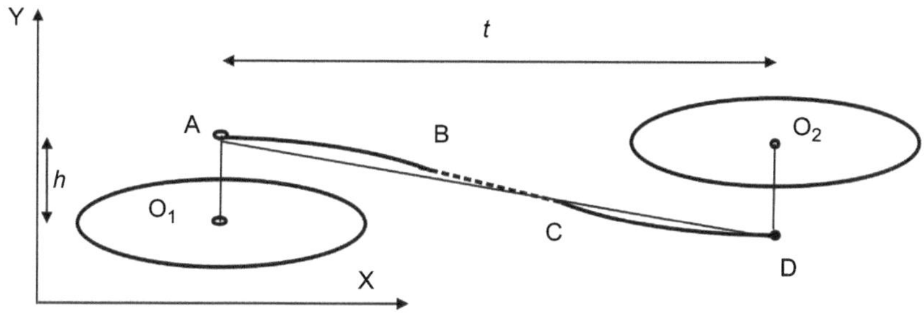

Figure 14.7 Approximation of the yarn length with a line in order to consider the crimp.

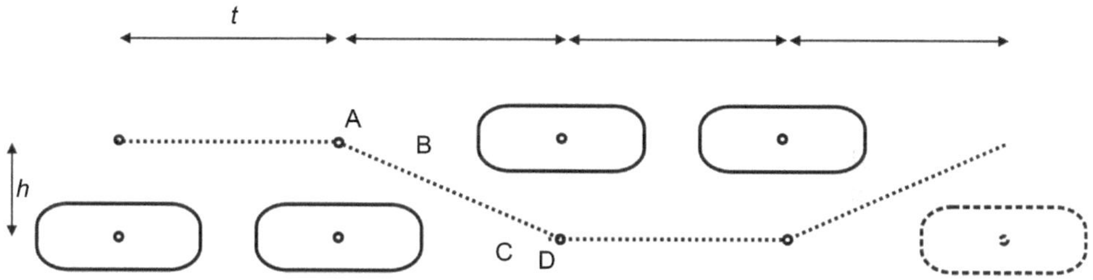

Figure 14.8 Geometric relations in one repeat of flat braid.

and, per unit cell, the length will then be:

$$L = 2 \cdot \sqrt{t^2 + h^2} \quad [14.9]$$

as the same geometry is repeated, albeit symmetrically.

In the case of a floating length of 2, each yarn goes over and under two other yarns, respectively. The floating over and under of the yarns is with length equal to the step size t, and so, according to Fig. 14.8, the yarn length in one unit cell will be:

$$L_2 = 2t + 2\sqrt{t^2 + h^2} \quad [14.10]$$

In the general case of a floating length of N, where each yarn floats over N cells and under N cells, the yarn length in the unit cell will be:

$$L_N = 2 \cdot (N - 1) \cdot t + 2\sqrt{t^2 + h^2} \quad [14.11]$$

From Equation [14.11], as well as from Figs 14.6 to 14.8, it is clear that in the case of very flat and thin cross-sections (as carbon or glass fibre tows) and for higher floating lengths, the influence of the yarn crimp over the total yarn length per unit cell is generally lower and less significant. For dense structures, where the step t is comparable with the yarn diameter – especially when braided with a floating length of 1 – the crimp cannot be neglected. A simple approximation of the yarn axis from Fig. 14.6 with a circular arc, where the free gap space between two yarns is exactly

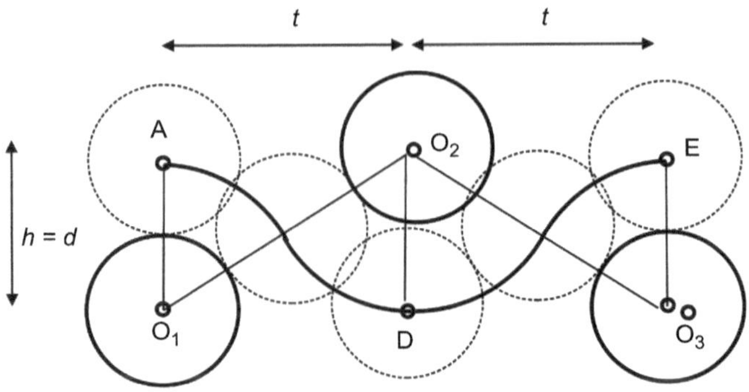

Figure 14.9 Geometric relations in one repeat of flat braid.

equal to the yarn diameter (Fig. 14.9), leads to a configuration where $O_2D = 2 \cdot O_1O_2$ and thus the angle s. There are in all four arcs, so the total yarn length AD + DE in one unit cell will be:

$$L_{1\max} = 4 \cdot d \cdot \frac{60}{180} \cdot \pi = d \cdot \frac{4}{3} \cdot \pi = 4.1886 \cdot d \qquad [14.12]$$

The projection of the yarn axis is:

$$2 \cdot t = 2 \cdot \sqrt{3} \cdot d = 2 \cdot 1.73 \cdot d = 3.46 \cdot d \qquad [14.13]$$

meaning that the relation between the length of the crimped yarn to its projection on the horizontal axis, defined here as crimp factor K_{crimp}, would be maximally:

$$K_{\text{crimp}} = \frac{(4/3) \cdot d \cdot \pi}{2 \cdot \sqrt{3} \cdot d} = \frac{2 \cdot \sqrt{3}}{9} \cdot \pi \approx 1.209 \qquad [14.14]$$

which in turn means that the calculated yarn length from the previous section has to be increased by about 20% to obtain the real length.

In less crimped cases for any floating length, the factor K_{crimp} can be calculated from the above equations as:

$$K = \frac{L_N}{L_{\text{Projection}}} = \frac{2 \cdot (N-1) \cdot t + 2\sqrt{t^2 + h^2}}{N \cdot t} \qquad [14.15]$$

14.2.4 Yarn length taking into account the yarn crimp

Correcting the yarn length in Equation [14.1) with the crimp factor from Equation [14.14] or [14.15] leads to:

$$L_{\text{yarn,crimped}} = K_{\text{crimp}} \cdot \frac{1}{\cos \alpha} \qquad [14.16]$$

where, as already demonstrated, $1.0 < K_{\text{crimp}} \leq 1.2$. For soft, voluminous yarns, prone to greater lateral compression, this factor can increase to values of >1.2, but it should remain less than 2 (as, in the braiding process, no beat-up forces are available, as in weaving.)

Of course, the crimp factor depends on the braiding angle, α, that is indirectly presented in the equations in the step size *t*. These relations are intentionally skipped here to keep the equations as simple for use as possible.

14.2.5 Total yarn length in the braid

The total yarn length required for a 1 m braid for all carriers can be then simply estimated from multiplying the yarn length, from Equation [14.16], by the number of the carriers N_{carriers}:

$$L_{\text{total}} = N_{\text{carriers}} \cdot K_{\text{crimp}} \cdot \frac{1}{\cos \alpha} \qquad [14.17]$$

For the braids with inlay yarns, (also called middle ends or, for the composites familiar reader, 0° layer for triaxial braids) and cores with the angle $\alpha = 0$ and in normal cases, the crimp can be neglected, as the middle yarns are compressed from both sides by braiding yarns. Their required length is then equal to the length of the braid.

14.3 Weight per metre

For these calculations, the linear density of the yarn material is required. For textile yarns, the most commonly used system should be the SI tex system, where the tex shows the weight in grams of 1 km of yarn:

$$1\,\text{tex} = \frac{1\,\text{g}}{1\,\text{km}} \qquad [14.18]$$

For yarns in fine medical and clothing applications, fineness definition in dtex can be found (1dtex= 0.1tex); at the other end of the spectrum – e.g. for thick ropes – fineness is measured in ktex (1 ktex = 1000 tex). There are also several legacy measurement systems, so yarns can be found in Nm, den, Ne, etc., presenting challenges for buying agents and product developers alike. Tables and equations for recalculation between the different systems are given in the Appendix of this book, but can be found as well in several places on the Web; here the tex system is used.

If the fineness of the braiding yarns in tex is Tt_{Braid}, 1 m of yarn will have weight:

$$m_{1m[g]} = \frac{Tt_{Braid[tex]}}{1000} \quad [14.19]$$

or, even easier, in milligrams:

$$m_{1m[mg]} = Tt_{Braid[tex]} \quad [14.20]$$

The linear density of wire and other non-textile materials can be calculated from their density and cross-section. If the wire has diameter d, given usually in millimetres, its cross-section is:

$$S_{[mm^2]} = \frac{\pi \cdot d^2_{[mm]}}{4} \quad [14.21]$$

which is:

$$S_{[cm^2]} = \frac{\pi \cdot d^2_{[mm]}}{4 \cdot 100} \quad [14.22]$$

The volume of 1 m of wire will be:

$$V_{[cm^3]} = 100_{[cm]} \cdot \frac{\pi \cdot d^2_{[mm]}}{4 \cdot 100} = \frac{\pi \cdot d^2_{[mm]}}{4} \quad [14.23]$$

and the mass of this 1 m is thus:

$$M_{[g]} = \frac{\pi \cdot d^2_{[mm]}}{4} \cdot \rho_{[g/cm^3]} \quad [14.24]$$

If the tex value is required, so that the following equations can be used for any material, then the above mass has to be multiplied by 1000 to get the weight in grams per kilometre:

$$Tt_{[tex]} = 1000 \cdot \frac{\pi \cdot d^2_{[mm]}}{4} \cdot \rho_{[g/cm^3]} \quad [14.25]$$

For the density, ρ, several tables can be found. For steel it is 7.87 g/cm^3, copper 8.96 g/cm^3, compared to polyester 1.38 g/cm^3, polypropylene 0.9 ÷ 0.91g/cm^3, etc., but these values depend as well on the temperature and on the exact material combination.

Having the length of all braiding yarns in 1 m braid from Equation [14.17], the weight of the braiding yarns will be:

$$m_{\text{Braiding\_yarns}} = N_{\text{carriers}} \cdot K_{\text{crimp}} \cdot \frac{1}{\cos\alpha} . Tt_{\text{Braid[tex]}} \qquad [14.26]$$

Core yarns and middle ends usually have different fineness from the braiding yarns; let these be defined as $Tt_{\text{Core[tex]}}$ and $Tt_{\text{Middle[tex]}}$.

The mass of the core yarn in 1 m braid will thus be:

$$m_{\text{Core\_yarns[mg]}} = Tt_{\text{Core[tex]}} \qquad [14.27]$$

and the mass of the middle ends in 1 m yarn:

$$m_{\text{MiddleEnd\_yarns[mg]}} = N_{\text{MiddleEnds}} \cdot Tt_{\text{Middle[tex]}} \qquad [14.28]$$

The total mass of 1 m braid will be:

$$M_{\text{Braid[mg]}} = m_{\text{MiddleEnd\_yarns[mg]}} + m_{\text{Core\_yarns[mg]}} + m_{\text{Braiding\_yarns[mg]}} \qquad [14.29]$$

or

$$M_{\text{Braid[mg]}} = N_{\text{carriers}} \cdot K_{\text{crimp}} \cdot \frac{1}{\cos\alpha} . Tt_{\text{Braid[tex]}} + Tt_{\text{Core[tex]}} + N_{\text{MiddleEnds}} \cdot Tt_{\text{Middle[tex]}} \qquad [14.30]$$

As the above equations are based on several simplified assumptions, the calculated values will differ from the real values, due to a combination of the longitudinal deformation of the yarns under tension, lateral deformation of the yarns under contact forces, elongation of the rope due to winding tension, and so on.

14.4 Time for preparation

14.4.1 Winding time

Given the length of the yarn required for 1 m of braid, and the length of the product that has to be produced, calculation of the winding time depends on the winding velocity and the number of the winding places of the machine. The time required per bobbin with length L is then:

$$T_{\text{[min]}} = \frac{L_{\text{[m]}}}{V_{\text{[m/min]}}} \qquad [14.31]$$

Depending on the material, the velocity can vary from few metres per minute for very fine, very thick or very sensitive materials, up to several hundred metres per minute for robust materials.

The total time required for winding of all bobbins for a braid will then be:

$$T_{[\text{min}]} = \frac{N_{\text{Carriers}}}{N_{\text{Winding\_Places}}} \cdot \left[\frac{L_{[\text{m}]}}{V_{[\text{m/min}]}} + T_{\text{bobbin\_change}} \right] \qquad [14.32]$$

where $N_{\text{Winding\_places}}$ is the number of the winding heads on one machine. In older or single-place machines, this number is just 1, while for specialized winding machines for braiding bobbins it can be 2 or 4. $T_{\text{bobbin\_change}}$ is the time taken to change the bobbin on the winding machine. In the case of automated winding machines, it can be a few seconds, but, if expensive and special materials have to be wound, it can be more than a minute per bobbin.

14.4.2 Machine preparation

Preparation time also has to be considered, especially for machines with a large number of bobbins. Usually for classical braiding machines for ropes, laces, etc. with up to 20–30 carriers, two workers can change the bobbins in less than an hour if the old bobbins are still on the machine and their yarns are still in working positions, as new yarns can be easily knotted onto the older ones. The time taken can increase two- or three-fold if the carriers are without bobbins, or if the yarns are already pulled out of some guides because the yarn has then to be inserted through all guides again (see Chapter 7 *Carriers for braiding machines*). For radial braiders with 96, 144, 288 or more carriers (or megabraiders with 600–800 carriers), it is clear that the time for the bobbin change will be longer than the braiding time. Due to this, some companies, especially in composites production, put two braiding machines in a line, so that while braiding/overbraiding takes place with one, the other can be prepared with bobbins (Henkel *et al.*, 2011).

14.5 Take-off speed, braiding time and productivity

The take-off speed for a given machine and material, the braiding angle, and the relationship between these can be found from:

$$V_{[\text{m/s}]} = \frac{L_{\text{Repeat[m]}}}{t_{\text{Repeat[s]}}} \qquad [14.33]$$

The time $t_{\text{Repeat[s]}}$ for one repeat is the time required for one carrier to go through a complete track (Fig. 14.10). For tubular braiding machines, this is the time taken for one revolution of the carrier around the machine centre (Fig. 14.11a); for flat braiding

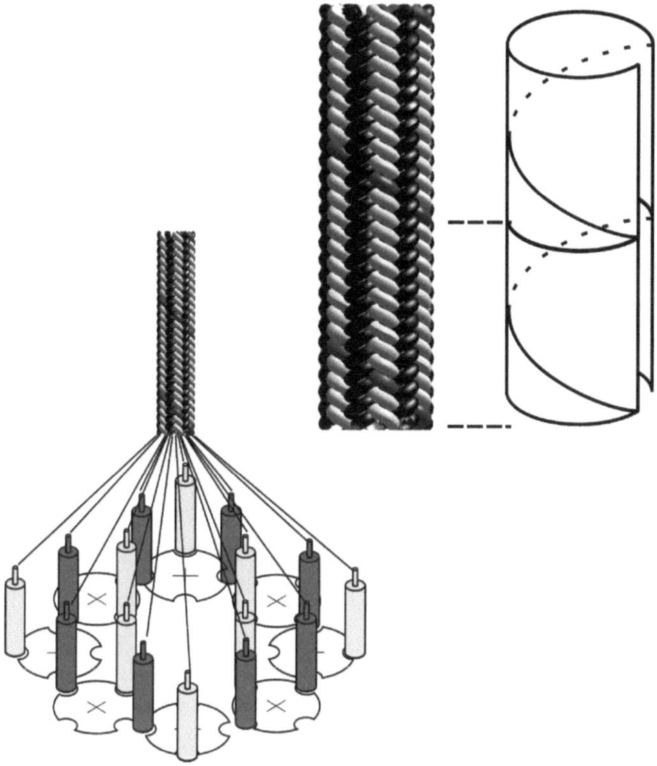

Figure 14.10 Following the track, for one way (counter-clockwise or opposite) one carrier lie the yarn for one repeat at the tubular braiding.

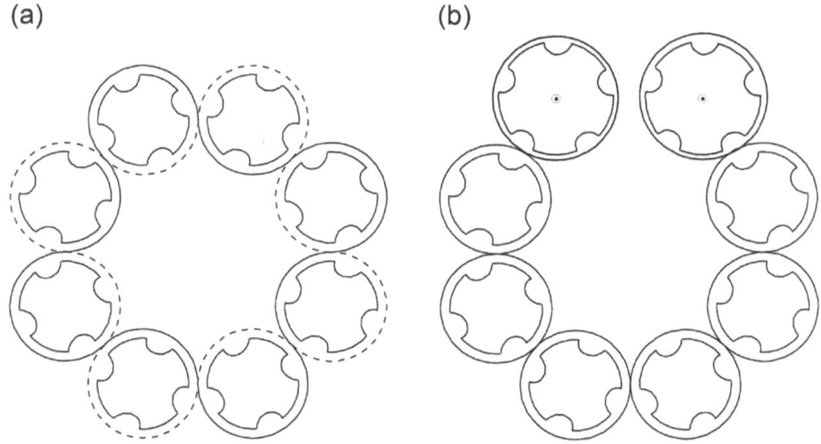

Figure 14.11 Following the track, for one way (counter-clockwise or opposite) one carrier succeeds in passing only half of its path and lie the yarn for half repeat for flat braiding (b) contrary to one carrier from tubular braiding (a), which track has the half-length.

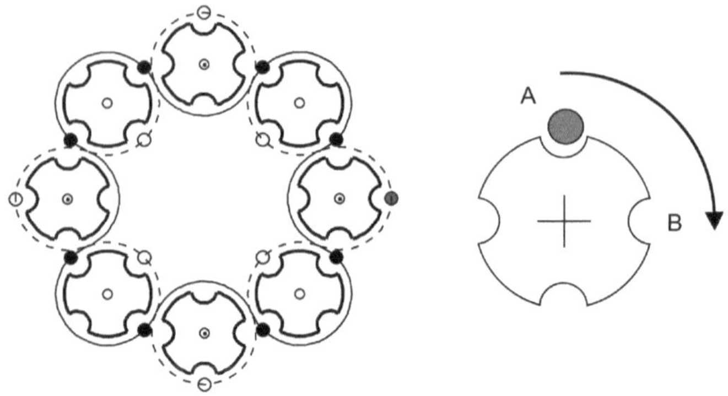

Figure 14.12 For calculation of the time, the path of one carrier for the motion from place A to place B is analysed. This motion corresponds to the rotation of the horn gear on angle, equal to the one slot.

machines, this time is twice as long, as one carrier has to go through the entire track in both directions (Fig. 14.11b). For machines with more tracks, the time for one repeat is the smallest common multiple of the times required by the carriers to fully go through each track.

To illustrate, let us consider one of the tracks in Fig. 14.12. One carrier following such a track will be carried by each of the eight horn gears, and will travel a distance equivalent to the length of two slots per track, 16 slots in all. Generally, with $N_{\text{HornGears}}$ number of horn gears, each with $N_{\text{Slots\_per\_horngear}}$ slots, there will be a track from N_{Slots}:

$$N_{\text{Slots}} = \frac{N_{\text{HornGears}} \cdot N_{\text{Slots\_per\_horngear}}}{2} \qquad [14.34]$$

For flat braiding machines, it has to be taken into account that the horn gears at both sides have a larger (or smaller) number of slots. The carrier goes through all of the slots, not around half of them as for tubular machines. If the normal horn gears have four slots each and both end horn gears (where the carrier turns the moving duration) have five slots, the equation will be:

$$N_{\text{Slots}} = N_{\text{4Slots\_HornGears}} \cdot 4 + 2 \cdot 5 \qquad [14.35]$$

The time to travel around one slot depends on the angular speed of the horn gears, $n_{\text{HornGears[rpm]}}$. Because one revolution leads to a rotation angle of $2 \cdot \pi$ radians, the angular velocity of the horn gear is:

$$\omega_{[s^{-1}]} = 2 \cdot \pi \cdot \frac{n_{[\text{rpm}]}}{60} = \frac{\pi \cdot n_{[\text{rpm}]}}{30} \qquad [14.36]$$

For one horn gear with $N_{\text{Slots\_per\_horngear}}$ the angle at which the horn gear has to rotate in order to move the carrier to the same place on the next slot is:

$$\varphi_{\text{Slot}} = \frac{2 \cdot \pi}{N_{\text{Slots\_per\_horngear}}} \quad [14.37]$$

and the time for moving one carrier this angle is

$$t_{[\text{s}]} = \frac{\varphi_{\text{Slot}}}{\omega_{[\text{s}^{-1}]}} = \frac{2 \cdot \pi / N_{\text{Slots\_per\_horngear}}}{\pi \cdot n_{[\text{rpm}]} / 30} = \frac{60}{N_{\text{Slots\_per\_horngear}} \cdot n_{[\text{rpm}]}} \quad [14.38]$$

The time required for one carrier to move along its total track with length N_{Slots}, i.e. for the production of one repeat, is:

$$t_{\text{Repeat}[s]} = \frac{N_{\text{Slots}} \cdot 60}{N_{\text{Slots\_per\_horngear}} \cdot n_{[\text{rpm}]}} \quad [14.39]$$

Equations [14.39], [14.33] and [14.3] give the relations between the take-off speed, depending on the machine and braid parameters:

$$V_{[\text{m/s}]} = \frac{L_{\text{Repeat}[\text{m}]}}{t_{\text{Repeat}[\text{s}]}} = \frac{2 \cdot w_{[\text{m}]} / \tan\alpha}{N_{\text{Slots}} \cdot 60 / N_{\text{Slots\_per\_horngear}} \cdot n_{[\text{rpm}]}} = \frac{w_{[\text{m}]} \cdot N_{\text{Slots\_per\_horngear}} \cdot n_{[\text{rpm}]}}{N_{\text{Slots}} \cdot 30 \cdot \tan\alpha} \quad [14.40]$$

For commonly used machines with four slots per horn gear, this equation can be simplified to the following for the tubular braiding machines (applying Equation [14.34]):

$$V_{[\text{m/s}]} = \frac{w_{[\text{m}]} \cdot n_{[\text{rpm}]}}{15 \cdot N_{\text{Horn\_Gears}} \cdot \tan\alpha} \quad [14.41]$$

and for flat braiding machines:

$$V_{[\text{m/s}]} = \frac{4 \cdot w_{[\text{m}]} \cdot n_{[\text{rpm}]}}{\left(N_{\text{4Slots\_HornGears}} \cdot 4 + 10\right) \cdot 30 \cdot \tan\alpha} \quad [14.42]$$

Of course, it is better to have the number of the machine or the number of the carriers in the equations, but these equations have to be used very carefully, because the possibility of errors in calculation by entering the number of the carriers at one partially occupied machine (instead of the maximal number of carriers on the machine) is very high. For a tubular machine with the *maximal number* of carriers N_{Carriers}

$$V_{[m/s]} = \frac{w_{[m]} \cdot n_{[rpm]}}{15 \cdot (N_{\text{Carriers}} / 2) \cdot \tan \alpha} \qquad [14.43]$$

and for a flat machine:

$$V_{[m/s]} = \frac{w_{[m]} \cdot n_{[rpm]}}{15 \cdot N_{\text{Carriers}} \cdot \tan \alpha} \qquad [14.44]$$

Multiplying the speed from the last equations by 60, 3600, etc. will give the length of the braid, produced for 1 min, 1 h, etc. These equations are valid for configurations that are producible, not accounting for the density of the yarns in the braid, cover factor, etc.

The time taken to braid a certain length of rope can be now calculated analogously to the winding time (Equation [14.31]), as the take-off speed is known. To calculate the productivity in kilograms per hour, the length of the produced braid has to be multiplied by the linear density (weight per metre according [14.30]) of the braid:

$$P_{[kg/h]} = V_{[m/s]} \cdot 3600_{[s]} \cdot \frac{M_{\text{Braid}[mg]}}{10^6} \qquad [14.45]$$

14.6 Calculation examples

14.4.1 Tubular braid

Assume a tubular braiding machine with horn gears of speed 200 RPM, 16 carriers with full occupation (1 Full 1 Empty), is to be used to braid a polypropylene rope of 110 tex, with diameter (width in compressed form) of 4 mm and braiding angle 45°. The repeat length in mm will thus be:

$$L_{\text{repeat}} = \frac{2 \cdot w}{\tan \alpha} = \frac{2 \cdot 4}{1} = 8 \qquad [14.46]$$

The take-off speed:

$$V_{[m/s]} = \frac{w_{[m]} \cdot n_{[rpm]}}{15 \cdot (N_{\text{Carriers}} / 2) \cdot \tan \alpha} = \frac{0.004_{[m]} \cdot 200_{[rpm]}}{15 \cdot (16/2) \cdot 1} = 0.0066_{[m/s]} \qquad [14.47]$$

is thus 0.399 m/min, or 23.9 m per hour. For 1 km of rope, 1000/23.9 = 41 h braiding is required. The length of one yarn per metre length of rope is thus:

$$L_{\text{yarn}} = \frac{1}{\cos 45^\circ} = \frac{1}{0.707} = 1.41_{[\text{m}]} \quad [14.48]$$

and, considering an approximated crimp factor $K = 1.1$, the real length required per metre braid will be:

$$L_{\text{yarn}} = 1.41_{[\text{m}]} \cdot 1.1 = 1.55_{[\text{m}]} \quad [14.49]$$

For 16 yarns, this gives 24.8 m yarn in a 1 m braid.

The weight of a metre of braid is:

$$M_{[\text{mg}]} = 16 \cdot 1.55_{[\text{m}]} \cdot 110_{[\text{tex}]} = 2480_{[\text{mg}]} = 2.48_{[\text{g}]} \quad [14.50]$$

or, for 1 km rope, 2.48 kg of yarn material is required.

The winding time, assuming an automatic winding machine with four heads and a winding speed of 120 m/min, is thus

$$\begin{aligned} T_{[\text{min}]} &= \frac{N_{\text{Carriers}}}{N_{\text{Winding\_Places}}} \cdot \left[\frac{L_{[\text{m}]}}{V_{[\text{m/min}]}} + T_{\text{bobbin\_change}} \right] \\ &= \frac{16}{4} \cdot \left[\frac{1000}{100} + 0.2 \right] = 4 \cdot 10.2 = 40.8_{[\text{min}]} \end{aligned} \quad [14.51]$$

On a machine with single winding head, this will take 160 min.

14.4.2 Tubular braid for thick rope

Assuming that tubular braiding machine with horn gear speed of 20 RPM, with 16 carriers with full occupation (1 Full 1 Empty) is used for braiding a polyester rope of 110 ktex, with a diameter (width in compressed form) of 80 mm and a braiding angle 45°.

The repeat length in mm will thus be:

$$L_{\text{repeat}} = \frac{2 \cdot w}{\tan \alpha} = \frac{2 \cdot 80}{1} = 160 \quad [14.52]$$

The take-off speed is thus calculated as follows:

$$V_{[\text{m/s}]} = \frac{w_{[\text{m}]} \cdot n_{[\text{rpm}]}}{15 \cdot (N_{\text{Carriers}} / 2) \cdot \tan \alpha} = \frac{0.080_{[\text{m}]} \cdot 20_{[\text{rpm}]}}{15 \cdot (16/2) \cdot 1} = 0.0133_{[\text{m/s}]} \quad [14.53]$$

giving 0.8 m/min, or 48 m per hour. For 1 km of rope, 1000/48 = 20 h braiding is required.

The length of one yarn per metre of rope is

$$L_{\text{yarn}} = \frac{1}{\cos 45^\circ} = \frac{1}{0.707} = 1.41_{[\text{m}]} \quad [14.54]$$

and, taking into account an approximated crimp factor $K = 1.1$, the real length required per metre braid will be:

$$L_{\text{yarn}} = 1.41_{[\text{m}]} \cdot 1.1 = 1.55_{[\text{m}]} \quad [14.55]$$

For 16 yarns, this gives 24.8 m yarn in a 1 m braid.

The weight of 1 m braid is:

$$M_{[\text{mg}]} = 16 \cdot 1.55_{[\text{m}]} \cdot 110000_{[\text{tex}]} = 2480000_{[\text{mg}]} = 2.48_{[\text{kg}]} \quad [14.56]$$

meaning that 2 480 kg yarn material is required to produce 1 km of rope.

The winding time of an automatic winding machine with one head and winding speed of 20 m/min is then:

$$\begin{aligned} T_{[\text{min}]} &= \frac{N_{\text{Carriers}}}{N_{\text{Winding\_Places}}} \cdot \left[\frac{L_{[\text{m}]}}{V_{[\text{m/min}]}} + T_{\text{bobbin\_change}} \right] \\ &= \frac{16}{1} \cdot \left[\frac{1000}{20} + 4 \right] = 16 \cdot 54 = 864_{[\text{min}]} = 14.4_{[\text{h}]} \end{aligned} \quad [14.57]$$

In practice, for very thick ropes, the required length may not fit on the bobbin, leading to multiple bobbin windings. The time required for this is not considered here.

14.4.3 Flat braid (lace)

Assume a flat braiding machine with horn gears speed 240 RPM, and 33 carriers with full occupation (1 Full: 1 Empty) is to be used for braiding polyester lace of 176 tex, with a width of 10 mm and braiding angle 60°.

The repeat length in mm will be:

$$L_{\text{repeat}} = \frac{2 \cdot w}{\tan \alpha} = \frac{2 \cdot 10}{\tan 60^\circ} = \frac{20}{1.73} = 11.5\,\text{mm} \quad [14.58]$$

The take-off speed is:

$$V_{[\text{m/s}]} = \frac{w_{[\text{m}]} \cdot n_{[\text{rpm}]}}{15 \cdot N_{\text{Carriers}} \cdot \tan \alpha} = \frac{0.010_{[\text{m}]} \cdot 240_{[\text{rpm}]}}{15 \cdot 33 \cdot 1.73} = 0.0028_{[\text{m/s}]} \quad [14.59]$$

which is 0.168 m/min or 10.1 m per hour. For 1 km of lace, 1000/10.1 = 99.1 h braiding is required.

The length of one yarn per metre of lace is:

$$L_{\text{yarn}} = \frac{1}{\cos 60^\circ} = \frac{1}{0.5} = 2_{[\text{m}]} \qquad [14.60]$$

and, considering an approximated crimp factor $K = 1.2$, the real length required per metre braid will be:

$$L_{\text{yarn}} = 2_{[\text{m}]} \cdot 1.2 = 2.4_{[\text{m}]} \qquad [14.61]$$

For 33 yarns this gives 79.2 m yarn in 1 m of braid.

The weight of 1 m of braid is:

$$\text{M}_{[\text{mg}]} = 33 \cdot 2.4_{[\text{m}]} \cdot 176_{[\text{tex}]} = 13939.2_{[\text{mg}]} = 13.9_{[\text{g}]} \qquad [14.62]$$

thus, 1 km lace requires 13.9 kg of yarn material.

The winding time, assuming an automatic winding machine with four heads and a winding speed of 220 m/min, is then:

$$T_{[\text{min}]} = \frac{N_{\text{Carriers}}}{N_{\text{Winding\_Places}}} \cdot \left[\frac{L_{[\text{m}]}}{V_{[\text{m/min}]}} + T_{\text{bobbin\_change}}\right] = \frac{33}{4} \cdot \left[\frac{1000}{220} + 0.2\right] = 8.25 \cdot 4.74 = 39.1_{[\text{min}]} \qquad [14.63]$$

or, on a machine with single winding head, 156 min.

14.7 Sources of further information and advice

Some good information about yarn parameters and braiding calculations can be found in the Handbook of Industrial Braiding (Ko *et al.*, 1989). Most such calculations can be performed with the CAB-Design software (August Herzog Textilmaschinenfabric GmbH & Ko KG, Oldenburg, 2013), while some of these may also be conducted with TexMind software (Kyosev, 2012), both of which are presented in Chapter 13.

There are many papers on the calculation of yarn crimp and exact length, from the first geometrical models for woven structures (Peirce, 1937; Kemp, 1958), to the basics of textile mechanics (Postle *et al.*, 1988) to modern computational solutions (Lomov *et al.*, 2000, 2002; Sherburn, 2007). The older, simpler, geometrical models are generally not accurate enough for day-to-day use, and the computational methods are too complex for simple calculations, requiring software implementation. Some production calculations can also be found in German-language books (Engels, 1994a, b; Lepperhoff, 1953) and (Goseberg *et al.*, 1981).

References

August Herzog Textilmaschinenfabric GmbH & Ko KG, Oldenburg, (2013). *CAB Design: Computer Aided Braided Design* [online]. Available from: http://www.herzog-online.com/_rubric/index.php?rubric=Products+Software+EN.

Engels, H. (1994a). *Handbuch der Schmaltextilien. Die Flechttechnologie. Teil I: Maschinen und Verfahren zur Erzeugung konventioneller Geflechte*. Mönchengladbach.

Engels, H. (1994b). *Handbuch der Schmaltextilien. Die Flechttechnologie. Teil II: Maschinen und Verfahren zur Erzeugung von Flechtprodukten mit speziellen physikalischen und chemischen Anforderungen*. Mönchengladbach.

Goseberg, F., Brabender, K. and Moeller, P. (1981). *Textiltechnik Maschinengeflechte: Ausbildungsmittel Unterrichtshilfen*. Frankfurt am Mai: Arbeitgeberkreis Gesamttextil.

Henkel, F., Brockmanns, K.-J. and Baumgart, G. (2011). Automatisierte Fertigung von Spantpreforms mittels Flechttechnologie. *In: 17. Nationales Symposium SAMPE Deutschland e.V.* Aachen: IKV-RWTH Aachen, 43.

Kemp, A. (1958). An extension of Peirce's cloth geometry to the treatment of non-circular threads. *Journal of the Textile Institute Transactions* [online], **49** (1), T44. Available from: http://www.tandfonline.com/doi/abs/10.1080/19447025808660119.

Ko, F.K., Pastore, C.M. and Head, A.A. (1989). *Handbook of Industrial Braiding*: Atkins & Peirce.

Kyosev, Y.K. (2012). *TexMind Braider* [online]. Mönchengladbach. Available from: www.texmind.com.

Lepperhoff, B. (1953). *Die Flechterei.* 3rd edn.: Eugen G. Leuze Verlag Saulgau/WTTBG.

Lomov, S. *et al.* (2000). Textile geometry preprocessor for meso-mechanical models of woven composites. *Composites Science and Technology* [online], **60** (11), 2083–2095. Available from: http://www.sciencedirect.com/science?_ob=GatewayURL&_origin=ScienceSearch&_method=citationSearch&_piikey=S0266353800001214&_version=1&_returnURL=&md5=86d8d0f28c4d93c992c704eb787a7b37.

Lomov, S.V. *et al.* (2002). Experimental and theoretical characterization of the geometry of two-dimensional braided fabrics. *Textile Research Journal*, **72** (8), 706–712.

Lomov, S.V. and *et al.*, (2012). *Wisetex* [online]. Available from: http://www.mtm.kuleuven.be/Onderzoek/Composites/software/wisetex.

Peirce, F.T. (1937). 5 – The geometry of cloth structure: Journal of the Textile Institute Transactions. *Journal of the Textile Institute Transactions* [online], **28** (3), T45. Available from: http://www.tandfonline.com/doi/abs/10.1080/19447023708658809.

Postle, R., Carnaby, G. and Jong, S. de (1988). *The Mechanics of Wool Structures*: Ellis Horwood Limited.

Sherburn, M. (2007). *Geometric and Mechanical Modelling of Textiles*. PhD. University of Nottingham.

Using MATLAB® for calculations in braiding

15

15.1 Introduction

"MATLAB® is a high-level language and interactive environment for numerical computation, visualization, and programming" (Mathworks, 2014). It is similar to many modern programming languages, but has a simplified syntax and a large set of toolboxes with functions for computing different kinds of numerical problems and visualizing large sets of data types. This chapter gives a short introduction to a minimal set of MATLAB® commands and toolboxes (MATLAB® is the registered trademark of The MathWorks, Inc.) which can be useful for speeding up the computation and analysis of braided fabrics, machines, and processes. Large computational models have revolutionized research since 2000, permitting the consideration of significantly more process parameters and more accurate modelling.

As textile engineers are seldom familiar with these models, the goal of this chapter is to demonstrate the simplicity of computing with MATLAB and to motivate the young (and the experienced) researchers and engineers to build their own computational models in the area of braiding. A short introduction of the basics of the MATLAB environment and language will be followed by an introduction to some plotting and computing functions, and the chapter ends with an example showing how yarn tension may be measured using the Data Acquisition Toolbox™. No prior knowledge of programming is assumed. Once the first steps are done, the learning curve using MATLAB is very small, and the extension of the commands and implementation of code to solve other tasks can be done in a very short time. Detailed tutorials and examples for all MATLAB toolboxes can be found in the documentation of the software (Mathworks, 2014), where step-by-step "Getting Started" sections for each toolbox are available.

While Mathworks provides academic pricing for both students and institutions, some researchers may still not have access to MATLAB; in this case, some open source products supporting the MATLAB language – such as FreeMat, Octave, Scilab – are usable alternatives; they have stable codebases, and contain the most important features of MATLAB, although they differ slightly in syntax and do not provide all the functions of MATLAB. As open source products, the quality of documentation is less detailed than that of MATLAB; while this may be initially daunting, perseverance will allow MATLAB code to be adjusted for these products.

Braiding Technology for Textiles. http://dx.doi.org/10.1533/9780857099211.4.357

15.2 MATLAB background

15.2.1 Environment

The MATLAB environment has changed in appearance since its first Windows version in several respects, but the main features remain the same (Fig. 15.1).

The main window consists of set windows with different purposes. In the "Command Window" (Fig. 15.1, Position 1) all commands are entered and their output is printed. Figure 15.2 demonstrates the output of the command:

```
ver
```

giving the current version of the software and the available licensed toolboxes. The list of all commands in the current session and former days over long period of time (adjustable in the settings) has been stored in the "Command History" (Fig. 15.1 Position 2). This history is especially useful if a sequence of commands has been tested; it also allows easy cutting and pasting of commands into new files. (Fig. 15.3).

The scripts and functions are the main elements of the MATLAB libraries, and these can be edited in the Editor (Fig. 15.1, Position 3).

The example file ch15_1.m consists of one line

```
x=0:0.5:10;
```

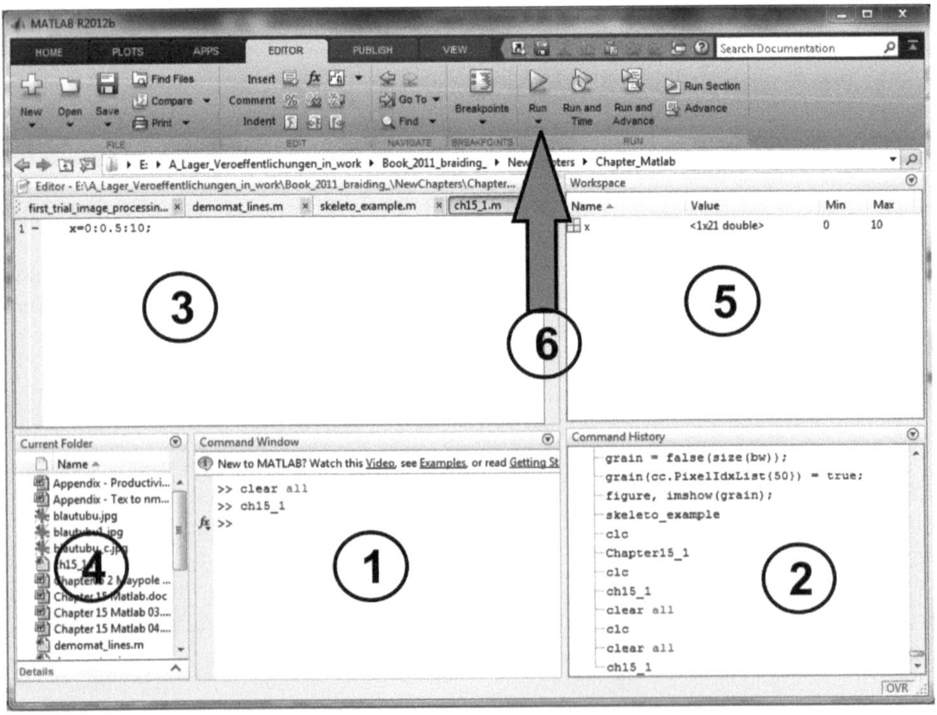

Figure 15.1 Main window of MATLAB Ver. 2012.

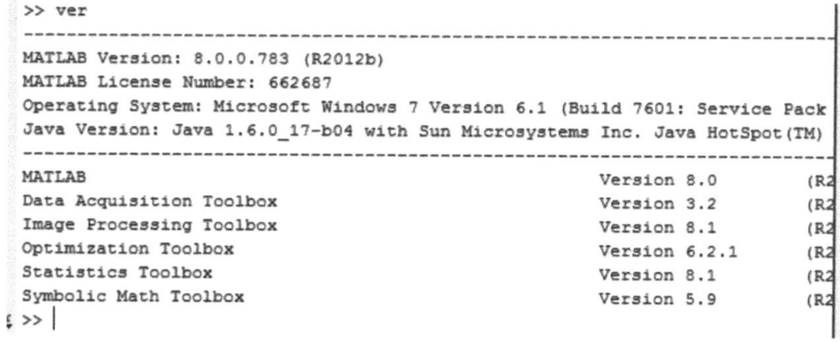

Figure 15.2 Command window with version information.

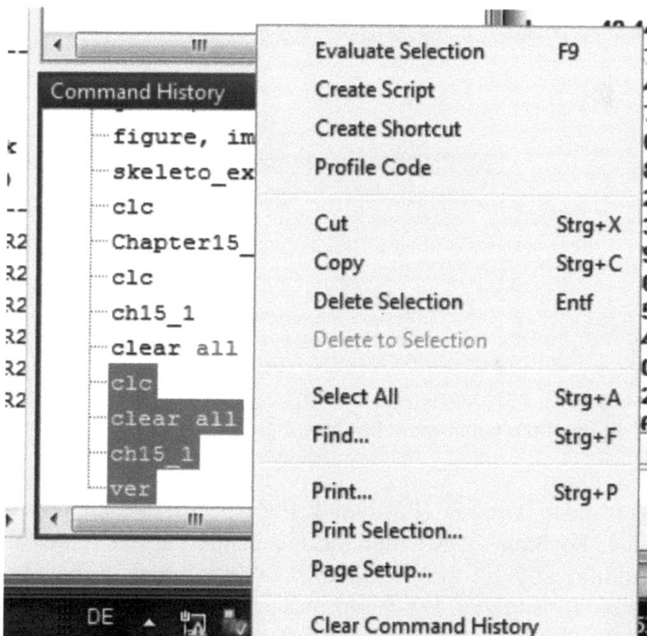

Figure 15.3 Commands from the history can be selected (Shift + Arrow) and with right mouse button a suitable option can be chosen, as for instance creating a script.

and does not require any input variables. This is a script file. All variables defined in the script remain valid and accessible in the workspace after the script has finished its work. The workspace is the current "memory" of the program, where variable data are stored and can be seen – and directly accessed – from the window Workspace (Position 5, Fig. 15.1). As can be seen, there is the variable x there, because the script ch15_1.m was executed. The running of single scripts can be done either by clicking the Run button (Position 6, Fig. 15.1) or entering the name of the script (or function), without the file ending, on the command line.

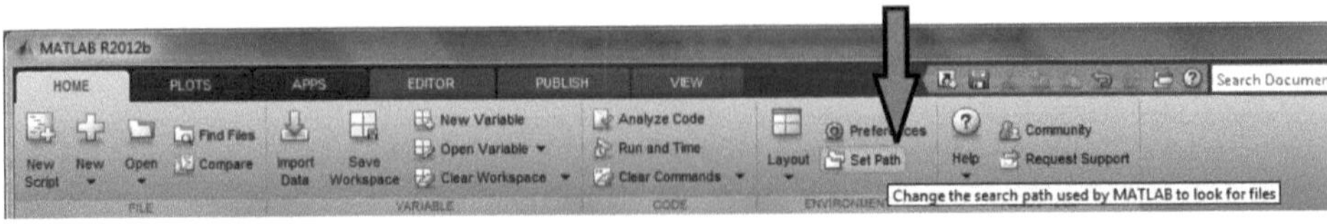

Figure 15.4 Setting the path to the (own) MATLAB files.

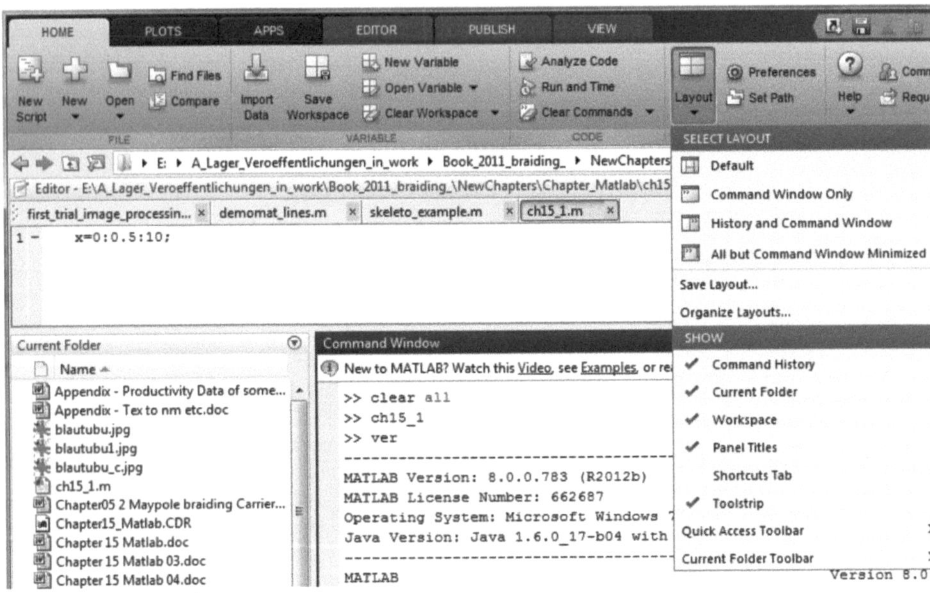

Figure 15.5 Selection of the windows to be shown.

The "Current folder" window (Position 4, Fig. 15.1) shows the files in the current folder (Fig. 15.1, Position 7). All files located in the current folder are accessible for running. Additionally, all files, located in folders stored in the MATLAB Path, are also accessible for running. For larger projects, it is sufficient for organizing the code into several subfolders. In order for all these files to be accessible, the folders and subfolders have to be stored in the MATLAB Path (Menu -> Home, Set Path, Fig. 15.4). If some of these windows are not visible, they may be selected from the layout settings (Fig. 15.5).

15.2.2 Functions and scripts

As mentioned above, scripts can contain a set of commands, but they do not require any input variables. Functions are more complex elements; they may have input and output variables, and all (normal) variables used within functions are accessible only within the function while the function is in use.

Let us consider the function `ch15_2.m` (Fig. 15.6).

```
function [x_, y_]=ch15_2(r,N)
fii=0:pi/N:pi;
x_=r * cos (fii);
y_=r * sin (fii);
```

Figure 15.6 Function ch15_2.m.

```
>> which beta
C:\Program Files\MATLAB\R2012b\toolbox\matlab\specfun\beta.m
```

Figure 15.7 Checking the existence of a function with name beta.

The function begins with the keyword function. The variables x_ and y_ are the local variables in the function, which will be outputted into the workspace after the function is completed. The function name – here, ch15_2 – has to begin with a character and be unique, while the file containing the function must be named after the function. The command which can establish whether a function of the same name already exists in the MATLAB Path. Figure 15.7 demonstrates how to check if there is an existing function named beta – as it shows that the function beta already exists, the name beta should not be used.

The function name in brackets is then given the input variables, in this case, r and h, representing radius and height. The variable fii defined in the function body will not be visible in the workspace after the function is executed; this can be checked in the Workspace window or by typing the command who. Figure 15.8 demonstrates the common call of the function with inputs $r = 10$ and $h = 20$. Before calling the function, it is recommended that all variables in the workspace are cleared with the command clear all, and that all open figures are closed with close all.

The function creates a matrix vector named fii, beginning with 0 and ending with pi (3.1415) in steps of pi/N; the individual elements of fii can be accessed using fii(k) where k is the index, starting from 1. The row x=r * cos(fii) then creates a vector x, multiplying each component of fii by the radius, while the function plot(x,y) takes the output variables from the function to create a 2D plot.

15.2.3 Function with global variables and 3D plot

If some variable should remain accessible both within a function and outside a function, it has to be defined as global. This is shown in Fig. 15.9 within the code of the function ch15_3, where both variables v (representing takeoff speed) and EndAngle (denoting the maximal angle for the angular coordinate of the points to be drawn) are defined as global. This means that the values of these variables have also to be declared as global in the calling function, which in this case is ch15_4, presented in Fig. 15.10.

The global variables here are normally constant during calculations assuming one machine, product or other object. In a practical scenario, there is a large set of variables that remain global, where it is more convenient to initially specify these as global. This allows the lists of input variables for the functions to be kept shorter.

```
>> clear all
>> close all
>> [x,y]=ch15_2(10,20);
>> plot(x,y)
>>
```

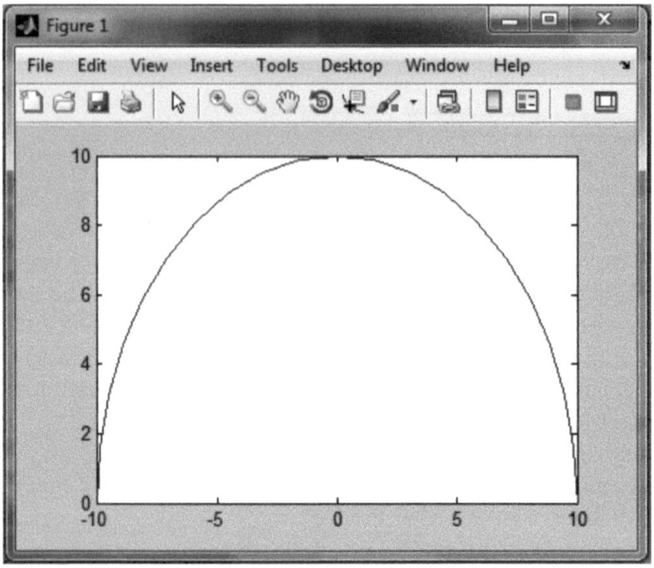

Figure 15.8 Running the function and the plotting result.

```
function [x_, y_, z_]=ch15_3(r,N)
% function for calculation of 3D coordinates

% Global variables
global v EndAngle

% Definition of fii for angle
fii=0:pi/N:EndAngle;
x_=r * cos (fii);
y_=r * sin (fii);

% Pre-allocate the variable for speed
z_=zeros(1,length(fii));

% fill the z_
for i=1:length(fii)
z_(i)=v*i;
end
```

Figure 15.9 Extended function ch15_3.m for using of global variables v and EndAngle.

```
% Script ch15_4.m
% plotting 3D

% Definition of the global variables
global v EndAngle
v=1;
EndAngle=20;

% definition of the local variables
r=10;
N=20;

% calling the function
[x, y, z]=ch15_3(r,N);

%plotting the results
figure(1);
plot3(x,y,z,'ro-');
```

Figure 15.10 Running the function and the plotting result.

The next new code in the function `ch15_3` is in the last lines; if a third coordinate `z_` has to be defined that increases at every step with value v, then one possible way is to set the single values within for –cycle and direct access to the elements (there are shorter ways to achieve this as well, for instance using z_=0:v:v*length(fii), but commands are more complicated), The command `length(fii)` returns the number of the elements in `fii`, required in order to specify the same number of elements for `z_`.

Figure 15.10 lists the code of the function ch15_4, where the global variables are defined again as global and their values are set. The variables `r` and `N` are assigned values (`r` = 10 and `N` = 20), but these values are not global. After the calculation the returned values from the function `ch15_3` will be stored in the variables x, y and z, which are used to plot a line with the function `plot3`. The characters after the three vector `ro-` define the colour (*r* = red), the marker (o – means circle) and the line type (– means solid line) of the line. All options for plotting can be seen using the "build it" help function, or by invoking `help plot` in the command window.

The resulting plot is shown in Fig. 15.11. For 3D rotation to be switched on, the corresponding button shown on the picture has to be clicked, after which the figure can be rotated with the mouse.

15.2.4 Carrier coordinates of spiral braiding machine

The motion of the carrier in the spiral braiding machine is explained in Chapter 11. The function `draw_hopy(k)` uses the equations provided there so as to plot the path

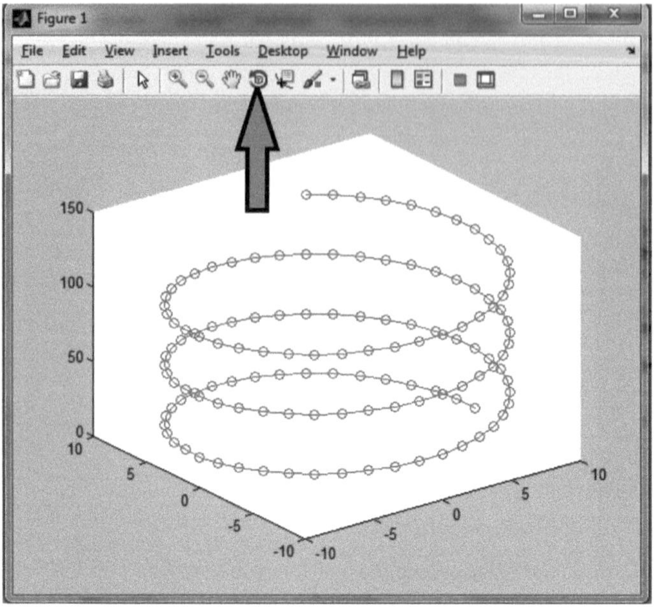

Figure 15.11 Example of 3D curve, plotted using the script ch15_4.

```
function draw_hypo(k)
%% Draws hypocycloide with k stopping points
%
% The radius of the horn gear r is defined as global variable, in order to
% be accessible inside of the function getcoord(), which calculates the
% coordinates at a single point.
global   r

R = 600; % mm %Larger Radius
r=R / k; % Radius of the horn gear
lambda=k-1;  % Lamda number
i=0;
for fi=0: 0.01: 2*pi
    i=i+1;
    [x(i),y(i)]=getcoord(fi, lambda)  ;
end
plot(x,y);hold on;
daspect([1 1 1])
axis off;

% function for calculation of the coordinates of one carrier
function  [x,y]=getcoord(fi, lambda)
global   r
x=lambda*r*cos(fi) + r * cos(lambda*fi);
y=lambda*r*sin(fi) - r * sin(lambda*fi);
```

Figure 15.12 Function for drawing hypocycloid for spiral drawing machine.

of one carrier during the work of the machine. Figure 15.12 presents the code of the function. This code demonstrates the nesting of functions within other functions; the function `getcoord` is a local function that is accessible only for the code inside the

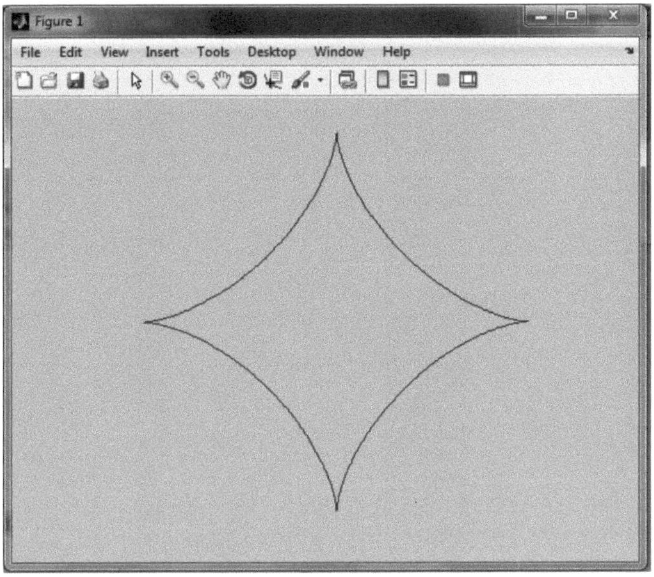

Figure 15.13 Resulting curve from draw_hypo(4).

function `draw_hypo. getcoord` calculates the coordinates of a point for given angular position `fi` and coefficient lambda, and is called as required for calculation of all points of the hypocycloid. Figure 15.13 shows the result from the calculation of the path of carrier with `k` = 4.

15.2.5 Calculation of the yarn length per repeat depending on the braiding angle

It is easier to understand the influence of a single variable over some product parameter if the relation between the input and output parameters is presented visually. If the relations become more complex and several equations are used, it is more convenient to draw such relations in MATLAB instead of in MS Excel or OpenOffice Calc. The calculation sheets can have equations, but they are not directly visible during the calculation (if the user does not use Visual Basic for Applications), and looking for errors can take long time. In the MATLAB code, the equations are visible one after another, and debugging (running the program step by step) allows for errors in programming to be found more rapidly.

The file `lflat.m` (Fig. 15.14) presents the code for some of the calculations, presented in Chapter 14 *Productivity calculations in braiding*. Here, the variable a stores the values for the braiding angle – between 20° and 80° – and the variable I is the index for the single values. The example demonstrates plotting of two curves on the same figure (Fig. 15.15) with different settings, setting the legend, x- and y-axis, and the switches to export the figures as jpeg files with no compression (`-jdpeg100` means 100% quality) and resolution 300 dpi (`-r300`), required for the figures in the book.

```
%% script lflat for Length per repeat of flat braided fabrics
w = 5; % mm width of the braid

%% Length per 1 m braid
% tan(alfa)=w/Lrepeat;  Lrepeat=w/tan(alfa)
% Number_Repeats_in_1m = 1000mm / Lrepeat (mm) = 1 * tan(alfa) / w
i=0;% Yarn Lengt per 1 m   = number_repeats * yarn_in Repeat

for a=20:2:80
   i=i+1;
   alfa(i)=a;  % store the angle as degrees for ploting
   Lyarn(i)=2*w/sin(a*pi/180);  % calculate the yarn length
   Lrepeat(i)=2*w/tan(a*pi/180);
   numb_repeats(i)=1000/Lrepeat(i);  %
   Lyarn1m(i)=Lyarn(i)*numb_repeats(i)/1000;
   Lyarn1mdirect(i)=1/cos(a*pi/180);
end

figure(1)
plot(alfa,Lyarn,'-v',alfa,Lrepeat,'-o','LineWidth',2,'Color',[0 0 0]);hold on;
legend('Yarn per repeat','Repeat length');
grid on;
xlabel('Braiding angle [degree]','FontSize',12,...
    'FontName','Arial Black');
ylabel('Yarn length per repeat, [mm]','FontSize',12,...
    'FontName','Arial Black');
title('Yarn length per repeat','FontSize',14,...
    'FontName','Arial Black')
print YarnPerRepeat -djpeg100 -r300

figure(2)
plot(alfa,Lyarn1m,'LineWidth',2,'Color',[0 0 0]);
grid on;
xlabel('Braiding angle [degree]','FontSize',12,...
    'FontName','Arial Black');
ylabel('Yarn length , [m]','FontSize',12,...
    'FontName','Arial Black');
title('Yarn length  per 1 m braid ','FontSize',14,...
    'FontName','Arial Black')
print YarnPer1m -djpeg100 -r300
```

Figure 15.14 Code for visualization of the yarn length per repeat for flat braids.

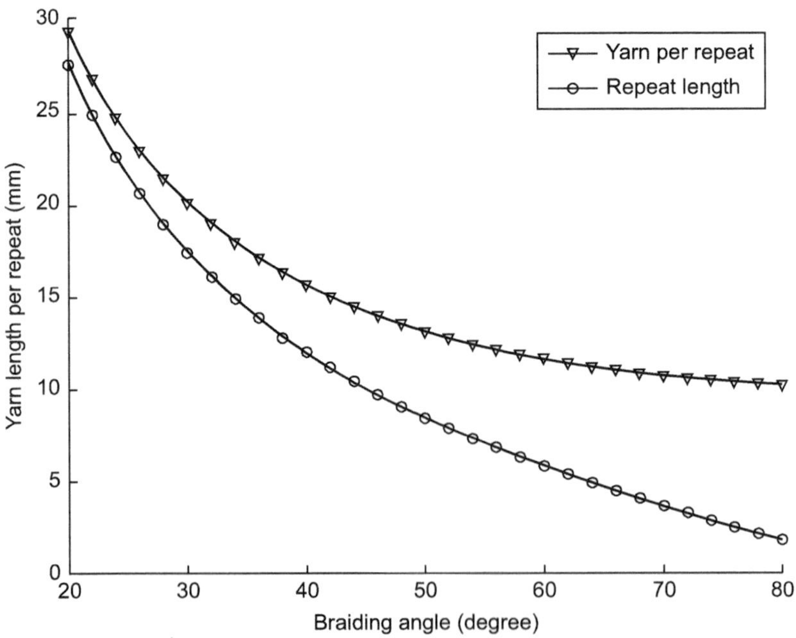

Figure 15.15 The figure shows the yarn length per repeat and repeat length for different braiding angles, created with the program given in Fig. 15.14.

```
% plot moments in the bobbin

%% Bobbin Radius
% empty bobbin with radius 0.5
rh=0.5;
rfull = 2; % cm
% empty bobbin with mass 20 g
m_empty_bobbin = 20; % g

r=rfull:-0.05:rh;  % Let the bobbin radius changes between 0.5 and 2 cm
ro=1.3; % Package density of 1.3 g/cm^3
g=9.8; % earth acceleration
h=40;  % bobbin hight = 40 cm
rfr=0.1;  % Friction radius. It is like friction coefficient, but for rotation

% Wound package volume
V=pi * (r.^2 - rh^2)*h;

% yarn mass
m_yarn= V *ro;

%mass bobbin with yarn
m=m_yarn + m_empty_bobbin;

% plot dependence between radius and mass
F=( r*pi*h*ro + (1./r)*(m_empty_bobbin - rh^2*pi*h*ro) )*g*rfr;

% let we supoose the friction is lower, you have good carrier
F1=( r*pi*h*ro + (1./r)*(m_empty_bobbin - rh^2*pi*h*ro) )*g*rfr *0.5;

% of you have a steel and the desnity is 7 time higher then PES
ro=7.8;
F2=( r*pi*h*ro + (1./r)*(m_empty_bobbin*5 - rh^2*pi*h*ro) )*g*rfr;

figure(2)
plot(r,F,'-ko','LineWidth',2); hold on;
plot(r,F1,'-kx','LineWidth',2);
legend('Some standard case','50% less bearing friction')
figure(3)
plot(r,F,'-ko','LineWidth',2); hold on;
plot(r,F2,'-kv','LineWidth',2);
legend('PES','Steel');
```

Figure 15.16 Calculation of the yarn tension, required to rotate the bobbin.

15.2.6 Yarn force during unwinding

The calculations required to calculate the yarn force necessary to rotate the bobbin during the braiding process was described in Chapter 8. Figure 15.16 presents the MATLAB code, where these equations are implemented and can be used for parametric analysis of yarn tension.

15.2.7 Horn gear form drawing

For several demonstration and animation processes, it is efficient to have a simplified representation of the horn gears with their slots. For the purposes of this book, a function to draw horn gears was written, so that – depending on the machine, and number of slots of the horn gear – appropriate drawings could be generated automatically. Figure 15.17 presents the first part of the code, where the function `hg=plot_horn_gear` creates a structure `hg` with all important properties, based on the horn gear

```
function hg=plot_horn_gear(D, xc, yc, nsec, alfa)
% Setting all parameters as a properties of the structure hg (horn gear)
hg.xc=xc; hg.yc=yc; hg.rfus=10; hg.D=D; hg.NPos=nsec; hg.dfi=2*pi/nsec;
hg.omega=1; hg.angles= alfa + (0:hg.NPos-1)*hg.dfi;
hg.x_pos=hg.xc + 0.5*hg.D*cos(hg.angles);
hg.y_pos=hg.yc + 0.5*hg.D*sin(hg.angles);

plot(hg.x_pos,hg.y_pos,'ro'); hold on; plot(hg.xc, hg.yc,'b+');
dr=10;  hgD=hg.D - 2* dr;

%   plot(hg.x_pos,hg.y_pos,'ro'); hold on;
if hg.omega >0
    plot(hg.xc, hg.yc,'k.','LineWidth',1,'MarkerSize',10);hold on;
    plot(hg.xc, hg.yc,'ko','LineWidth',1,'MarkerSize',10);
else
    plot(hg.xc, hg.yc,'kx','LineWidth',1,'MarkerSize',10); hold on;
    plot(hg.xc, hg.yc,'ko','LineWidth',1,'MarkerSize',10);
end
daspect([1 1 1]); axis off; h=gcf; hold on
```

Figure 15.17 Horn gear drawing function, part 1 – initialization and angles of the slots.

```
%% Plot Foot Arc
% Central Angle pro Carrier Foot
beta_halb=asin(hg.rfus/(0.5*hgD));
%
% Start and End angle per Foot
start_angl =hg.angles - beta_halb;
end_angl =  hg.angles + beta_halb;
%
% Start Point of the Foot arc
x_pos_start=hg.xc + 0.5*hgD*cos(start_angl);
y_pos_start=hg.yc + 0.5*hgD*sin(start_angl);

% End Point of the Foot Arc
x_pos_end=hg.xc + 0.5*hgD*cos(end_angl);
y_pos_end=hg.yc + 0.5*hgD*sin(end_angl);

for i=1:hg.NPos   %
    hg_x_pos(i)=hg.xc + 0.5*hgD*cos(hg.angles(i));
    hg_y_pos(i)=hg.yc + 0.5*hgD*sin(hg.angles(i));
end

for i=1:hg.NPos   % norm has to be per vector...
    %local OY
    loc_oy=[hg.xc - hg_x_pos(i) hg.yc  - hg_y_pos(i) 0];
    oye=loc_oy/norm(loc_oy);
    %local OX
    % vector product: ox * oy = oz ;  oy * oz = ox
    oz=[0 0 1];
    loc_ox=cross(oye,oz);
    oxe=loc_ox/norm(loc_ox);  % norm per vector  % eigentlich unnötig
    oxe=oxe(1:2);  % only the first two componentx
    oye=oye(1:2);

    iarc=0;
    d_fi=abs((pi - beta_halb) ) / 10;
    for filoc=beta_halb/2 : d_fi : pi - beta_halb/2
        iarc=iarc+1;
        xy=[hg_x_pos(i) hg_y_pos(i)] + ...
            hg.rfus * oxe * cos(filoc) + ...
            hg.rfus * oye * sin(filoc);
        xfa(iarc)=xy(1);
        yfa(iarc)=xy(2);
    end  % for the Foot_Arc - Points
    plot(xfa,yfa,'k','LineWidth',2,'MarkerSize',1); hold on;
    clear xfa; clear yfa
end  % for the Positions _ Foot Arc calculations
```

Figure 15.18 Horn gear drawing function, part 2 – drawing the arc of the slots.

```
%% plot Horn Part
% plotet each arc separately, starting from the first end to
% the second begin
for is=1:hg.NPos-1
    ii=0;
    d_angl=abs(start_angl(is+1)-end_angl(is))/10;
    for horn_angle=end_angl(is): d_angl : start_angl(is+1)
        ii=ii+1;
        xho(ii)=hg.xc + 0.5*hgD*cos(horn_angle);
        yho(ii)=hg.yc + 0.5*hgD*sin(horn_angle);
    end
    plot(xho, yho,'k','LineWidth',2,'MarkerSize',1); hold on;
end

% last one sector
ii=0;
d_angl=abs(start_angl(1)+2*pi-end_angl(hg.NPos))/10;
for horn_angle=end_angl(hg.NPos): d_angl : start_angl(1)+2*pi
    ii=ii+1;
    xho(ii)=hg.xc + 0.5*hgD*cos(horn_angle);
    yho(ii)=hg.yc + 0.5*hgD*sin(horn_angle);
end
plot(xho, yho,'k','LineWidth',2,'MarkerSize',1); hold on;
end
```

Figure 15.19 Horn gear drawing function, part 2 – drawing the arc between the slots.

diameter `D`, the centre with coordinates (`xc`, `yc`), number of slots `nsec` and starting angle `alfa`. It is not necessary to add parameters structure `hg` in the form:

```
hg.xc;
hg.yc;
```

but to do so would allow these data to be used and stored after finishing the drawing; for instance, in an array for all horn gears, this can be useful for simulation of the braiding process. This is how it is used in the software TexMind Braider Geo3D, albeit implemented in C++.

The variable `dr` is the difference between the radius of the track and the radius of the horn gear. Normally, these are almost equal, but here the horn gears are drawn with a smaller radius for better visibility and clarity of the pictures.

After initialization of all the data, the second part of the code (Fig. 15.18) plots the arcs for the carrier feet – the slots. Some vector operations are used for the right orientation to be found; an arc, consisting of a set of short lines between the points, is then drawn around the slot centre. Finally, the curves between the slots are drawn as presented in Fig. 15.19, and the results from these single steps lead to the final drawing in Fig. 15.20. Using this function, any braiding machine can be drawn within a parametric cycle, or step by step in adding single drawings to the same figure.

15.3 Data Acquisition Toolbox™ and yarn tension measurement

The Data Acquisition Toolbox™ (DAQ) is a separate library giving MATLAB the ability to read or write signals from/to external devices. A small set of commands – which

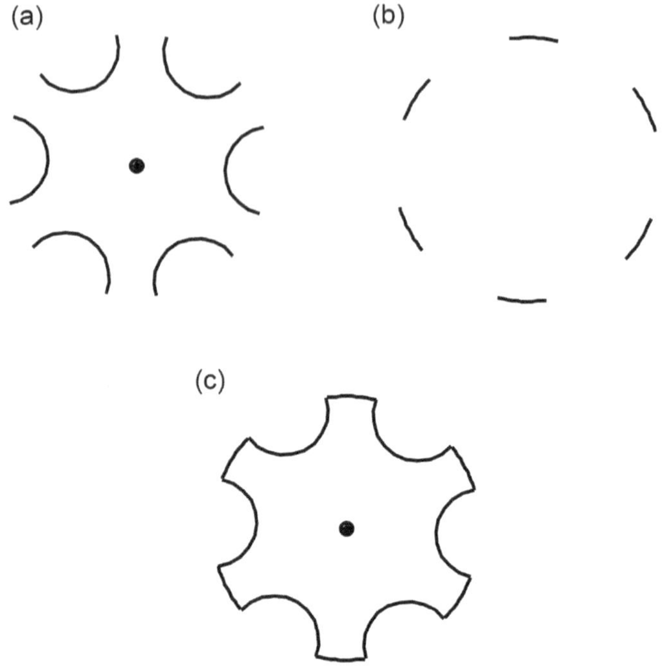

Figure 15.20 Steps of drawing horn gear (a) slot arcs (b) arcs between slots (c) final results.

can be used for measurement of yarn tension – will be demonstrated. The yarn tension is not measured during machine activity, as such a device would require both an electricity source and data transmission or storage abilities on the carrier. Here, the yarn tension was measured on a stopped machine, in order to analyse the state of the carrier using the most important characteristic for braiding – yarn tension. Another direct application could be the measurement and control of tension during yarn winding. The principle can be applied to any kind of measurement where the investigated value is received in some electric signal.

15.3.1 Equipment

The equipment mentioned here is given only to ensure the integrity of the information. The measurements can be performed with any other device compatible with the available software. For measuring tensile force, the measurement element manufactured by BTSR (BTSR Yarn feeding control, 2013) for 500 cN is currently used. This is an analogue device, which gives an output voltage between 0 and 10 V. This voltage can be transferred to the computer using an analogue digital converter (ADC). Currently used is one of the analogue inputs of the National Instruments device NI USB-6211 (National Instruments, 2013). Before attempting measurement, the driver for the ADC has to be installed on the computer; it should be determined whether the

```
>> daq.getDevices

ans =

ni: National Instruments USB-6211 (Device ID: 'Dev1')
   Analog input subsystem supports:
      4 ranges supported
      Rates from 0.1 to 250000.0 scans/sec
      16 channels ('ai0' - 'ai15')
      'Voltage' measurement type

   Analog output subsystem supports:
      -10 to +10 Volts range
      Rates from 0.1 to 250000.0 scans/sec
      2 channels ('ao0','ao1')
      'Voltage' measurement type

   Digital subsystem supports:
      8 channels ('port0/line0' - 'port1/line3')
      'InputOnly','OutputOnly' measurement types

   Counter input subsystem supports:
      Rates from 0.1 to 80000000.0 scans/sec
      2 channels ('ctr0','ctr1')
      'EdgeCount','PulseWidth','Frequency','Position' measurement types

   Counter output subsystem supports:
      Rates from 0.1 to 80000000.0 scans/sec
      2 channels ('ctr0','ctr1')
      'PulseGeneration' measurement type
```

Figure 15.21 Getting the devices plugged into the computer and recognized by DAQ of MATLAB.

device drivers are compatible with both the operating system on the computer and the MATLAB toolbox.

15.3.2 Setting the environment and creating measuring session

The first step in MATLAB is to check which devices are recognized for measurement.

Figure 15.21 demonstrates the results of checking the devices on the current computer with the command `daq.getDevices`. The computer shown has only one device – `ni` – with sixteen analogue channels. The device ID is `Dev1`, the first channel is named ai0, and they are ready for voltage measurement. These data are required for the next step, as the other channels are not used in the current example.

The DAQ toolbox has two types of interface – session-based and legacy-based. These are not supported for all devices and operating systems. The session-based interface works on "sessions," for NI devices, multiple devices and measurement types including temperature, current, and accelerometer, and works as well on the

```
>> s=daq.createSession('ni');
>> s

s =

Data acquisition session using National Instruments hardware:
   Will run for 1 second (1000 scans) at 1000 scans/second.
   No channels have been added.

Properties, Methods, Events
```

Figure 15.22 Creating new session.

```
>> s.addAnalogInputChannel('Dev1','ai0','Voltage')
>> s

s =

Data acquisition session using National Instruments hardware:
   Will run for 1 second (1000 scans) at 1000 scans/second.
   Number of channels: 1
      index Type Device Channel MeasurementType      Range            Name
      ----- ---- ------ ------- --------------- ---------------- ----
      1     ai   Dev1   ai0     Voltage (Diff)  -10 to +10 Volts
```

Figure 15.23 Adding Analog Session for the channel ai0, device Dev1 and type Voltage.

64 bit MATLAB system. The legacy interface can be used for all other types of devices, but was at this time (MATLAB 2012b, Year 2013) not available on 64 bit MATLAB.

For the session interface, a new session has to be created with the command `s=daq.createSession('ni')`, where `s` is the session object, and `ni` is the device name, taken from the previous result (Fig. 15.22).

The channel that receives the signal is currently the first one (`ai0`)– this is determined from the place, where the cable from the sensor is plugged into the ADC. This channel has to be added to the session with the command `daq.addAnalogInputChannel` where the device ID, the channel name and the type of the signal have to be specified (Fig. 15.23).

Additionally, the type of sensor connection (single-ended or double-ended) and the time for the recording has to be set, as presented in Fig. 15.24.

If some other properties have to be changed, as for instance sampling rate, etc., the properties of the sessions can be set in the same manner as the `DurationInSeconds` in Fig. 15.24.

15.3.3 Measurement

In the current examples, the following command set was used for the measurement:

```
clear data; data=s.startForeground(); plot(data);
```

```
>> s.Channels.InputType='SingleEnded';
>> s

s =

Data acquisition session using National Instruments hardware:
   Will run for 1 second (1000 scans) at 1000 scans/second.
   Number of channels: 1
      index Type Device Channel   MeasurementType         Range          Name
      ----- ---- ------ ------- ------------------- ---------------- ----
      1     ai   Dev1   ai0     Voltage (SingleEnd) -10 to +10 Volts

 s.DurationInSeconds=10;
```

Figure 15.24 Setting the single-ended type of the sensor connection and 10 s for the measuring time.

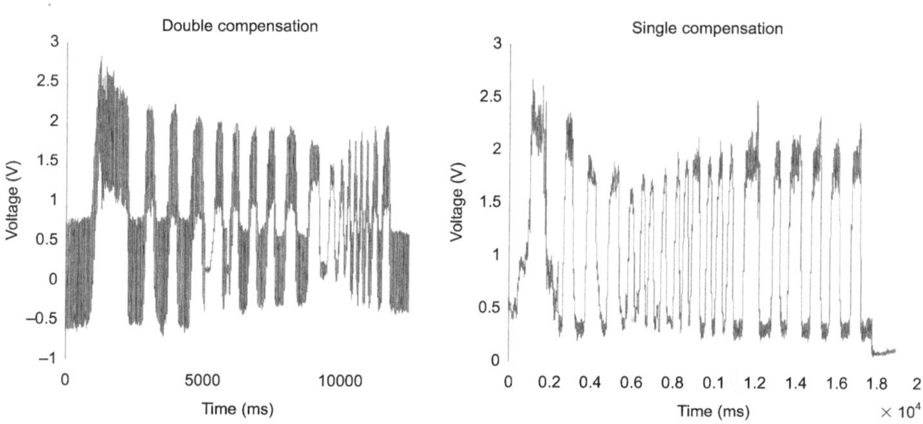

Figure 15.25 Measured yarn tension for double and single compensated carrier.

The first command clears any variable data still present from previous measurements. Normally, the data will be overwritten, but the clean start ensures that the variable data will be reinitialized to null values. The second command begins measurement in the foreground, and stores all data in the vector. The third command plots all single values on the current figure.

15.3.4 Examples

Figure 15.25 shows the measurement of yarn tension on one carrier in cases of double and single compensation (see Chapter 7 *Carriers for braiding machines*), where the yarn was taken up and down slowly so that the lever swung around its middle position. This corresponds to the situation where the carrier is "dancing" around the horn gears, and compensates for the distance differences between the carrier and braiding point. The measurement shows that, in the case of double compensation, there were

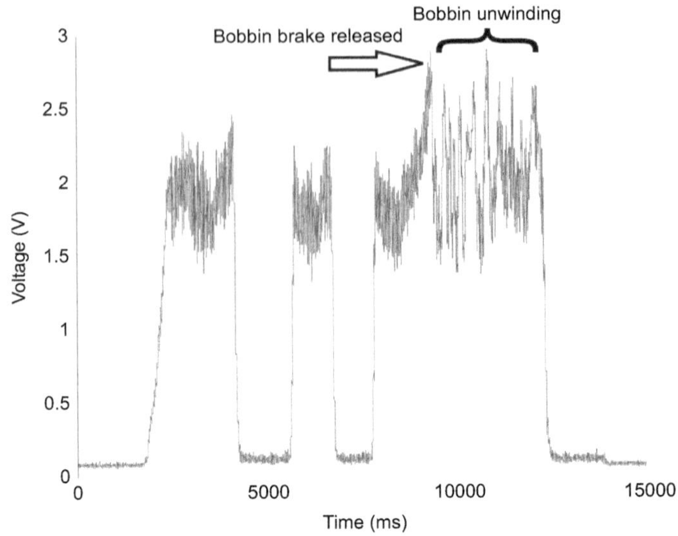

Figure 15.26 Yarn tension during two motion cycles of the lever and bobbin unwinding.

more fluctuations in yarn tension around the upper and lower tension values compared to single compensation. This is because the yarn goes through two more guides in double compensation than in single compensation; because of this, there are more places where the stick–slip effect from the friction between the yarn and guide influences the yarn tension.

In both cases, the yarn tension swings between two values: the higher one during the tension period, where the lever is travelling up and more yarn has to be given; and the lower one – nearly zero – where the available yarn length is greater than the distance between the carrier and braiding point, and the superfluous length is moving back into the compensation area. The yarn tension can be recalculated using the principle that the sensor gives 10 V for 500 cN; however, more exact calculation requires calibration of the device with some known weight(s).

Figure 15.26 illustrates the yarn tension after two cycles of bidirectional lever swinging; it is pulled further, so that the bobbin starts to rotate and unwind more yarn. It can be seen that the maximal yarn tension is reached at the moment before the bobbin starts rotating and, after that, during the unwinding, where the yarn tension fluctuates around values identical to these during the loading cycle during the normal braiding process.

15.4 Conclusions

The given examples demonstrate the power of the MATLAB environment; with a few lines of code, it is possible to measure and plot the signal for the yarn tension to be measured and plotted, or draw a 3D path. The code samples from this chapter are freely available from the MATLAB Central webpage – at http://www.mathworks.com/matlabcentral/fileexchange/47310-braiding-technology-fo-textiles--elsevier – as well as on the author's homepage www.kyosev.com.

References

BTSR Yarn feeding control (2013). *Constant Tension Feeders* [online], www.btsr.com. Available from: http://www.texgilles.de/fournisseure_btsr.html (Accessed 2013).

Mathworks (2014). *MATLAB, www.mathworks.com* [online]. Available from: www.mathworks.com.

National Instruments (2013). *National Instruments* [online]. Available from: www.ni.com.

Index

FSC
www.fsc.org
MIX
Papier | Fördert gute Waldnutzung
FSC® C083411

Kolibri 360 GmbH
Ferdinand-Jühlke-Straße 7
99095 Erfurt, Deutschland
produktsicherheit@kolibri360.de